AP

物理 1&2

AP Physics 1&2

赵 杰 编著

中国人民大学出版社
· 北京 ·

Preface

With the rapid growth of the Chinese economy, China's education reform has been confronted with a challenge of cultivating talent. The traditional education system cannot satisfy the huge demand of qualified talents for the development of the market economy. Chapter 16 of *National education reform and development of long-term planning programs* (*2010–2020)* clearly states that we need to cultivate talents who have international insights, a good understanding of international rules, and the ability to participate in international affairs. The internationalization of education entails teaching resources, environment, outcome and concept. Since 2003, Beijing Royal School (BRS) has implemented an international education along with constant research and trials in methodology and pedagogy. Additionally, BRS has established the A-level, AP and AEOP courses in which the AP course has been the most preferred.

AP (Advanced Placement), initiated by College Board, aims at helping high school students adapt to college education in advance. American universities consider AP grades as important criteria for assessing the students' overall qualities and research capabilities. They are also used to select students who can meet the standard of college admissions.

In 2006, BRS became the first school in China to offer AP courses. Because of our outstanding teaching achievements, BRS is the first high school in China to be admitted as a member of the College Board, and has been authorized as the only AP teaching demonstration school in China. Since the introduction of AP courses, we have been committed to the promotion of AP research. In 2010, BRS hosted the First

AP Teaching Forum along with the College Board. Some of our teachers have been granted the "AP Demonstration Teacher" certificates.

In order to help our students adapt to international education and to better integrate with creative learning concepts, BRS has sent many teachers of different disciplines to the United States, the United Kingdom, Korea and Hong Kong for further study, experience and training on a regular basis. As a result, our teachers have accumulated great teaching experience. We sincerely hope that this book can benefit students in China.

This set of books is based on the outlines of AP testing and includes key points, previous test questions and glossaries of the tests in hopes that students will have a better understanding of the AP tests. All the authors are accomplished and experienced AP teachers who specialize in AP tests. They are highly praised by experts from the College Board for applying new methodology, such as understanding-based and critical teaching. Our books can not only bring joy to students learning AP courses but can also lay a firm foundation for the reform of high school education here in China.

Due to time constraints, errors are inevitable. Any suggestions and corrections are welcome and appreciated.

AP 辅导教材序言

中国经济快速发展，人才培养给我国教育体制改革也带来了更大的挑战。传统教育模式不能满足我国市场经济对于高素质人才的巨大需求。《国家中长期教育改革和发展规划纲要（2010—2020 年）》在第十六章“扩大教育开放”中明确提出，要适应国家经济社会对外开放的要求，培养大批具有国际视野、通晓国际规则、能够参与国际事务和国际竞争的人才。教育国际化需要的是教育资源、教育环境、教育成果和教育理念。北京王府学校自 2003 年成立以来，一直奉行教育国际化的理念，在教育教学、教法和教材的改革上不断进行尝试和摸索，开设了 A-level，AP，AEOP 等课程。其中，尤以 AP 课程为广大学生所青睐。

AP（Advanced Placement），中文译为美国大学先修课程，由美国大学理事会（College Board）举办，旨在使高中学生提前适应大学教育，在高中阶段以美国大学标准挑战自我。美国大学将 AP 成绩看作衡量学生学习综合素质和研究能力的重要指标，也用于帮助筛选符合大学入学标准的优秀学生。

北京王府学校 2006 年成为中国首所引进美国 AP 课程的学校。由于教学成果出色，美国大学理事会吸收为中国首个会员单位，被美国大学理事会首选为“AP 教学示范学校”。自开设 AP 课程以来，王府学校一直致力于 AP 课程深度教科研活动的推广。2010 年，王府学校作为主办单位，同美国大学理事会共同成功举办了“首届（2010）中国 AP 教学与学习交流观摩论坛”。王府学校部分骨干教师成为中国首批获得“AP 示范教师”证书的教师。

王府学校教师受益于长期的教科研经验积累，学校鼓励和支持更多的学生加入到挑战自我的队伍中来，为让中国的高中学生早日适应先进的教育培养模式，学校各专业教师多次赴美国、英国、韩国及我国香港等地进行学习、交流经验、培训，通过学习收获教学经验积累，同 AP 课程教师和学生创造性学习的教育理念相融合，我们希望根据学生教学要编写能够结合 AP 课程精髓和中国学生思维方式的 AP 教材，以飨莘莘学子。

本套 AP 教材以 AP 考试大纲为基本框架，辅以对知识点、考题和专业词汇的缜密解析，旨在帮助学生更快地掌握 AP 知识架构，了解 AP 考试特点和突破专业词汇障碍。负责本教材编写工作的都是学校教学一线的 AP 教师，具有较丰富的教学经验，针对 AP 试题有着深入的研究，应用新教学方法，引领、理解教学，领导教学，比较教学，批判教学，得到了美国大学理事会等著名大学专家学者的认同。本套 AP 教材的编写，不仅能带给学生学习 AP 的乐趣，还能为国内高中教材教学改革抛砖引玉。

时间仓促，书中难免有不妥之处，望同行、读者不吝赐教。

Preface 前言

改革开放以来，我国已有几十万学生出国留学。很多人学成归来为我国的现代化建设做出重要贡献。随着改革开放步伐的加大，国强民富，我国留学生的数量逐年增多。从出国读硕士、博士，逐渐扩展到读本科甚至高中。尤其是自 2009 年美国大学本科向我国开放以来，我国高中毕业生申请到美国读书的人数激增。美国大学对非英语国家的考试，如 TOFEL、SAT 成为留学必备的考试。除此之外，美国的 AP 考试也在我国广泛流行。仅 2011 年，北京新增 AP 考点就有 10 个，其他省市的考点也迅速增长。2014 年 10 月，美国大学理事会在北京王府学校召开了 AP 峰会，把中国的 AP 教育推向新高峰。

为什么中国中学生喜欢参加 AP 物理考试呢？因为如果 AP 物理考试成绩能达到 4 分或 5 分，美国多数的大学会承认，并可获得大学学分，甚至有些大学也接受 3 分的成绩。这就意味着，你虽然还没进大学，就已经拿到大学的学分。这样，你在大学读书就可以省很多的时间和学费，何乐而不为呢？再说，即便 AP 物理考试不能拿到学分，这也足以证明你在中学已经学习了高等学校的课程，这对于大学入学有重要的参考价值。2014 年北京《新京报》采访北京王府学校时，王府学校执行校长指出，AP 成绩反映学生的学术能力。对于中国考生，AP 成绩对申请多数美国大学也有参考作用。

AP 物理自 2013 年开始改革，2015 年 5 月首次按改革以后的新方案考试。改革的重点是，将原来的 AP 物理 B，改成 AP 物理 1 和 2，原来一年的教学时间改为两年。改革的重点是提出六大物理思想（6 Big Ideas），重视物理概念教学。另外，加强物理实验的训练，提出加强物理七大科学实践 (7 Scientific Practices) 方法。为了应对这种形势，我们在原有的《美国 AP 物理》一书的基础上，编写了这本书，希望能对中国准备参加 AP 物理 1 和 2 考试的学生有所帮助。

本书主要介绍了美国 AP 物理 1 和 2 的考试内容、学科比例、考题类型和考试时间。

书中针对中国学生的特点，对应试策略提出了建议，同时也帮助学生制定学习计划。本书重点讲解了 AP 物理 1 和 2 涉及的 18 个知识点，对每个知识点的要点进行了剖析，这部分内容增加了中文解释。每个知识点附有练习题，特别增加了 AP 物理 1 和 2 的新题型。为了使中国学生通过练习对题型有深刻理解，练习答案采用中文表述，力求指出题目的内涵。书中附有近年来 AP 物理 1 和 2 的模拟题。希望学生能深入理解，做到举一反三。书后的附录，除了列出 AP 物理常用公式、数表以外，还包括常用物理词汇。为了使学生容易查找和记忆，物理词汇按知识点编排，希望对不熟悉物理英语词汇的考生有所帮助。

该书编写时间紧，而且新的 AP 物理 1 和 2 考试刚刚开始，参考资料不多，该书内容是仅在对美国大学理事会的改革精神的理解和多年 AP 物理教学经验的基础上编写的，因此错误在所难免，希望读者指正。

目录 Contents

Part I

Introduction of AP Physics

第一篇 AP物理简介

1 Why would you like to join the examination of AP Physics 为什么要参加 AP 物理考试

1954 年，美国大学理事会（College Board）首次推出 AP 物理考试。从此以后，很多学生参加 AP 物理考试。2009 年 5 万多名学生参加了 AP 物理 B 考试，2 万 5 千多名学生参加了 AP 物理 C 考试。这个数字还在逐年增加，特别是自从 2009 年美国大学本科向中国中学生开放以来，中国学生参加 AP 考试的人数激增，仅 2011 年，北京新增 AP 考点 10 个，其他省市的考点也迅速增长。

2013 年美国大学理事会提出对 AP 物理 B 的改革方案，将 AP 物理 B 改为 AP 物理 1 和 2，教学时间加长，突出对物理概念的理解，增强对学生实验技能的培养。2015 年已经采用改革后的 AP 物理 1 和 2 的考试取代原来 AP 物理 B 的考试。

为什么美国中学生喜欢参加 AP 物理考试呢？因为如果你的 AP 物理考试成绩能达到 4 分或 5 分，美国大多数的大学会承认你已经获得学分，甚至有些大学也接受 3 分的成绩。这就意味着，虽然还没进大学，你已经拿到大学的学分。这样，在大学读书就可以省很多的时间和学费，何乐而不为呢？另外，AP 物理成绩标志着一个学生的物理学术能力，对于申请高水平大学具有重要的参考价值。再说，即便你的 AP 物理考试成绩不能置换学分，这也足以证明你在中学已经学习了高等学校的课程，给大学物理学习打下坚实的基础。

2 What are the differences between AP Physics 1 & 2 and C AP 物理 1、2 和 C 有何区别

AP 物理 1 和 2：

AP 物理 1 和 2 是一门基础课，正如美国大学理事会公布的 AP 物理 1 和 2 大纲指出的，它涵盖的范围较宽，包括力学、热学、电学、磁学、光学和原子物理的基本知识。其中数学知识较浅显，只涉及代数方程，我国高中数学足以达到考试中的数学要求。由于

AP 物理 1 和 2 涵盖范围很宽，根据你今后在大学选什么专业，可以对应选择 AP 物理 1 或 AP 物理 2 的考试，即便你选择的专业今后不直接应用物理，AP 物理 1 和 2 也是相关知识的基础，对你今后的学习很有帮助。

AP 物理 C：

AP 物理 C 是在 AP 物理 1 和 2 的基础上，为今后准备选大学理工科专业的学生设置的。AP 物理 C 包括两部分：一是力学，二是电磁学。尽管 AP 物理 1 和 2 也包含这两部分内容，但 AP 物理 C 讲得更深，更细。学了它，你会对物理有更深层次的理解。在解决 AP 物理 C 问题时，需要微积分的知识，因此，学生应具备微积分基础。我国很多重点中学已经学习了微积分，因此，对于这些学校的学生，和那些已经自学了微积分的学生，参加 AP 物理 C 的考试也不成问题。

改革以后的 AP 物理给学生更灵活的选择，可以根据今后在大学选择的专业，自由选择学习 AP 物理 1、AP 物理 2、AP 物理 C 力学、AP 物理 C 电磁学这四门独立的考试中的一门或多门考试，获得相应的学分。

3 The test form and schedule of AP Physics
AP 物理考题的结构和考试时间

AP 物理考题量及考试时长见表 1—1。

表 1—1

AP Physics	Section	Number of questions	Time of test
AP Physics 1 & 2	Multiple Choice	50	90 minutes
	Free Response & Short Answer	4 or 5	90 minutes
AP Physics C Mechanics	Multiple Choice	35	45 minutes
	Free Response	3	45 minutes
AP Physics C E & M	Multiple Choice	35	45 minutes
	Free Response	3	45 minutes

AP 物理 C 力学考题涉及知识点占比情况见表 1—2。

表 1—2 **AP Physics C Mechanics**

	Contents	Percentage
1	Kinematics	18%
2	Newton's Laws	20%
3	Work, Energy and Power	14%
4	Liner Momentum	12%
5	Rotational Motion	18%
6	Oscillations and Gravitation	18%

AP 物理 C 电磁学考题涉及知识点占比情况见表 1—3。

表 1—3 AP Physics C Electricity and Magnetism

	Contents	Percentage
1	Electrostatics	30%
2	Conductors, Capacitors, Dielectrics	14%
3	Electric Circuits	20%
4	Magnetic Fields	20%
5	Electromagnetism	16%

4 The type of questions in AP Physics AP 物理的题型

传统的 AP 物理考题的类型分两大类：选择题（Multiple Choice）和计算题（Free Response）。2015 年改革以后，选择题里除了原来的单选题以外，增加了多选题，即在 A、B、C、D 四个答案中，选两个正确答案。在计算题里，增加了简答题（Short Answer Questions），每题 7 分，不但考查学生的计算能力，还要考查学生的科技语言表达能力。在简答题里，考查学生对物理概念、定义的叙述，对物理现象的解释。这对中国学生是有相当难度的。因此，要求教 AP 物理的老师在平时的教学中培养学生科技英语的表达和写作能力。除此之外，AP 物理 1 和 2 特别增加了对实验相关问题的考查，主要包括简单的实验设计和实验数据的处理这两个方面。因此，AP 物理老师平时要增加实验课时，以对实验能力进行培养。

A. 选择题：选择题分为单选题和多选题

在单选题中，概念题考查对基本概念的理解；公式题考查对基本物理定律、公式的理解和记忆；计算题考查简单计算能力等。

在多选题中，A、B、C、D 四个选项有两个是正确的，这就大大增加了难度，也能考查出学生对物理概念、规律的理解程度。

多选题中必须选出两个答案，只选一个不得分！

B. 计算题

每道计算题一般给 10～15 分钟的时间，获 15 分左右，不但考查对公式、概念的理解，还考查计算能力。计算题和原来 AP 物理 B 的要求相同。计算时可以参考试题给的公式表，同时可以用规定的计算器计算。因此，学生要熟悉自己的计算器，能熟练准确地计算，特别注意数量级别搞错。

C. 简答题

简答题是改革以后，在 AP 物理 1 和 2 中出现的新题型，一般有 2～3 题，要求在较短时间内回答问题，例如，每题学生在 13 分钟左右答完，获 7 分。注重对概念的理解，对现象的解释，因此英语叙述很重要。请看解题说明：

Directions:

The short free-response questions that require about 13 minute each to answer and are worth 7 points each. Show your work for each part in the space provided after that part.

例如：

(7 points，suggested time about 13 minutes)

The figure (图 1—1) shows part of a system consisting of a block at the top of an inclined plane that rests on a table，which is located on the Earth. The block and plane are at rest when the block is released. In trial 1 there is no friction between the block and the plane or between the plane and the table. In trial 2 the plane is fixed to the table so it cannot move，but there is still no friction between the block and the plane.

Indicate which the speed of the block relative to the table when the block reaches the bottom of the plane is greater in trial 1 or trial 2. Justify your answer in a clear，coherent，paragraph-length explanation.

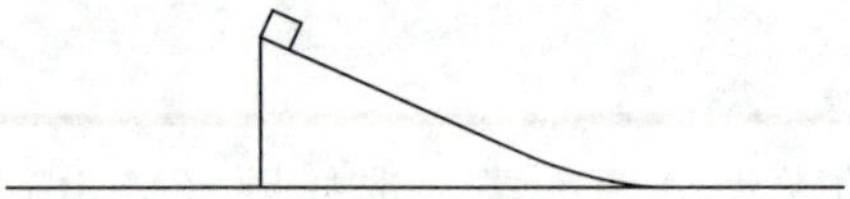

图 1—1

由于第一种情况物块和斜面、斜面和桌面都没有摩擦，因此，当物块释放以后，物块和斜面都会运动，即物块的势能转换成物块和斜面的动能。而第二种情况，斜面是固定在桌面上的，所以物块的势能全部转换成物块的动能，因此，后者速度大。注意，以上内容要求用英语表达，参考答案如下：

The speed of the block is greater in trial 2. Since gravity is the only force doing work in both trials，energy is conserved. The potential energy of the block at the top of the plane is converted into kinetic energy. In trial 1 the plane is free to move，so if the block ends up going to the right，the plane has to move to the left to conserve momentum. [*Alternate*：*In trial 1 the plane is free to move*，*so if the plane pushes the block to the right*，*the block pushes the plane to the left*.] Since the plane moves，the original potential energy of the block must be divided between the plane and the block. In trial 2 the plane doesn't move，so the block gets all of the available energy. More kinetic energy means a greater speed.

5 Make your plan 制订你的学习计划

在决定参加 AP 物理考试后，要对自己有充分的信心。要有重点地复习，对那些自己觉得困难，而 AP 物理考试又一定会考的内容要加倍努力，对自己熟悉并有把握得分的内容放在次要地位，重点放在能增加得分的内容上。AP 物理考试，是对学生大学物理基本知识的检验，不是物理竞赛，不是智力测验，因此，不必花很多时间去攻难题。

下面，分别对参加 AP 物理 1 和 2 及 AP 物理 C 的学生提出如下建议：

(1) How to make your plan for AP Physics 1 & 2 students?（参加 AP 物理 1 和 2 考试的学生如何备考?）

AP 物理 1 和 2 涵盖的知识范围较宽，不要面面俱到，要有重点。当制订学习计划时，首先要试做几套练习题，从而判断哪些内容自己已经很熟悉，哪些内容自己不熟悉，要重点复习自己不熟悉的内容。改革以后的物理 1 和 2，特别重视对物理概念的理解，加强了对学生实验技能的考核。因此，考生要注意对基本概念、定理、定律的书写，对简单结果的解释，多做物理实验，并熟悉实验结果的分析方法，同时，要有一定的实验设计能力。

注意，AP 物理 1 和 2 不涉及高等数学，在做计算题时，不要把问题想得过难，这些问题只用高中学的代数、几何、三角知识就能解决。

(2) How to make your plan for AP Physics C students?（参加 AP 物理 C 考试的学生如何备考？）

首先，注意 AP 物理 C 不是独立的，它是在 AP 物理 1 和 2 的基础上加以延伸，并分别向力学和电磁学两个方向发展的。因此，参加 AP 物理 C 考试的考生一定要有AP 物理1 和 2 的坚实基础。

其次，AP 物理 C 要求有高等数学微积分作基础，但应该看到，AP 物理 C 是物理而不是数学。它不是考查学生的微积分运算能力，而是考查学生能否运用微积分知识解决物理问题。什么是运用微积分的理念解决物理问题呢？这包括两方面：其一，运用微积分的必要性。例如，AP 物理 1 中，我们定义力做功的概念为：功等于力和沿着力的方向的位移的乘积，即：$W = FS$，但在这个公式里有两点必须强调，一是力和位移方向一致，二是力是恒力。当力是变力时，我们要用微积分的概念，功是力沿位移的积分。其二，要理解微积分的概念和图示表示方法。要知道，积分表示函数曲线和自变量轴间包含的图形的面积。例如，速度曲线下包含的面积是位移。导数是函数曲线上某一点的切线的斜率。另外，要学会从图线中读取数据，将数据绘制成曲线。这正是中国学生的弱项。对于有微积分知识基础的考生，可以根据自己今后的学科方向，选择学习 AP 物理 1 加 AP 物理 C 力学，或 AP 物理 2 加 AP 物理 C 电磁学。这样可以起到事半功倍的效果。

6 Find your weakness and overcome it
找到自己的弱点克服它

参加考试的考生要做到知彼知己才能百战不殆。考生要了解自己的优势和弱点，参加考试时，要发挥自己的优势，克服自己的弱点。例如，有的考生知识面宽，善于作选择题，但计算能力差，这样的考生就要迅速、准确地完成选择题，留出比较充裕的时间做计算题。反之，计算能力强的考生，应该先迅速准确地完成计算题，留出较充裕的时间做选择题。

7 How to do the multiple choice questions
怎样应对选择题

选择题概括起来有几种：考查概念的题、定律题、公式题、计算题等，我们应对的方法有：概念公式判断法、排他法、计算法、注意“否定”的选择、注意多选题的选择等。下面举例说明：

A. 概念公式判断法

例如：

When a light wave passes from air into glass, quantities that remain constant include which of the following?

Ⅰ. frequency

Ⅱ. wavelength

Ⅲ. speed

(A) Ⅰ only

(B) Ⅱ only

(C) Ⅰ and Ⅲ only

(D) Ⅱ and Ⅲ only

本题考查光在不同介质中传播时哪些量保持不变。如果知道光在空气、水、玻璃中不会变颜色，就得出频率保持不变。因此，你会立即选择（A）。如果对概念不清楚，就会错选速度、波长不变的选项。

又如：

A force of constant magnitude F and fixed direction acts on an object of mass m that is initially at rest. If the force acts for a time interval Δt over a displacement Δx, what is the magnitude of the resultant change in the linear momentum of the object?

(A) $F\Delta t$

(B) $F\Delta x$

(C) $F\Delta t/m$

(D) $F\Delta x/m$

这题是考查学生对动量定理的理解。如果知道动量定理是说物体线动量的变化等于作用于物体合力的冲量：$\Delta mv = F\Delta t$，就会很顺利地选择（A）。如果不知道什么是力的冲量，及冲量和动量的关系，就会错选答案。

B. 排他法

有的问题列出几个答案，只有一个是正确的，这时可以逐一排除错误的，留下正确的答案。

例如：

Which of the following experimental observations provides the best support for the statement "light behaves like a wave"?

(A) Light can be reflect by a mirror

(B) Light is scatted when passing through smoke

(C) Monochromatic light forms bright and dark bands after passing through two narrow slits

(D) White light can be broken into component colors by a prism

（A）是光的粒子性表现，可排除。（B）中光作为粒子和烟雾颗粒碰撞，也是粒子性的表现，可排除。（D）表现光在不同介质界面折射，可排除。只剩下正确答案（C），双缝干涉是波动性的最好佐证。

又如：

Which of the following is an expression of conservation of charge?

(A) Kirchhoff's loop rule

(B) Kirchhoff's junction rule

(C) Ohm's law

(D) Snell's law

答案是（B）。因为（A）是描述电路回路的。（C）是欧姆定律，涉及一段电路电压、电流和电阻的关系。（D）是光学定律。只有（B），Kirchhoff 节点定律，反映了电荷守恒的规律。或者说，节点定律是电荷守恒定律在电路中的应用。

C. 计算法

例如：

$^{238}_{92}U$ undergoes a series of reaction in which it emits eight $^{4}_{2}He$ nuclei and six electrons. What is the isotope that results from this series of reactions?

(A) $^{222}_{66}Dy$

(B) $^{206}_{81}Ti$

(C) $^{206}_{82}Pb$

(D) $^{207}_{82}Pb$

核反应方程要遵从质量数守恒和电荷数守恒两个规律。这里$^{238}_{92}U$辐射出 8 个$^{4}_{2}He$核后，质量数应变为：$238-4\times8=206$，只能选（B）或（C），电荷数变为：$92-2\times8+6=82$，因此正确答案为（C）。

又如：

A ball swings freely back and forth in an arc from point Ⅰ to point Ⅳ, as shown in the figure（图 1—2）. Point Ⅱ is the lowest point in the path, Ⅲ is located 0.5 meter above Ⅱ, and Ⅳ is 1 meter above Ⅱ. Air resistance is negligible.

The speed of the ball at point Ⅱ is most nearly

(A) 3.0 m/s

(B) 9.8 m/s

(C) 4.5 m/s

(D) 14 m/s

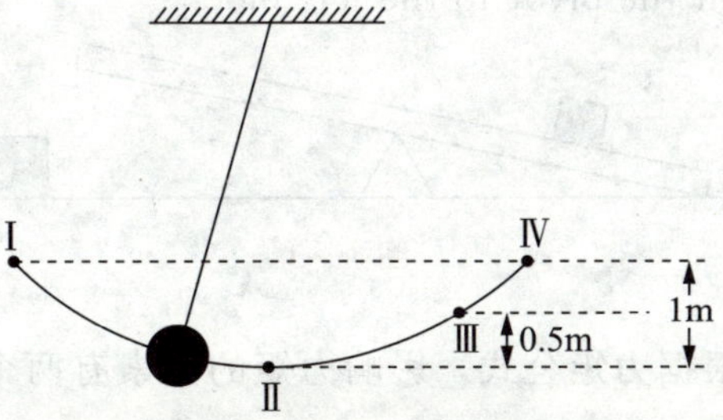

图 1—2

选（C）。根据能量守恒定律，在点Ⅰ的势能全部转化为在点Ⅱ的动能。所以，可以计算出在点Ⅱ处的速度近似为 4.5 m/s。

D. 注意"否定"的选择

例如：

Which of the following statements about electric and magnetic fields is FALSE?

(A) A charge moving along the direction of an electric field will experience a force, but a charge moving along the direction of a magnetic field will not experience a force

(B) All charges experience a force in an electric field, but only moving charges can experience a force in a magnetic field

(C) A positive charge moves in the direction of an electric field; a positive charge moves perpendicular to a magnetic field

(D) All moving charges experience a force parallel to an electric field and perpendicular to a magnetic field

选择（C）。因为正电荷在磁场中受力的方向不一定是运动方向。这是一个概念考查题，注意它是否定命题。

E. 注意多选题的选择

多选题是从 A、B、C、D 4 个选项中选出 2 个正确答案，因此，难度较单选题大，也很难得到"几率"分。多选题的内容，大多是考查对概念、定律的理解，涉及计算问题的题目很少。因此，考生只要有坚实的物理基础，对基本概念、定理、定律理解全面，仔细阅读题意，全面考虑，就不难得到正确结论。

例如：

A uniform plank is placed with a pivot at its center. A block is placed on the plank to the left of the pivot, as shown in the figure（图 1—3）. A student is asked to place a second block of greater mass on the plank so it will balance when horizontal. Which of the following quantities are needed to determine where the second block should be placed? Select two answers.

(A) The mass of the plank

(B) The mass of each block

(C) The length of the plank

(D) The distance from the pivot to the left block

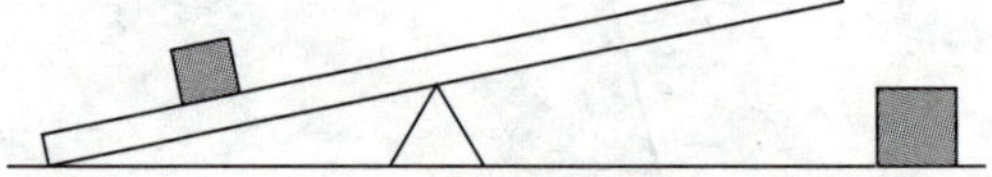

图 1—3

选择（B）和（D）。根据力矩公式，影响力矩的因素有两个，力和力到支点的垂直距离。只选一个选项不得分。

8 How to deal with the free response questions 怎样应对计算题

应对计算题，首先要清楚地了解考试的要点，透彻地了解题型。我不赞同某些教师给学生押题的做法，这种做法使考生存在侥幸心理，一旦押错题，全盘皆输。所以，掌握知识是基础，分析题型是方法。这样做，无论考题怎样变化，考生都能应对。例如，物理 1 或 2 的计算题，每年基本上有 4～5 题，包括力学、流体力学、热学、电学、磁学、光学、原子物理等内容。由于 AP 物理考试，不是物理竞赛，考查学生是否全面掌握了物理知识，大纲涉及的内容都会考。因此，我们只要掌握每一种可能的题型，不管题目怎样变化，都能顺利解决它们。

在改革后的 AP 物理 1 和 2 中，增加了一种简答题（Short Answer Questions），这种问题注重对概念、定律的理解，因此回答问题要用简明流畅的英语叙述。这对中国学生是一个难点。平时老师要重视培养学生的科技英语书写能力。

9 How to deal with the questions related with experiments 怎样回答和实验相关的问题

AP 物理考试的另一个特点是将实验内容插入考题，这种和实验相关的考题在逐年增加。在改革后的 AP 物理 1 和 2 中，加强了对实验的考查。在 4～5 道计算题中，总有一题是关于实验的。一方面，可能给出实验数据，要求学生根据数据绘图，由曲线斜率计算相关数值。另一方面，可能考查学生组织实验的能力，给出一些实验仪器和材料，供学生选择，并设计简单的物理实验。因此，要求教师平时在教学中加强对学生实验能力的培养。这正是中国考生的弱项，因此要特别引起注意，加强训练。

10 How to overcome the difficulties 如何应对难点

AP 物理考试中有少数难点，这些难点主要包括考查学生的知识面、计算能力等等。因此，学生除了课本上学的东西之外，要扩大知识面，多参与科技创新活动。例如，有的考题列出 5 个近代物理实验，然后让考生选择其中一个反映物质粒子性的实验。这就要求考生对物质的波粒二象性有所了解，对验证物质粒子性、波动性的实验有所了解，以做出正确判断。

11 The curricula of AP Physics 1 & 2 AP 物理 1 和 2 大纲要求

我们以 AP 物理 1 大纲为例，说明改革以后的 AP 物理 1 和 2 的大纲要求和主要改革重点，以供参加 AP 物理考试的师生参考（表 1—4）。

表 1—4 **Curricular Requirements of AP Physics 1**

	Curricular Requirements	Page
CR1	Students and teachers have access to college-level resources including college-level textbooks and reference materials in print or electronic format.	4
CR2a	The course design provides opportunities for students to develop understanding of the foundational principles of kinematics in the context of the big ideas that organize the curriculum framework.	4
CR2b	The course design provides opportunities for students to develop understanding of the foundational principles of dynamics in the context of the big ideas that organize the curriculum framework.	4
CR2c	The course design provides opportunities for students to develop understanding of the foundational principles of gravitation and circular motion in the context of the big ideas that organize the curriculum framework.	4
CR2d	The course design provides opportunities for students to develop understanding of the foundational principles of simple harmonic motion in the context of the big ideas that organize the curriculum framework.	4
CR2e	The course design provides opportunities for students to develop understanding of the foundational principles of linear momentum in the context of the big ideas that organize the curriculum framework.	5
CR2f	The course design provides opportunities for students to develop understanding of the foundational principles of energy in the context of the big ideas that organize the curriculum framework.	5
CR2g	The course design provides opportunities for students to develop understanding of the foundational principles of rotational motion in the context of the big ideas that organize the curriculum framework.	5
CR2h	The course design provides opportunities for students to develop understanding of the foundational principles of electrostatics in the context of the big ideas that organize the curriculum framework.	6
CR2i	The course design provides opportunities for students to develop understanding of the foundational principles of electric circuits in the context of the big ideas that organize the curriculum framework.	6
CR2j	The course design provides opportunities for students to develop understanding of the foundational principles of mechanical waves in the context of the big ideas that organize the curriculum framework.	5
CR3	Students have opportunities to apply AP Physics 1 learning objectives connecting across enduring understandings as described in the curriculum framework. These opportunities must occur in addition to those within laboratory investigations.	8
CR4	The course provides students with opportunities to apply their knowledge of physics principles to real world questions or scenarios (including societal issues or technological innovations) to help them become scientifically literate citizens.	9

续前表

	Curricular Requirements	Page
CR5	Students are provided with the opportunity to spend a minimum of 25 percent of instructional time engaging in hands-on laboratory work with an emphasis on inquiry-based investigations.	6
CR6a	The laboratory work used throughout the course includes a variety of investigations that support the foundational AP Physics 1 principles.	7
CR6b	The laboratory work used throughout the course includes guided-inquiry laboratory investigations allowing students to apply all seven science practices.	8，9
CR7	The course provides opportunities for students to develop their communication skills by recording evidence of their research of literature or scientific investigations through verbal，written，and graphic presentations.	7，10
CR8	The course provides opportunities for students to develop written and oral scientific argumentation skills.	9，10

AP Physics 1 is equivalent to most college-level introductory physics courses with a focus on the following topics：Kinematics，Dynamics，Universal Law of Gravitation，Simple Harmonic Motion，Momentum，Work and Energy，Rotational Motion，Electrostatics，Electric Circuits and Wave and Sound.

The content for the course is based on six big ideas：

Big Idea 1：Objects and systems have properties such as mass and charge. Systems may have internal structure.

Big Idea 2：Fields existing in space can be used to explain interactions.

Big Idea 3：The interactions of an object with other objects can be described by forces.

Big Idea 4：Interactions between systems can result in changes in those systems.

Big Idea 5：Changes that occur as a result of interactions are constrained by conservation laws.

Big Idea 6：Waves can transfer energy and momentum from one location to another without the permanent transfer of mass and serve as a mathematical model for the description of other phenomena.

从以上部分 AP 物理 1 的大纲可以看出，改革以后的 AP 物理，突出以下几个特点：

(1) 整个教学内容紧扣六个主导思想（6 Big Ideas）：

主导思想 1：世界上的物体和系统都具有质量和电荷，物体系统具有内部结构。

主导思想 2：空间存在“场”的物质形态，用场可以解释物体的相互作用。

例如，在空间存在引力场、电场和磁场，自然界的引力、电磁相互作用就是借助引力场和电磁场实现的。

主导思想 3：一个物体和其他物体之间的相互作用，就是“力”。

主导思想 4：系统之间的相互作用，可以导致系统的变化。

主导思想 5：系统的变化遵从守恒定律。

例如，系统质量变化遵从质量守恒，电荷变化遵从电荷守恒。

主导思想 6：波可以从一个地方到另一个地方传输能量和动量，而不传输质量。波可以作为描述物理现象的数学模型。

例如，机械波和声波只传输能量，而不传输质量，电磁波只传输能量和动量，不传输电荷。机械波、声波、电磁波都可以用共同的数学模型描述，都存在干涉、衍射等现象。

（2）实验教学大大加强，约占教学总量的 25%，不但要注意实验室内的实践活动，还要和社会实践结合，始终贯彻 7 大实践（7 Science Practices）。具体内容请参见美国大学理事会官网。

下面将逐一讨论 AP 物理 1 和 2 涉及的主要知识点。

Part II

The Key Points of Knowledge

第二篇　知识要点

本书第二部分将复习 18 个关于 AP 物理 1 和 2 的知识要点，有些知识点是 AP 物理 1 和 2 共有的，例如静电场和电势，我们把它归并在物理 1 的范围内。有些内容是具有深刻联系的，例如，波的概念，光的波动性，我们把这部分内容归并在物理 2 的范围内，以免重复。

Chapter 1

Vector and scalar

第一章 矢量和标量

需要掌握的要点是：矢量和标量的区别；用几何法计算矢量的和、矢量差；用矢量分量法计算矢量的和与差。了解矢量的点乘和叉乘。

Main points（主要知识点）

- 矢量和标量
- 矢量的运算：加、减、乘、除
- 矢量在坐标系的分量及其运算
- 矢量的点乘和叉乘

1 The definition of vector and scalar 矢量和标量的定义

1. Scalars（标量）

Scalars are numbers that have a magnitude but no directions, such as speed, temperature etc.

2. vectors（矢量）

Vectors have both magnitude and direction. The direction of vector also is called the orientation of a vector.

If the answer to a free-response question is the vector quantity, be sure to state both the magnitude and the direction.

2 Graphic representation of vectors 矢量的图示

Vectors are drawn as arrows. The length of the arrow corresponds to the magnitude of the vector. The direction in which the arrow points represents the direction of the vector.

The figure（图 2—1—1）shows a few examples：

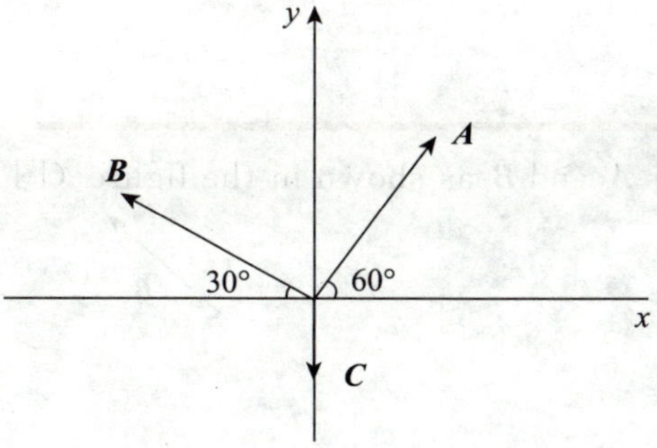

图 2—1—1 the direction of vectors

Vector **A** has a magnitude of 4 meters. Its direction is 60° degrees above the positive x-axis. Vector **B** also has a magnitude of 4 meters and its direction is 150° degree from x-axis in anticlockwise turning. Vector **C** has the magnitude of 2 meters and its direction is in the negative y-axis direction.

3 Vector components and basis vectors 矢量的分量和基矢

Any vector can be broken into its x- and y-components in a xOy coordinate system.

The horizontal vector is always considered as a scalar multiple of what is called horizontal basis vector $\boldsymbol{i}$ and the vertical vector is scalar multiple of the vertical basis vector $\boldsymbol{j}$. Both of those special vectors have a magnitude of 1, and for this reason, they are called unit vectors.

For instance, the vector **A** in the figure（图 2—1—2）is the sum of the horizontal vector $\boldsymbol{B}=3\boldsymbol{i}$ and the vertical vector $\boldsymbol{C}=4\boldsymbol{j}$.

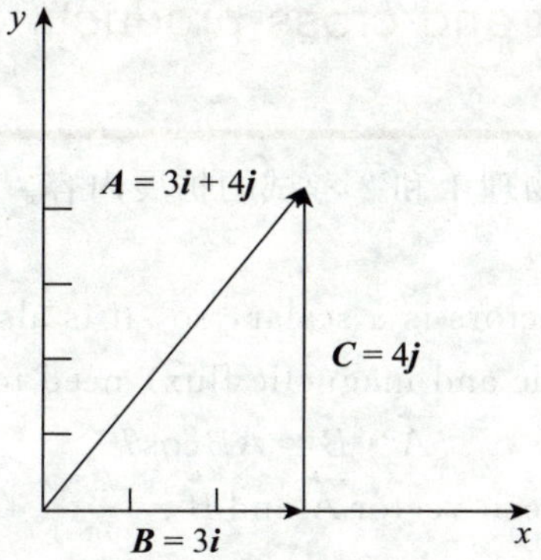

图 2—1—2 the sum of vectors using components

The vector **B** and **C** are called vector components of **A**, and the scalar multiples of $\boldsymbol{i}$ and $\boldsymbol{j}$ which give **A**, 3 and 4 are called scalar components.

4 Adding vectors 矢量和

Let's take two vectors $\boldsymbol{A}$ and $\boldsymbol{B}$ as shown in the figure（图 2—1—3）.

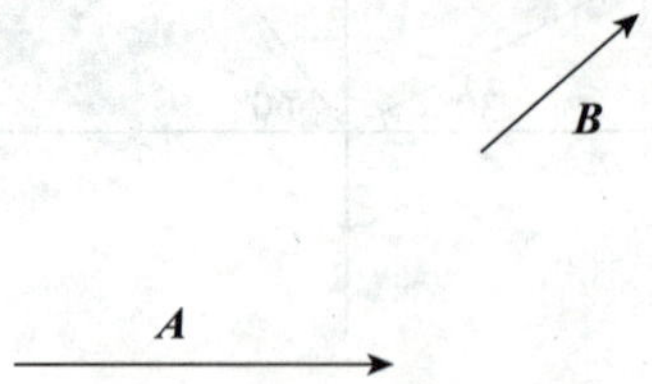

图 2—1—3　two vectors

Now we place them on a coordinate plane. We will move them around so that they line up head-to-tail. The vector sum of $\boldsymbol{A}$ and $\boldsymbol{B}$ is the vector that starts at the origin of vector $\boldsymbol{A}$ and ends at the end of vector $\boldsymbol{B}$. This is shown in the figure（图 2—1—4）.

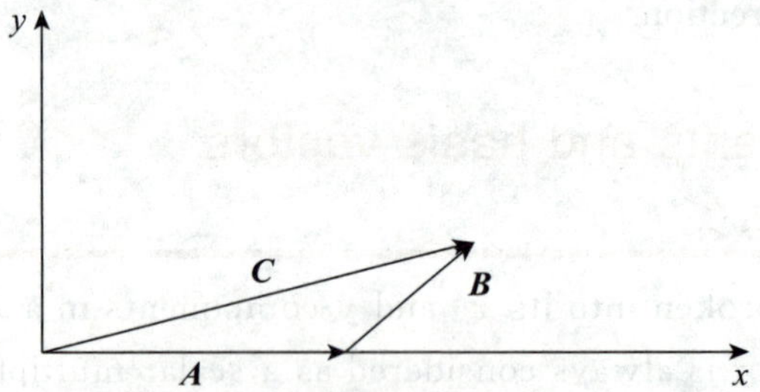

图 2—1—4　the triangle rule

Physicists called the vector sum the "resultant vector". This method of finding the sum of vectors is called triangle principle.

5 Vector's dot product and cross product 矢量的叉乘和点乘

这部分内容是为参加 AP 物理 1 和 2 考试的扩展内容。

1. The dot product（点乘）

The dot product of two vectors is a scalar, so, it is also called scalar product. Several physics concepts（work, electric and magnetic flux）need to use dot product.

$$\vec{A} \cdot \vec{B} = AB\cos\theta$$

Where θ is the angle between vector $\vec{A}$ and $\vec{B}$.

2. The cross product（叉乘）

The cross product of two vectors is a new vector.

Its magnitude is: $|\vec{A} \times \vec{B}| = AB\sin\theta$

The direction of new vector is determined by the right hand rule: Point your index finger in the direction of first vector, $\boldsymbol{A}$, then point your middle finger in the direction of the second vector，$\boldsymbol{B}$，and your thumb points in the direction of the cross product，$\boldsymbol{A}\times\boldsymbol{B}$.

6 Practice problems
练习题

Multiple-Choice（选择题）

Single-Select（单选题）

1. The work can be considered as a dot product of the following two vectors:

(A) force and velocity

(B) force and displacement

(C) force and momentum

(D) torque and acceleration

2. The force $\vec{F}$ exerted on a moving charge $(q\vec{v})$ in a magnetic field $\vec{B}$ is

(A) $F = qvB$

(B) $\vec{F} = \vec{B} \times q\vec{v}$

(C) $\vec{F} = q\vec{v} \times \vec{B}$

(D) $F = \vec{B} \cdot q\vec{v}$

Multi-Select（多选题）

3. In cross product of vectors，$\vec{C} = \vec{A} \times \vec{B}$，vector $\vec{C}$ is

(A) perpendicular to $\vec{A}$ only

(B) perpendicular to $\vec{B}$ only

(C) perpendicular to both $\vec{A}$ and $\vec{B}$

(D) $\vec{A}$ and $\vec{B}$ can be perpendicular to each other or not

Free-Response（计算题）

4. If $\vec{A} = -2\boldsymbol{i} + 4\boldsymbol{j}$ and $\vec{B} = 6\boldsymbol{i} + B_y\boldsymbol{j}$，find the value of B_y such that vector $\vec{A}$ and $\vec{B}$ will be perpendicular to each other.

题解

选择题

单选题

1. 选（B）。功定义为力和沿力方向的位移的乘积，即力矢量和位移矢量的点积。

2. 选（C）。在磁场中，作用于运动电荷的力为：$\vec{F}=q\vec{v}\times\vec{B}$。这里叉积的结果是矢量，不能选（D）。叉积不满足交换律，不能选（B）。只能选（C）。

多选题

3. 选（C）和（D）。两个矢量叉乘产生的新矢量一定和两个叉乘的矢量垂直，但叉乘的两个矢量不一定互相垂直。

计算题

4. 如果 A，B 两个矢量互相垂直，则 $B_y=-3$。

Chapter 2

Kinematics

第二章 运动学

运动学需要掌握的主要内容有：描述物体运动的物理量，位移、速度、加速度；匀变速运动的五个基本公式、运动的合成和分解；抛体运动。

Main points（主要知识点）

- 运动学的基本概念：包括位移、速度、加速度等
- 匀变速运动的 5 个基本公式
- 位移是矢量，路程是标量
- 速度（velocity）是矢量，速率（speed）是标量
- 加速度是速度的变化率
- 抛体运动是水平方向的匀速直线运动和竖直方向的匀变速直线运动的合成。两个分运动相互独立

1 Displacement, velocity and acceleration 位移、速度和加速度

1. Displacement（位移）

Displacement is the vector equivalent of distance. It has both magnitude and direction.

2. Velocity and speed（速度和速率）

Average speed（平均速度）is the total displacement you travel in a given time divided by the time it takes you to travel that distance. Expresses as $\bar{v}=\dfrac{\Delta x}{\Delta t}$, the average speed is a scalar but average velocity is a vector.

Instantaneous speed（瞬时速度）is the speed at any moment: $v=\lim\limits_{\Delta t\to 0}\dfrac{\Delta x}{\Delta t}=\dfrac{dx}{dt}$, the instantaneous velocity is the vector which direct the direction of the moving.

The unit of velocity is m/s（米/秒）.

3. Acceleration（加速度）

Average acceleration（平均加速度）is the ratio of the change of the velocity and the time interval.

$$\bar{a} = \frac{\Delta v}{\Delta t}$$

Because the velocity is a vector，usually the instantaneous acceleration（瞬时加速度）can be expressed as：

$$\vec{a} = \lim_{\Delta t \to 0} \frac{\Delta \vec{v}}{\Delta t} = \frac{d\vec{v}}{dt}$$

The unit of acceleration is m/s^2（米/秒²）.

2 Big five Equations in AP Physics 1 AP 物理 1 中匀变速运动的 5 个公式

If the acceleration is constant，the motion will obey the Big Five Equations：

Big Five Equations	missing variable
$\Delta x = \bar{v}t$	a
$v_f = v_0 + at$	x
$\Delta x = x_0 + v_0 t + \frac{1}{2}at^2$	v
$\Delta x = x_0 + vt - \frac{1}{2}at^2$	v_0
$v_f^2 = v_0^2 + 2a\Delta x$	t

Where v_0 is initial velocity

v_f is final velocity

x is displacement

a is acceleration

t is time interval

3 Problem-solving steps in kinematics 运动学解题步骤

第一步：弄清楚哪些变量已知，哪些未知。

第二步：如果已知量为 3 个以上，进行第三步，否则再想其他方法。

第三步：选择匀变速运动的 5 个方程，代入已知量，解出未知量。

第四步：标出正确的单位。一般情况用国际单位（SI）表示。

Example：

A car begins at rest and accelerates at a constant rate up to a velocity of 100 m/s. If it takes 5 seconds for the car to accelerate from rest to 50 m/s，how long does it take for the car to reach 100 m/s and how far does it travel in total?

Solution:

First let us analyze this problem. We can use the Big Five Equations to solve this problem because the problem states the acceleration is constant. It asks us to find two values: time and distance. Based on the information in the problem, we know that the time needed for the car to reach 100 m/s should be greater than 5 seconds because it takes 5 seconds for the car just to reach 50 m/s. Further, we can estimate that the car will travel several hundred meters in total.

Step 1: What we know, the car begins at rest, so $v_0 = 0$. The final velocity of the car is 100 m/s. What we need to find out? The time and the distance. The problem does not provide the acceleration. How to find it ?

Step 2: To find the acceleration. Based on the information the problem provided, we know that it takes 5 seconds for the car to accelerate from rest to the velocity of 50 m/s. So, according to the definition of acceleration, we can find out the acceleration as follows:

$$a = \frac{\Delta v}{\Delta t} = \frac{50\text{m/s}}{5\text{s}} = 10\text{m/s}^2$$

Step 3: Using the Big Five Equations to solve the time and distance:

Based on $v_f = v_0 + at$

$$100\text{m/s} = 0 + (10\text{m/s}^2)\,t$$

Solve out: $t = 10\text{s}$

Based on $\Delta x = v_0 t + \frac{1}{2}at^2$

Put the numbers in: $\Delta x = \frac{1}{2}(10\text{m/s}^2)(10\text{s})^2 = 500\text{m}$

Step 4: Unit. Pay attention to the unit. We should use SI unit system to describe any quantity in the problem.

4 Freefall 自由落体

In solving freefall problem, we should pay attention to two problems:

First, the acceleration is g which is the acceleration caused by gravity near the Earth's surface, and the value is about 10 m/s^2.

Second, choose the positive direction and solve the problem using displacement. For instance, if you choose the "up" as the positive, the acceleration g should be -10 m/s^2.

Example:

You are standing on a cliff, 30 m above the valley. You throw a watermelon vertically upward at a velocity of 3 m/s. How long does it take until the watermelon hits the valley floor?

Solution:

Step 1: Make sure the variables: For the motion of the watermelon, we know the initial velocity v_0, displacement -30m, and the acceleration $-10\ \text{m/s}^2$.

Step 2: Set "up" as the positive direction.

Step 3:

方法 1：用位移求解。

Using the displacement to solve it based on Big Five Equation:

$$x - x_0 = v_0 t + \frac{1}{2}at^2$$

Put the numeric information into this equation:

$$-30\text{m} = (3\text{m/s})\ t + \frac{1}{2}\ (-10\text{m/s}^2)\ t^2$$

This is a quadratic equation. With the help of the formula: $x = \dfrac{-b \pm \sqrt{b^2 - 4ac}}{2a}$

We can get the solution: $t = 2.8\text{s}$

注意：运用位移解运动学问题时，一定要确定位移的正方向。

方法 2：我们也可以避免直接用位移公式，先用 $v_f^2 = v_0^2 + 2a(x - x_0)$ 求出西瓜落到地面的末速度，然后代入 $v_f = v_0 + at$，求出时间。

方法 3：考虑西瓜经过三段路程：从起始点到最高点，再从最高点落回到起始点，然后再自由落体。这样分析，运用考虑路程，可以避免位移的正负规定，但比较繁琐。

5 Projectile motion
抛体运动

The projectile motion is a two dimensional motion; the motion in one dimension does not affect its motion in any other dimension.

No matter what the problem looks like, the rules to solve it are as follows:

- The vertical component of velocity: $v_y = v\sin\theta$
- The horizontal component of velocity: $v_x = v\cos\theta$, where θ is the angle between the direction of velocity v and the horizontal.
- The horizontal velocity is constant vector.
- The vertical acceleration is g; it directs downward.

6 Motion graphs
运动图线

用图线描述运动是运动学的重要内容之一。特别是对中国学生，作图的练习相对国外学生少得多，这是一个弱点，要引起特别注意。

通常有位移对时间的图线，代表位移和时间的函数关系，位移图线的变化率为速度。速度图线代表速度和时间的函数关系，速度的变化率为加速度，图线下面的面积为位移。

还要特别注意学会从图中读取数据，或用已知数据绘图。

Example:

Object A is released from rest at height h. At the same instant, object B is thrown downward from the same location. Which of the following graphs of speed v as a function of time t is correct for the two objects?

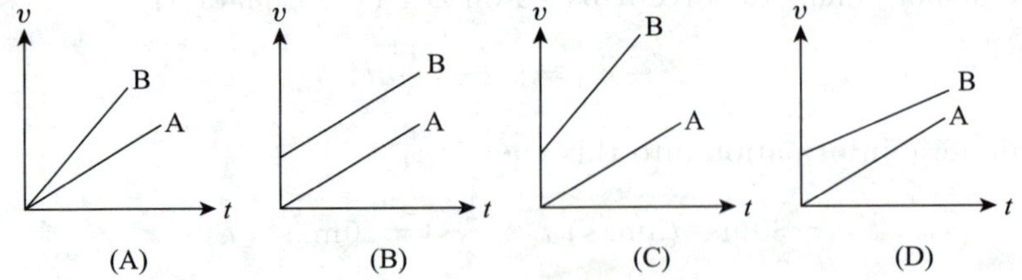

motional graphs

Solution:

The correct answer is B, because these are velocity-time graph. Object A is freefall motion, the initial velocity is zero, but object B is thrown downward with the initial velocity. Both have the same acceleration g. Therefore, the two lines should be parallel and A from origin, B starts with an initial velocity.

Note: Please pay attention to the units. For instance, when you calculate slope, you are dividing the unit of vertical axis variable by the unit of the horizontal axis variable. If you take the slope of a distance-time graph, the slope should have unit of "meters divided by second" or m/s. Similarly, The area under velocity-time graph will give you an answer with unit of (m/s) · (s) which equal to m. So, the area under velocity-time graph will give you a distance.

7 Practice problems
练习题

Multiple-Choice（选择题）

Single-Select（单选题）

1. The graph（图 2—2—1）shows the velocity versus time for an object moving in a straight line. At what time after $t=0$ does the object again pass through its initial position?

(A) Between 0 and 1s (B) 1s

(C) Between 1 and 2s (D) 2s

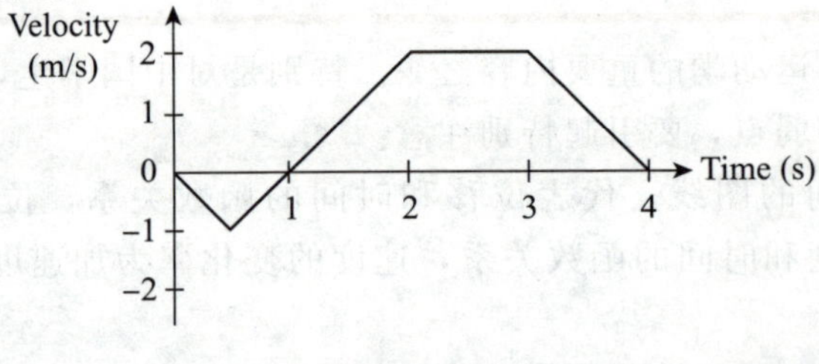

图 2—2—1

2. The displacement x of an object moving along the x axis is shown in the figure（图 2—2—2）as a function of time t. The acceleration of this object must be

(A) zero　　(B) constant but not zero

(C) increasing　　(D) decreasing

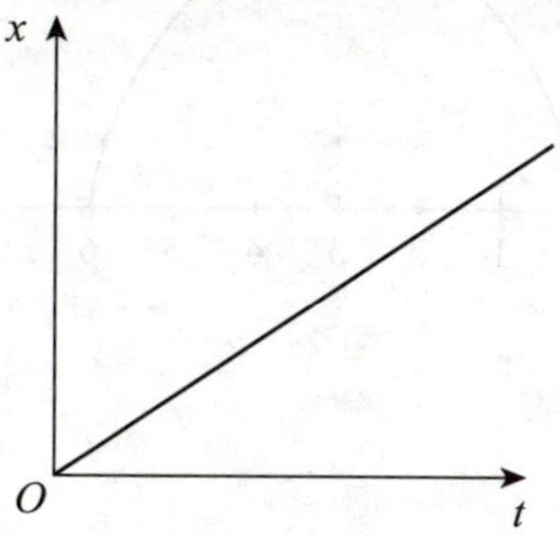

图 2—2—2

3. A 2-kilogram block rests at the edge of a platform that is 10 meters above level ground. The block is launched horizontally from the edge of the platform with an initial speed of 3 meters per second. Air resistance is negligible. The time it will take for the block to reach the ground is most nearly

(A) 0.3s　　(B) 1.0s

(C) 1.4s　　(D) 2.0s

4. A diver initially moving horizontally with speed v dives off the edge of a vertical cliff and lands in the water a distance d from the base of the cliff. How far from the base of the cliff would the diver have landed if the diver initially had been moving horizontally with speed $2v$?

(A) d　　(B) $\sqrt{2}d$

(C) $2d$　　(D) $4d$

5. In which of the following situations would an object be accelerated?

Ⅰ. It moves in a straight line at constant speed.

Ⅱ. It moves with uniform circular motion.

Ⅲ. It travels as a projectile in a gravitational field with negligible air resistance

(A) Ⅰ only　　(B) Ⅲ only

(C) Ⅰ and Ⅱ only　　(D) Ⅱ and Ⅲ only

Multi-Select（多选题）

6. A ball is thrown straight up by a student at rest on the surface of the Earth. A graph of the position y as a function of time t, in seconds, is shown in the figure（图 2—2—3）. Air resistance is negligible. Two of the following statements describe the acceleration of the ball correctly are

(A) It is downward and constant from 0 to 6 s

(B) It is downward and increases in magnitude from 0 to 3 s, and then decreases

(C) It is downward and decreases in magnitude from 0 to 3 s, and then increases

(D) It is downward and kept a constant from 0 to 3 s, and then from 3 to 6 s

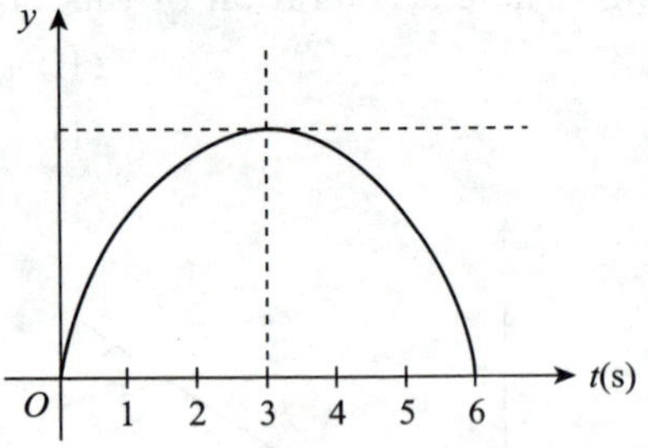

图 2—2—3

Free-Response (计算题)

7. This question concerns the motion of a car on a straight track. The car's velocity as a function of time is plotted in the figure (图 2—2—4).

(a) Describe what happened to the car at time $t=0$ and $t=1$s.

(b) How does the car's average velocity between time $t=0$ and $t=1$s compare to its average velocity between time $t=1$s and $t=5$s?

(c) What is the displacement of the car from time $t=0$ to time $t=7$s?

(d) Plot the car's acceleration during the time interval as a function of time.

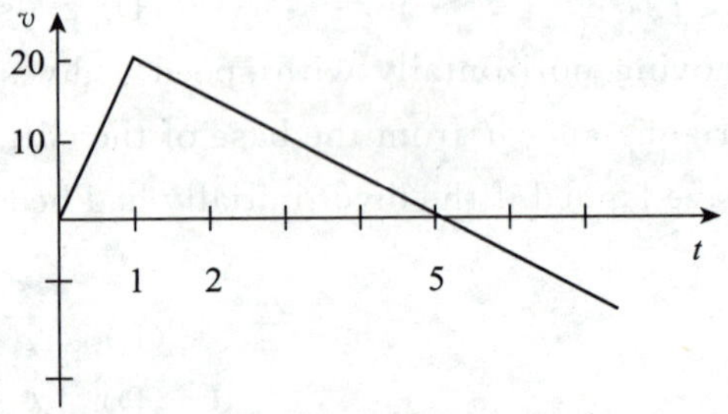

图 2—2—4

题解

选择题

单选题

1. 选 (C)。速度—时间曲线下的面积为位移，该物体第一秒速度先减小再增大，位移为负，从第一秒末到第二秒末，位移为正，其中有一点位移为 0。

2. 选 (A)。位移曲线的斜率为速度，等于常数，因此加速度为零。

3. 选 (C)。物体在竖直方向做自由落体运动，利用公式：$s=0.5gt^2$，可以算出 $t=1.4$s.

4. 选 (C)。平抛运动物体的水平位移取决于水平速度，水平速度增大一倍 ($2v$)，水平位移也增大一倍，即 $2d$。

5. 选 (D)。情况Ⅱ：物体做匀速率圆周运动，必定有向心加速度；情况Ⅲ：物体做抛体运动，竖直方向为自由落体运动，有加速度。

多选题

6. 选（A）和（D）。物体做竖直上抛和自由落体运动的加速度都是方向向下，大小保持常数。

计算题

7.（a）从 $t=0$ 时开始加速运动，在 $t=1s$ 处，由加速变减速。

（b）汽车在 $t=0$ 到 $t=1s$ 的时间间隔里，平均速度为 10m/s；在 $t=1s$ 到 $t=5s$ 时间间隔里，平均速度也为 10 m/s。

（c）从 $t=0$ 到 $t=5s$ 的时间间隔内汽车位移为速度曲线和 t 轴之间的面积，即三角形面积 $=\frac{1}{2}$（5s）（20 m/s）=50m，从 $t=5$ 到 $t=7s$，为 $-10m$，因此，总位移为 40m。

（d）从 $t=0$ 到 $t=1s$ 之间，为平行 t 轴的直线，在加速度轴上为 $+20m/s^2$，从 $t=1s$ 到 $t=7s$，也为平行 t 轴的直线，在加速度轴上为 $-5m/s^2$。如图 2—2—5。

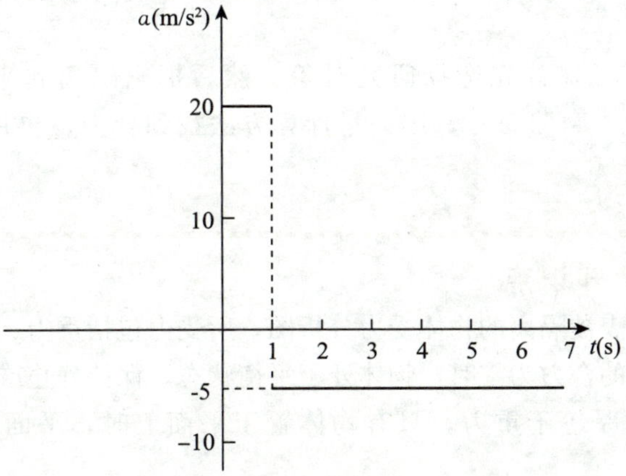

图 2—2—5

Chapter 3

Free-body analysis

第三章　隔离体方法受力分析

隔离体方法的要点是，首先选择研究对象，然后把它隔离起来，考虑其所受的各种力，例如重力、支撑力、摩擦力等，用矢量计算方法找到合力，再用牛顿第二定律列出方程，计算出未知量。

Main points（主要知识点）

- 隔离体图代表被隔离的物体受力分析图，所受力包括重力、支撑力、摩擦力等
- 当物体受到的合力为零时，物体处于平衡状态，保持静止或匀速直线运动状态
- 支持力不总是等于重力，只有物体置于平面上时，平面对物体的支持力才等于重力
- 绳或弦作用在物体上的力为张力，张力在绳、弦上处处相等
- 物体和表面接触时两个表面间产生摩擦，摩擦力分为静摩擦力和动摩擦力
- 物体在斜面上，一般选取沿斜面和垂直于斜面的方向为坐标轴，计算较方便
- 当一个力矩作用于物体时，会使物体发生转动

1 What is a free-body diagrams? How to draw this diagram? 什么是隔离体图？怎样画隔离体图？

（1） A free-body diagram is a picture that represents one or more objects, along with the forces acting on those objects.

（2） When you draw a free-body diagrams, first you have to choose one body as the object which you will consider as the object to be analyzed, and then isolated it from other bodies. Draw all forces exerted on that body as vectors and add these vectors to get the resultant force.

注意：我们把“free-body”翻译成隔离体，原因是，当你在分析几个物体组成的体系

时，为了使问题简化，隔离某个物体作为研究对象，对其进行受力分析。

Example:

In the figure (图 2—3—1), two objects m_1 and m_2 hang across a fixed pully by an unchangeable string. Find out the tension T on the string and the acceleration of any object.

Solution:

Let choose one object m_1 as an isolated object, and analyze all forces acting on it as shown in the figure (b). Choose the direction of weight m_1g as the positive direction, according to Newton's second law, we have: $m_1g - T = m_1a$. Then we choose another object m_2 to draw a free-body diagram too, we have another equation: $T - m_2g = m_2a$. Solve them,

We have : $a = \dfrac{m_1 - m_2}{m_1 + m_2}g$, and $T = \dfrac{2m_1m_2}{m_1 + m_2}g$

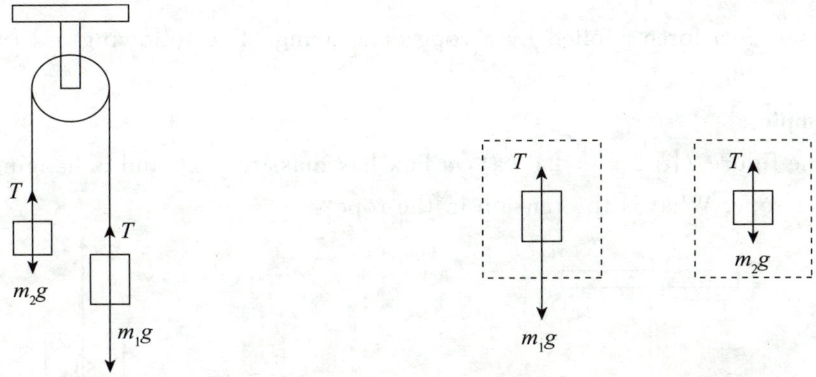

(a)　　(b) free-body diagram

图 2—3—1

2 About equilibrium
平衡

● The condition of equilibrium (平衡条件):

When the net force on an object equals zero, that object is in equilibrium. At equilibrium state, an object is either at rest or moving with a constant velocity, but it is not accelerating. An object in equilibrium is obeying Newton's first law.

● How to solve the problems when an object is in equilibrium (解决平衡问题的方法):

Follow the steps below, it is an easy way to solve equilibrium problems:

1. Draw a free-body diagram.

2. Establish a proper coordinate system and break the force vectors into x-and y-components.

3. Write an expression for the vector sum of forces in x direction. Then write an expression for the vector sum of forces in y direction. Set each of these expression equal to zero.

4. Solve the resulting algebraic equations.

3 The various forces
各种力的分析

Usually we should consider gravity, normal force, tension, friction when you deal with the equilibrium questions. The followings are some examples:

1. Normal force (支持力)

A box is sitting on the table. The force of gravity pulls downwards labelled as weight (or mg). Because the box is at rest, there must be some forces exerted on the box oppose the downwards pull of gravity. This force is called normal force, and it is usually abbreviated F_N.

注意：支持力不总是等于重力，只有物体置于平面上时，平面对物体的支持力才等于重力。

2. Tension (张力)

Tension is a force applied by a rope or a string. The following are two examples on tension.

Example:

In the figure (图 2—3—2) (a), a box has mass of 5 kg and is hanging from the ceiling with a rope. What is the tension in the rope?

图 2—3—2

Solution:

We can solve this problem by 4 steps:

Step 1: Draw a free-body diagram as shown in the figure (b).

Step 2: Set up coordinate: Choose the up direction as positive y.

Step 3: Set the equations: $F_{net,y} = T - mg = 0$.

Step 4: solve the equation: $T = 50\text{N}$.

Example:

A box has mass of 5kg and is hanging with two ropes. One makes a 45° angle with the ceiling, the other makes a 30° angle with the ceiling shown in the figure (图 2—3—3 (a)). What is the tension in each rope?

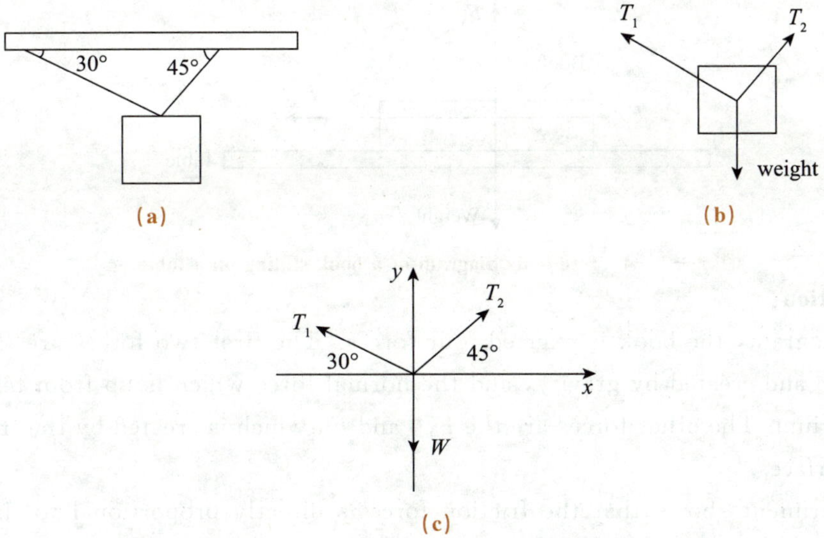

图 2—3—3

Solution:

Step 1: Draw a free-body diagram as shown in the figure (b).

Step 2: Set up coordinate: Choose the horizontal direction as x and the perpendicular direction as y.

Step 3: Set the equations:

For x direction: $F_{net,\ x} = T_{2x} - T_{1x} = 0$

$\because T_{1x} = T_1 \cos 30°$ and $T_{2x} = T_2 \cos 45°$

$\therefore T_2 \cos 45° - T_1 \cos 30° = 0$

For y direction: $T_{net,\ y} = (T_{1y} + T_{2y}) - mg = 0$

$\because T_{1y} = T_1 \sin 30°$ and $T_{2y} = T_2 \sin 45°$

$\therefore T_1 \sin 30° + T_2 \sin 45° - 50\text{N} = 0$

Step 4: Solve the equations, we have: $T_1 = 37\text{N}$, and $T_2 = 45\text{N}$.

3. Friction(摩擦力)

摩擦力产生的条件:

Friction is only found when there is contact between two surfaces.

Definition of friction(摩擦力的定义)

Friction is a force which acts parallel to two surfaces in contact. Its direction is always opposite the direction of the motion.

Example:

If you slide a book at a constant speed across a table(图 2—3—4)and the table is not frictionless, the table will exert a friction force on the book opposite its direction of motion. What is the friction exerted on the book? (Using free-body analysis)

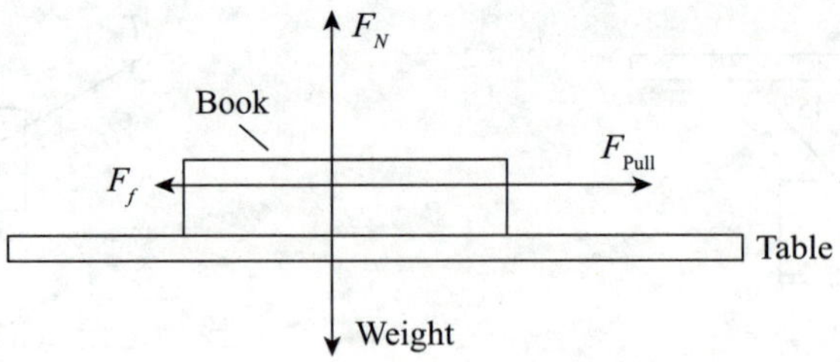

图 2—3—4　free-body diagram of a book sliding on a table

Solution:

In this diagram, the book is exerted four forces. The first two forces are the weight which is down and created by gravity, and the normal force which is up from table; they are in equilibrium. The other forces are the F_{pull} and F_f which is created by the friction between two surfaces.

The experiment shows that the friction force is directly proportional to the normal force. The constant is called coefficient of friction, usually denoted by μ.

$$F_f = \mu F_N$$

It is a dimensionless number that describes how big the force of friction is between two objects. It is found experimentally that the coefficient of friction differs for every combination of materials.

Pay attention to the cases in which the normal force is not equal to the weight because some other forces act on the object.

Static and kinetic friction（静摩擦力和滑动摩擦力）

There are two frictions which describe the two situations in the motion: static and kinetic friction. If the objects are stationary, we use the coefficient of static friction. If the objects are moving, we use the coefficient of kinetic friction. The equation for the force of friction is essentially the same in either case: $F_f = \mu F_N$. However, we have to pay attention to that the coefficient of static friction is the maximum value. It is equal to the active force in magnitude if the active force less than that maximum value.

注意：公式计算的静摩擦力是指最大静摩擦力，当小于最大静摩擦力时，静摩擦力等于主动力。

Think about this for a situation: If a book just sits on a table, it does not need any friction to stay in place. But if you apply very small horizontal force on it, the static friction can act on the book. Only the direction of it is opposite to your force. To find the maximum coefficient of static friction, find out how much horizontal pushing force will just cause the book to move using $F_f = \mu F_N$.

4 The application of free-body diagram 隔离体力图的应用

Inclined Planes（斜面）

If there is the friction between the block m and the inclined plane（图 2—3—5）, we can draw a free-body diagram as follows (图 2—3—6). Break the weight vector into components parallel and perpendicular to the plane.

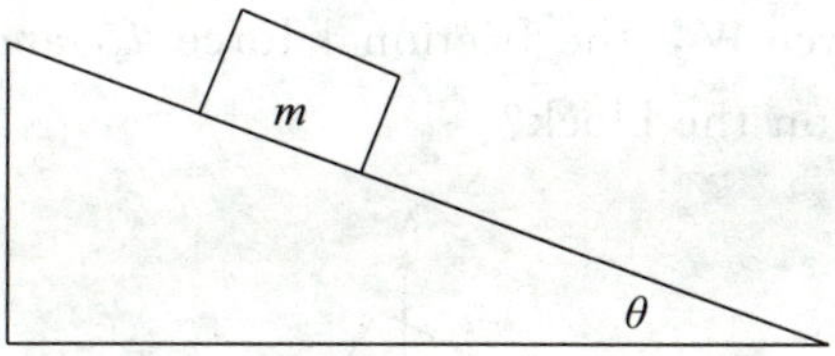

图 2—3—5 Inclined-plane situation

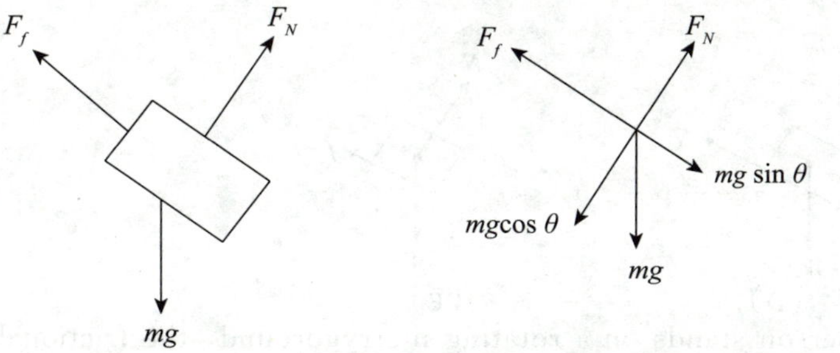

图 2—3—6 free-body diagram of a body on inclined plane

5 Torque
力矩

The torque exerted on an object equals the force F on that object multiplied by the distance between where the force is applied and the fulcrum (d) as long as the force acts perpendicular to the object. That is $\tau = Fd$, where the fulcrum is the point about which an object rotates.

The unit of torque is the newton · meter (牛顿 · 米).

6 Practice problems
练习题

Multiple-Choice (选择题)

Single-Select (单选题)

1. A ball of mass m is suspended from two strings as shown in the figure (图 2—3—7). The tensions T_1 and T_2 in the strings must satisfy which of the following relations?

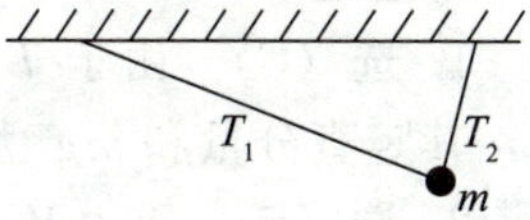

图 2—3—7

(A) $T_1 = T_2$

(B) $T_1 > T_2$

(C) $T_1 < T_2$

(D) $T_1 + T_2 = mg$

2. A 2 kilogram block slides down a 30° incline as shown in the figure 2—3—8. Which of the following diagrams best represents the gravitational force W, the frictional force f, and the normal force N that act on the block?

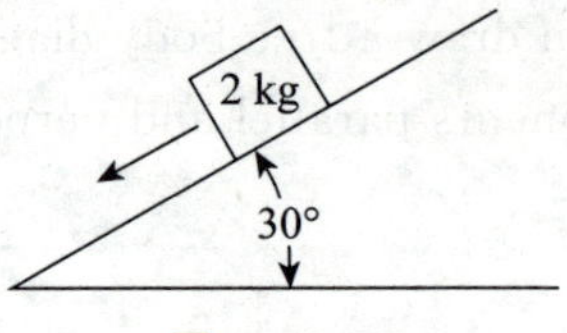

图 2—3—8

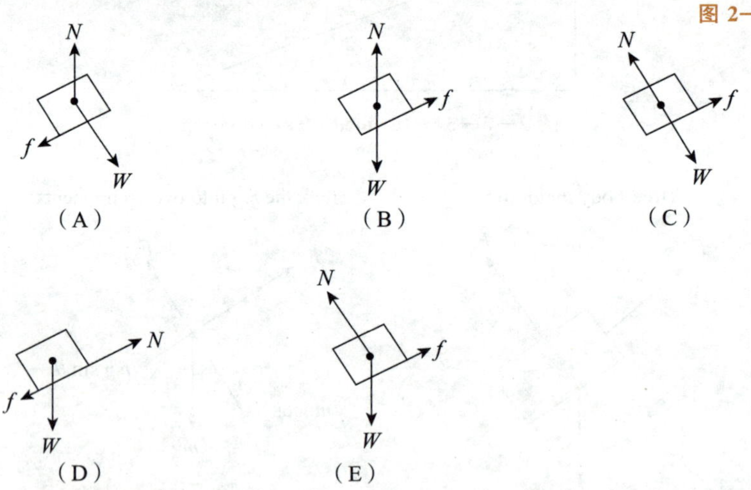

3. When a person stands on a rotating merrygoround, the frictional force exerted on the person by the merrygoround is

(A) greater in magnitude than the frictional force exerted on the merrygoround by the person

(B) opposite in direction to the frictional force exerted on the merrygoround by the person

(C) directed away from the center of the merrygoround

(D) zero if the rate of rotation is constant

Multi-Select (多选题)

4. Free-body diagram could help students

(A) to analyze the net force exerted on one object

(B) to break the vectors to components in a coordinate system

(C) to change the internal force to external force when study a multi-object system

(D) to calculate the velocity of the object

题解

选择题

单选题

1. 选（C)。由于 T_1 和 T_2 在竖直方向分量的和与重力 mg 平衡，T_2 和竖直方向夹角小，其竖直分量和 mg 大小接近，因此，$T_1 < T_2$。

2. 选（E)。重力 W 应垂直向下，支持力 N 应垂直斜面向上，摩擦力和运动方向相反。

3. 选（B)。转盘作用在人上的摩擦力和人作用在转盘上的摩擦力大小相等，方向相反。

多选题

4. 选（A）和（C)。Free-body diagram 可以帮助学生分析物体受力情况，找到合力，在研究几个物体形成的体系（连体运动）时，可以化内力为外力。

Chapter 4

Newton's laws of motion

第四章 牛顿运动定律

该知识点介绍牛顿运动定律，重点讨论牛顿第二定律。应用牛顿第二定律时要特别注意用合力，不要用分力；用隔离体法分析，注意考虑各种不同的力。

Main points（主要知识点）

- 运用牛顿第二定律解题时应注意，力是合力，在 AP 物理中用 F_{net} 表示
- 注意用隔离体方法解题时的方法和要点
- 两个物体直接接触，注意相互之间的压力（Normal force）和接触面垂直；当两个物体用绳子连接时，绳中的张力（Tension）处处相等
- 运用牛顿第三定律时，要特别注意，虽然作用力和反作用力大小相等，方向相反，但分别作用在不同物体上，它们不能抵消

Relative equations（相关公式）

Newton's second law: $F = ma$

Friction force: $F_{\mu} = \mu N$

1 Newton's first law and inertia 牛顿第一定律和惯性

Newton's first law of motion: If the net force acting on an object is zero, that object maintains its state of rest or constant velocity.

Newton's first law also puts forward the concept of inertia: The ability of an object to resist changes to its motion is a fundamental property of all matter called inertia（惯性）. Inertia tends to keep a stationary object at rest or a moving object in motion in a straight line at a constant speed. Thus, the first law of motion is often called the law of inertia（惯性定律）.

2 Newton's second law 牛顿第二定律

Newton's second law of motion: If the external net force on an object is not zero, the object accelerates in the direction of the net force. The acceleration is directly proportional to the net force and inversely proportional to the object's mass.

$$\sum \vec{F} = m\vec{a}$$

运用牛顿运动定律的解题步骤：

第一步：选择研究对象。例如，连体问题要隔离选择一个物体为研究对象，不是整个体系。

第二步：对隔离的物体进行受力分析，画出隔离体受力图。AP 物理中称为：free-body diagram。

第三步：根据隔离体受力图，建立合适的坐标系。例如，斜面问题一般沿斜面向下为 x 的正向，垂直斜面向上为 y 的正向。

第四步：将隔离体的受力，分解到已建立的坐标系中，沿各个坐标方向分解力，用牛顿第二定律列出分量方程。

第五步：解分量方程。注意，AP 物理 1 和 2 中，用到的是代数方程，AP 物理 C 中用到的是微分方程。

注意：在参加考试时，一定要按以上步骤解题，不要只写计算结果。因为一旦计算结果出错，如果解题步骤没错，思路明确，你仍然可以得分。如果没有步骤，就不会得分！

3 The application of Newton's laws 牛顿定律的应用

1. Inclined plane（斜面问题）

Example:

A block of mass m (2kg) is placed on a plane inclined at an angle θ (30°)（图 2—4—1）. The coefficient of friction between the block and the plane is μ. What is the acceleration of the block down the plane?

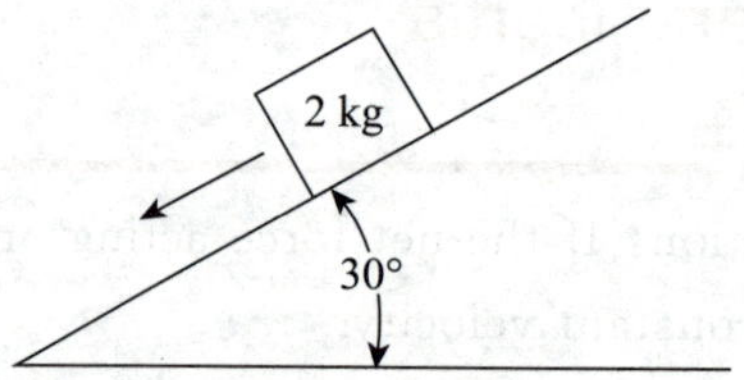

图 2—4—1

Note: In this question, only variables are given. It just means that our answer should be in variables. Only the given variables—in this case m, θ and μ and constants such as g can be used in the solution. And you should not plug in any numbers (such as 10 m/s^2 for g) even if you know them.

Solution：

斜面问题的解题步骤：

第一步：选择斜面上的物体为研究对象。

第二步：对斜面上的物体进行受力分析，画出隔离体受力图（free-body diagram）。

该物体受地心引力 mg，斜面的支撑力 F_N（不是物体对斜面的正压力，该力作用在斜面上！）和斜面作用在该物体上的摩擦力 F_f（图 2—4—2）。

第三步：建立坐标：沿斜面向下为 x 轴正方向，垂直斜面向上为 y 轴正方向。

第四步：应用牛顿第二定律，在每个坐标方向建立分量方程。

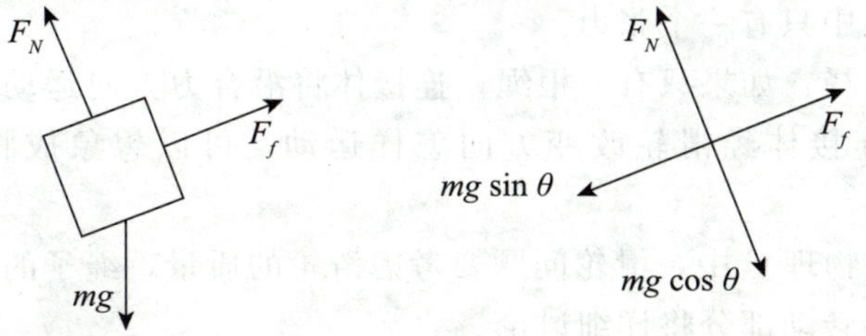

图 2—4—2 free-body diagram in the inclined plane

注意：在垂直斜面方向，物体处于平衡状态，因此，合力为零：

用 $F_{net,\ perpendicular}$ 表示沿垂直于斜面方向的合力：

$$F_{net,\ perpendicular}=F_N-mg\ \cos\theta=0$$

$$mg\ \cos\theta=F_N \quad ①$$

沿斜面向下方向的合力为：$F_{net,\ down\text{-}the\text{-}plane}$

$$F_{net,\ down\text{-}the\text{-}plane}=mg\ \sin\theta-F_f=ma_{down\text{-}the\text{-}plane} \quad ②$$

第五步：解方程。

考虑到摩擦力应为正压力乘以摩擦系数：$F_f=\mu F_N=\mu mg\ \cos\theta$

代入方程②

$$ma_{down\text{-}the\text{-}plane}=mg(\sin\theta-\mu\cos\theta)$$

得到物体下滑的加速度： $a_{down\text{-}the\text{-}plane}=g(\sin\theta-\mu\cos\theta)$

我们可以看到，下滑加速度都是用已知量，即重力加速度 g，摩擦系数 μ 和斜面倾角 θ 表示的。另外，$\cos\theta$，$\sin\theta$ 都是无量纲的量，因此是合理的。再进一步分析，当 θ 为 90° 时，物体为自由落体，加速度变为 g 。所以，此解是正确的。

For the particular incline, what coefficient of friction would cause the block to slide with constant speed?

Constant speed means $a=0$. The solution for F_N in the perpendicular direction is the same as before ：$F_N=mg\ \cos\ \theta$. But in the down-the-plane direction, no acceleration means that $F_f=mg\ \sin\theta$, Because $\mu=\dfrac{F_f}{F_N}=\dfrac{mg\sin\theta}{mg\cos\theta}$，

So，$\mu=\tan\theta$

You might note that neither this answer nor the previous one includes the mass of the block, so on the same plane, both heavy and light masses move in the same way.

2. Pulley problem（滑轮问题）

在讨论滑轮问题之前，我们先要了解以下几点：

(1) 滑轮通过连接两个或两个以上物体的运动，可以改变物体的运动方向和受力的方向。因此，滑轮可以和其他机械，如斜面、轮轴等结合。

(2) 解决滑轮问题，首先要选择其中一个物体为研究对象，用隔离体法（free-body）进行受力分析，这样就将绳子的张力从系统内力转化为外力。每个研究对象可以根据牛顿第二定律列出一个方程，产生方程组，求得未知量。

(3) 在 AP 物理 1 和 2 的滑轮问题中，一般不考虑绳子的质量，认为绳子的张力在绳中处处相等，每根绳中只有一个张力。

(4) 坐标系的选择：如果只有一根绳，连接体将沿合力方向运动，因此选连接体运动方向为正向。不管连接体经滑轮改变方向怎样运动，可以想象成将连接体拉直的一维运动。

(5) P 物理 1 和物理 C 中，滑轮问题要考虑滑轮的质量，绳子的张力形成合力矩，引起滑轮转动，我们在转动部分将详细讨论。

Example:

Three blocks of masses 1.0, 2.0, and 4.0 kilograms are connected by massless strings, one of which passes over a frictionless pulley of negligible mass, as shown in the figure（图 2—4—3）. Calculate each of the following.

a. The acceleration of the 4 kilogram block.

b. The tension in the string supporting the 4 kilogram block.

c. The tension in the string connected to the 1 kilogram block.

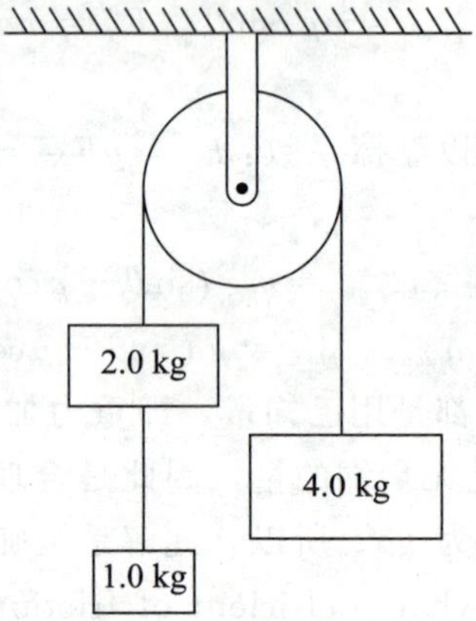

图 2—4—3　pully problem

Solution:

第一步：分别选择连接体 M (4.0 kg) 和 m_1 (2.0 kg)，m_2 (1.0 kg) 为研究对象。

第二步：对每一个研究对象画出受力的隔离体图（Free-body diagram），如图 2—4—4 所示：

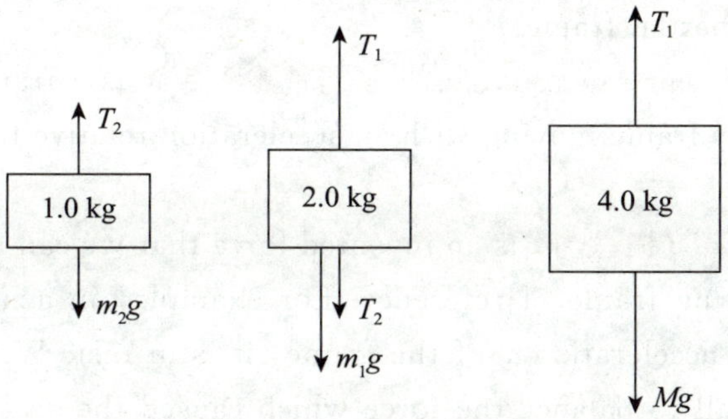

图 2—4—4 the free-body diagram in pully problem

第三步：因为 M 大于 m_1+m_2，规定连接体 M 的向下运动为正向。

第四步：根据牛顿第二定律对 M 和 m_1，m_2 分别列出方程：

Block M：$F_{net}=Mg-T_1=Ma$

Block m_1：$F_{net}=T_1-T_2-m_1g=m_1a$

Block m_2：$F_{net}=T_2-m_2g=m_2a$

注意到绳子是不可伸缩的，即没有弹性。因此，连接体加速度相同，绳子的张力相同。

第五步：解方程组。这是一个三元一次代数方程组，可以很容易解出张力 T_1 和 T_2 以及加速度 a.

消去张力，解出加速度：

$$Mg-m_1g-m_2g=(M+m_1+m_2)a$$

$$a=\frac{(M-m_2-m_1)g}{(M+m_1+m_2)}$$

代回方程，解出张力：$T_1=M(g-a)$

$$T_2=m_2(g+a)$$

如果需要，可以代入数据，得到最后结果。

4 Newton's third law 牛顿第三定律

Newton's third law of motion: For every action force, there is a simultaneous reaction force equal in magnitude but opposite in direction.

在应用牛顿第三定律时要特别注意，牛顿第三定律中的作用力和反作用力是作用在不同物体上，因此不能抵消！

5 Inertia and noninertia frame 惯性参照系和非惯性参照系

(1) An inertial frame of reference（惯性参照系）is a frame in which the law of inertia and other physics laws are valid. Any frame moving at a constant velocity relative to the

first frame is also an inertial frame.

(2) A noninertial frame of reference（非惯性参照系）is a frame in which the law of inertia is not valid. Any frame moving with an acceleration relative to an inertial frame is an noninertial frame.

(3) Fictitious force（惯性力）is an invented force that we can use to explain observed motion in an accelerating frame of reference. For example, in a noninertial frame a ball moving with the same acceleration a of the frame, it is in rest. We can invent a fictitious force exerted on the ball to balance the force which caused the acceleration. The fictitious force is in the opposite direction to the acceleration of the noninertial frame.

在运用牛顿定律处理问题时，要特别注意选取惯性参照系，惯性参照系是相对研究物体静止或做匀速直线运动的参照系，在惯性参照系里，牛顿运动定律成立。例如，在一个做匀速直线运动的列车上，从列车顶端用细线悬挂一个质量为 m 的小球。我们在列车上，看到悬挂小球的细线垂直向下，和我们在静止参照系看到的一样。这时，重力 mg 和绳子的张力 T 作用在小球上，互相抵消，使小球处于平衡状态，如图 2—4—5（a）所示。但是，如果列车以加速度 a 向右行驶，我们如果站在静止系看问题，通过 free-body diagram 分析，会看到小球在重力 mg 和绳的张力 T 作用下，产生向右的水平合力 F_a（图（b）），使小球做匀加速直线运动。如果我们跳上火车，在火车的非惯性参照系，看到小球是静止的，但由于绳子的偏斜，仍然看到小球受到水平向右的力 F_a，这时，小球对火车仍然静止，牛顿第二定律已经不适用了。为了让牛顿定律仍然成立，我们引进一个惯性力 F_a' 向左作用在小球上，与水平向右的合力 F_a 平衡，因此，小球相对列车静止。

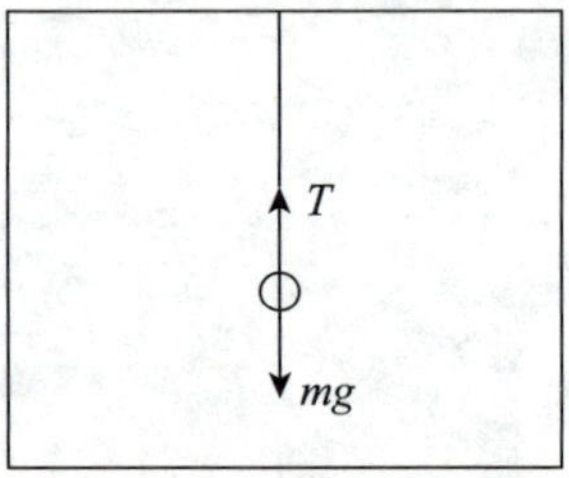

列车静止或做匀速直线运动，小球重力mg和绳的张力T平衡

(a) 列车静止或做匀速直线运动

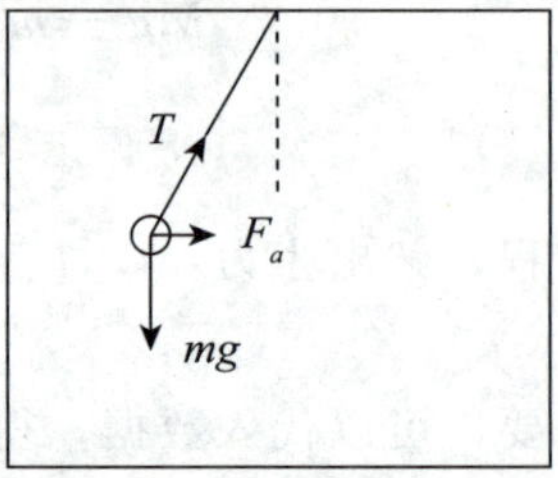

列车向右做匀加速速直线运动，小球重力mg和绳的张力T的合力F_a提供加速力

(b) 列车向右做匀加速直线运动

图 2—4—5

6 Practice problems 练习题

Multiple-Choice（选择题）

Single-Select（单选题）

1. When the frictionless system shown in the figure（图 2—4—6）is accelerated by an applied force of magnitude F, the tension in the string between the blocks is

(A) $2F$ (B) F

(C) $\frac{2}{3}F$　　(D) $\frac{1}{3}F$

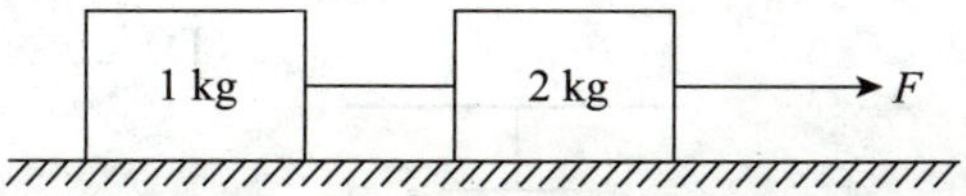

图 2—4—6

2. A block of weight W is pulled along a horizontal surface at a constant speed v by a force F, which acts at an angle of θ with the horizontal, as shown in the figure (图 2—4—7). The normal force exerted on the block by the surface has magnitude

(A) $W - F\cos\theta$　　(B) $W - F\sin\theta$

(C) W　　(D) $W + F\sin\theta$

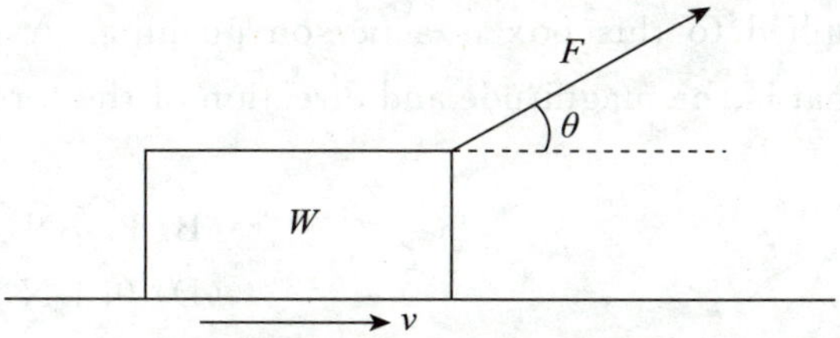

图 2—4—7

3. An object weighing 120 N is set on a rigid beam of negligible mass at the distance of 3 m from a pivot, as shown in the figure (图 2—4—8). A vertical force is applied to the other end of the beam, a distance of 4 m from the pivot, to keep the beam at rest and horizontal. What is the magnitude F of the force required?

(A) 10 N　　(B) 30 N

(C) 90 N　　(D) 120 N

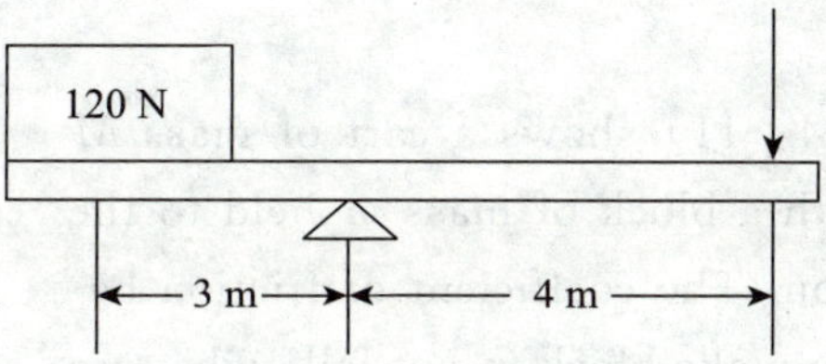

图 2—4—8

4. A car slides up a frictionless inclined plane. How does the normal force of the incline on the car compare with the weight of the car?

(A) The normal force must be equal to the car's weight

(B) The normal force must be less than the car's weight

(C) The normal force must be greater than the car's weight

(D) The normal force must be zero

5. Two blocks of wood, each of mass 2 kg, are suspended from the ceiling by strings of negligible mass, as shown in the figure (图 2—4—9), What is the force exerted on the

upper block by the lower string?

(A) zero

(B) 10 N upward

(C) 20 N upward

(D) 20 N downward

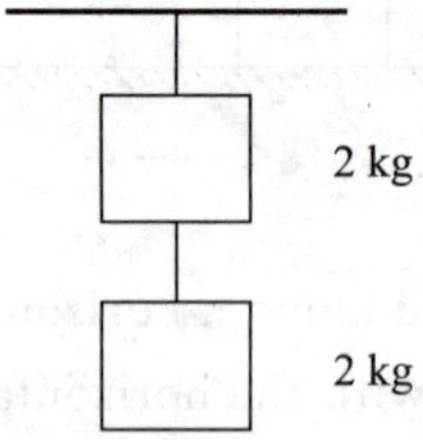

图 2—4—9

6. The velocity-time graph in the figure (图 2—4—10) represents the motion of a 5kg box. The only force applied to this box is a person pushing. Assuming that the box is moving to the right, what is the magnitude and direction of the force applied by the person pushing?

(A) 2.0 N, right

(B) 2.0 N, left

(C) 0.4 N, right

(D) 0.4 N, left

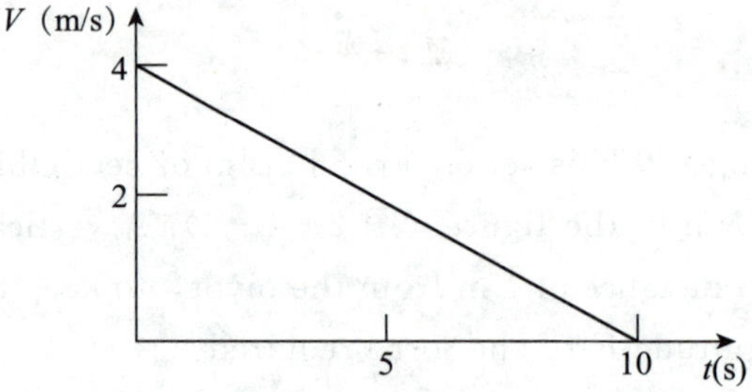

图 2—4—10

Multi-Select (多选题)

7. The figure (图 2—4—11) shows a cart of mass M accelerating to the right with a block of mass m held to the front surface only by friction. The coefficient of friction between the surfaces is μ. To make block m not fall, what are the possible ways?

(A) increase the acceleration a

(B) decrease the acceleration a

(C) decrease the coefficient of friction μ

(D) increase the coefficient of friction μ

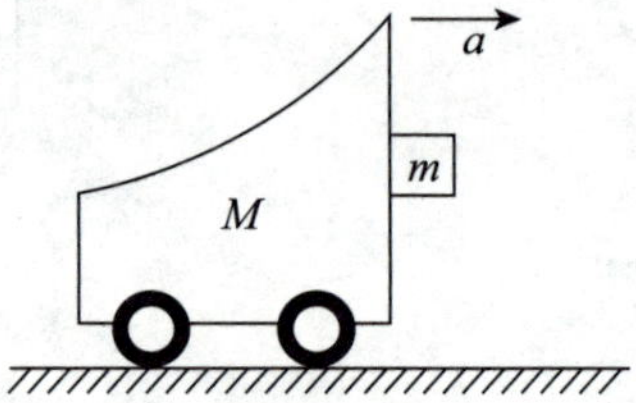

图 2—4—11

Free-Response (计算题)

8. A 2 kg block and a 5 kg block are connected as shown (图 2—4—12). The coefficient of friction between the 5kg block and the flat surface is $\mu = 0.2$.

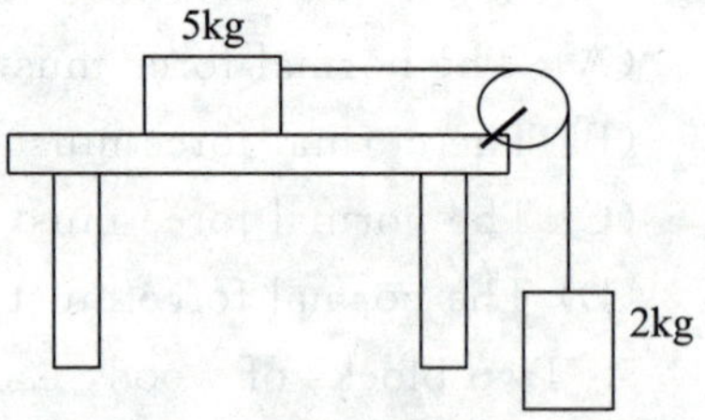

图 2—4—12

(a) Calculate the magnitude of the acceleration of the 5kg block.

(b) Calculate the tension in the rope connecting the two blocks.

Short-Answer（简答题）

9. Bert，Ernie，and Oscar are discussing the gas mileage of cars. Specifically，they are wondering whether a car gets better mileage on a city street or on a freeway. All agree (correctly) that the gas mileage of a car depends on the force that is produced by the car's engine—the car gets fewer miles per gallon if the engine must produce more force.

Whose explanation is completely correct?

Bert says：Gas mileage is better on the freeway. In town the car is always speeding up and slowing down because of the traffic lights，so because $F_{net}=ma$ and acceleration is large，the engine must produce a lot of force. However，on the freeway，the car moves with constant velocity，and acceleration is zero. So the engine produces no force，allowing for better gas mileage.

Ernie says：Gas mileage is better in town. In town，the speed of the car is slower than the speed on the freeway. Acceleration is velocity divided by time，so the acceleration in town is smaller. Because $F_{net}=ma$，the force of the engine is smaller in town getting better gas mileage.

Oscar says：Gas mileage is better on the freeway. The force of the engine only has to be enough to equal the force of air resistance—the engine does not have to accelerate the car because the car maintains a constant speed. Whereas in town，the force of the engine must often be greater than the force of friction and air resistance in order to let the car speed up.

题解

单选题

1. 选（D）。由于力 F 使整个系统产生的加速度为$F/3$，所以两个物块之间的绳子张力为 $ma=F/3$。

2. 选（B）。拉力 F 的竖直向上的分力为 $F\sin\theta$，由于重力 W 压在桌面，而桌面向上的支持力为 W，因此总的向上的支持力为 $W-F\sin\theta$。

3. 选（C）。为了保持杠杆平衡，杠杆另一端的力应为 90N。

4. 选（B）。斜面对汽车的支持力为 $mg\cos\theta$，θ 是斜面的倾角。如果 θ 大于 0，则 $\cos\theta$ 小于 1，所以该支持力一定小于重力。

5. 选（D）。绳子的张力为$T=$（2.0kg）g。

6. 选（B）。因为速度—时间曲线的斜率为加速度，数值为-0.4m/s^2，因此合外力为 $F_{net}=ma$，即 $5\text{ kg}\times0.4\text{ m/s}^2=2.0\text{ N}$，方向向左。

多选题

7. 选（A）和（D）。增加加速度 a，就增加了作用在 m 上的压力，从而增加了摩擦力。增大摩擦系数，也可以增大摩擦力。

计算题

8. 在 5kg 物块上有向左的摩擦力，正压力为 50N，所以摩擦力为 $\mu F_N=10$ N。

选择向下为正方向，分别列出两个物体的隔离体方程：

$$(2\text{kg})g - T = (2\text{kg})a$$
$$T - F_f = (5\text{kg})a$$

(a) 两个方程联立解出加速度为 1. 4 m/s^2。

(b) 将加速度带回任意一个方程，可得绳子的张力为 17N。这比无摩擦时的 14N 大，多余的力用于克服摩擦。

简答题

9. Bert and Oscar's explanation is correct. On the freeway, the car kept constant speed, nearly no acceleration, so the resultant force is very small. The force provided by the engine just overcomes the air resistance.

Ernie's answer is wrong. In town, the speed of the car is small, however, the acceleration is bigger. The force provided by the engine is to accelerate the car.

Chapter 5

Momentum and its conservation

第五章 动量和动量守恒

主要掌握动量、冲量的概念，掌握动量定理，动量守恒定律，弹性碰撞和完全非弹性碰撞。

Main points (主要知识点)

- 力的冲量为力和力作用时间的乘积，对于变力，冲量为变力对时间的积分
- 动量定理表述为：物体动量的变化等于合外力的冲量
- 由于物体碰撞时，它们相互之间的作用力为内力，外力为零，所以碰撞前后，体系的动量守恒
- 弹性碰撞前后不但动量守恒，动能也守恒。但非弹性碰撞，碰撞前后的动量守恒，但动能不守恒

Relative equations (相关公式)

The definition of momentum (动量的定义)：$p=mv$

The momentum theorem (动量定理)：$\Delta p=F\Delta t$

The conservation of momentum (动量守恒)：$\Delta p=0$，或 $p_1=p_2$

1 Momentum and impulse 动量和冲量

1. The definition of momentum (动量的定义)

The momentum of a body is the product of the mass and the velocity of the object.

$$\text{momentum}=mv$$

The unit of momentum (动量的单位)：kg · m/s，或 N · s.

2. The definition of impulse (冲量的定义)

Impulse is the product of the force which acts on an object and the time interval over which that force was applied.

$$\text{impulse} = F\Delta t$$

The definition of impulse explains why airbags are used in cars and why hitting someone with a pillow is less dangerous than hitting him or her with a cement block. The key is the Δt term.

3. The momentum theorem（动量定理）

$$F \cdot \Delta t = \Delta p$$

2 Conservation of momentum 动量守恒

Based on the momentum theorem, if the net external force is equal to zero, we have:

$$\Delta p = 0, \text{ and } p_1 = p_2$$

Momentum in an isolated system, where no net external forces act, is always conserved.

注意：应用动量定理解题的关键是要牢记动量是矢量！在解决二维问题时，要将动量矢量分解到各个坐标方向，动量的各个分量守恒。

Example:

A satellite floating through space collides with a small UFO. Before the collision, the satellite was traveling at 10 m/s to the right, and the UFO was traveling at 5 m/s to the left. If the satellite's mass is 70 kg, and the UFO's mass is 50 kg, and assuming that the satellite and the UFO bounce off each other upon impact, what is the satellite's final velocity if the UFO has a final velocity of 3 m/s to the right?

Solution:

Let us begin by drawing a picture: (The reader draws the picture, pay attention to the direction.)

Momentum is conserved, so we write

$$p_{satellite} + p_{UFO} = p'_{satellite} + p'_{UFO}$$

$$(m_s)(v_s) + (M_{UFO})(v_{UFO}) = (m_s)(v_s)' + (m_{UFO})(v_{UFO})'$$

The tick marks on the right side of the equation mean "after the collision". We know the momentum of each space traveler before the collision, and we know the UFO's final momentum. So we solve for the satellite's final velocity. (Note that we must define a positive direction. Because the UFO is moving to the left, its velocity is plugged in as negative.)

$$(70\text{kg})(+10\text{m/s}) + (50\text{kg})(-5\text{m/s}) = (70\text{kg})v'_s + (50\text{kg})(+3\text{m/s})$$

$$v'_s = 4.3\text{m/s}, \text{ to the right}$$

Now, what if the satellite and the UFO had stuck together upon colliding? We can solve for their final velocity easily:

$$p_{satellite} + p_{UFO} = p'_{satellite+UFO}$$

$$(m_s)(v_s) + (M_{UFO})(v_{UFO}) = (m_{s+UFO})(v'_{UFO+s})$$

$$(70\text{kg})(+10\text{m/s}) + (50\text{kg})(-5\text{m/s}) = (70\text{kg} + 50\text{kg})v'_{s+UFO}$$

$$v'_{s+UFO} = 3.8 \text{ m/s}, \text{ to the right}$$

3 Elastic and inelastic collisions
弹性碰撞和非弹性碰撞

1. Elastic collision: A collision in which kinetic energy is conversed

注意：弹性碰撞动量和动能均守恒。

2. Inelastic collision: A collision in which kinetic energy is NOT conserved

注意：非弹性碰撞动量守恒但动能不守恒！

The extreme case of an inelastic collision is called a perfectly inelastic collision.

3. Perfectly inelastic collision: The colliding objects stick together after impact

注意：完全非弹性碰撞动量守恒但动能不守恒！

AP 考试中只涉及弹性碰撞和完全非弹性碰撞两种情况。

4 Two dimensional collisions
二维碰撞

The key to solving a two dimensional collision problem is to remember that momentum is a vector and as a vector it can be broken into x and y components. Momentum in the x-direction is always conserved, and momentum in the y-direction is always conserved. In general, the component in one coordinate is always conserved.

注意：解决二维碰撞问题时，可以利用动量的分量守恒建立方程。

Example:

Two objects of masses $M_1=1$ kilogram and $M_2=4$ kilograms are free to slide on a horizontal frictionless surface. The objects collide and the magnitudes and directions of the velocities of the two objects before and after the collision are shown in the figure (图 2—5—1). ($\sin 37°=0.6$, $\cos 37°=0.8$, $\tan 37°=0.75$)

(a) Calculate the x and y components (p_x and p_y, respectively) of the momentum of the two objects before and after the collision.

(b) Show, using the data that you listed in the table, that linear momentum is conserved in this collision.

(c) Calculate the kinetic energy of the two object system before and after the collision.

(d) Is kinetic energy conserved in the collision?

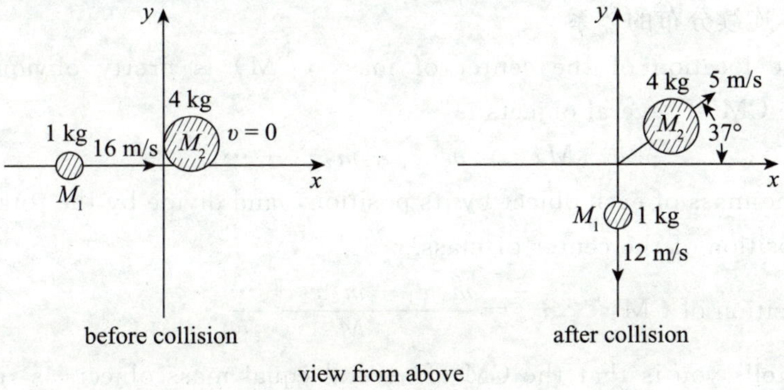

图 2—5—1

Solution：

（a）对于 $M_1=1\text{kg}$，碰撞前：$p_x=16\text{kgm/s}$，$p_y=0$；碰撞后：$p_x=0$，$p_y=-12\text{kgm/s}$

对于 $M_2=4\text{kg}$，碰撞前：$p_x=0$，$p_y=0$；碰撞后：$p_x=16\ \text{kgm/s}$，$p_y=12\ \text{kgm/s}$.

注意：碰撞后 $M_1=1\text{kg}$ 的物体的 y 方向动量为负值！

（b）对于 M_1，M_2 系统而言，碰撞前后动量均为 16 kg m/s，因此动量守恒。

（c）碰撞前动能为：$\frac{1}{2}\times1\times16^2=128$（J）

碰撞后动能为：$\frac{1}{2}\times1\times12^2+\frac{1}{2}\times4\times5^2=122$（J）

（d）说明动能不守恒，不是弹性碰撞。

对于二维碰撞问题，处理时一定要先建立坐标系，写出动量在不同坐标方向的分量。

5 Motion of the center of mass 质心运动

1. The center of mass of a system of objects obeys Newton's second law

Two common examples might illustrate the point：

（1）Imagine that an astronaut on a space walk throws a rope around a small asteroid, and then pulls the asteroid toward him. Where will the asteroid and the astronaut collide?

Answer：at the center of mass. Since no forces acted except due to the astronaut and asteroid，the center of mass must have no acceleration. The center of mass started at rest, and stays at rest all the way until the objects collide.

（2）A toy rocket is in projectile motion，so that it is on track to land 30 m from its launch point. While in the air，the rocket explodes into two identical pieces，one of which lands 35 m from the launch point. Where does the other piece land?

Answer：25 m from the launch point. Since the only external force acting on the rocket is gravity，the center of mass must stay in projectile motion，and must land 30 m from the launch point. The two pieces are of equal mass，so if one is 5 m beyond the center of mass's landing point，the other piece must be 5 m short of that point.

2. Finding the center of mass（质心的求解方法）

（1）质量不连续分布的体系

Usually the location of the center of mass（CM）is pretty obvious. The formal equation for the CM of several objects is

$$Mx_{cm}=m_1x_1+m_2x_2+\cdots$$

Multiply the mass of each object by its position，and divide by the total mass M，and you have the position of the center of mass.

So，the position of CM：$x_{cm}=\dfrac{m_1x_1+m_2x_2+\cdots}{M}$

What this tells you is that the CM of several equal mass objects is right in between them；if one mass is heavier than the others，the CM is closer to the heavier mass.

（2）质量连续分布的体系

Sometimes in AP Physics C, you will be asked to find the center of mass of a continuous body using calculus. The formula is

$$x_{cm}=\frac{\int r\,\mathrm{d}m}{\int \mathrm{d}m}$$

6 Practice questions 练习题

Multiple-Choice（选择题）

Single-Select（单选题）

1. Which of the following is true when an object of mass m moving on a horizontal frictionless surface hits and sticks to an object of mass M ($M>m$), which is initially at rest on the surface?

(A) The collision is elastic

(B) All of the initial kinetic energy of the less massive object is lost

(C) The momentum of the objects that are stuck together has a smaller magnitude than the initial momentum of the less massive object

(D) The speed of the objects that are stuck together will be less than the initial speed of the less massive object

Question 2 and 3:

A ball of mass M is caught by someone wearing a baseball glove. The ball is in contact with the glove for a time t, the initial velocity of the ball (just before the catcher touches it) is v_0.

2. If the time of the ball's collision with the glove is doubled, what happens to the force necessary to catch the ball?

(A) It doesn't change

(B) It is cut in half

(C) It is cut to one fourth of the original force

(D) It quadruples

3. If the time of collision remains t but the initial velocity is doubled, what happens to the force necessary to catch the ball?

(A) It doesn't change

(B) It is cut in half

(C) It is cut to one fourth of the original force

(D) It doubles

4. Two balls, of mass m and $2m$ (图 2—5—2), collide and stick together. The combined balls are at rest after the collision. If the ball of mass m was moving 5 m/s to the right before the collision, what was the velocity of the ball of mass $2m$ before the collision?

(A) 2.5 m/s to the right　　(B) 2.5 m/s to the left

(C) 10 m/s to the right　　　　(D) 10 m/s to the left

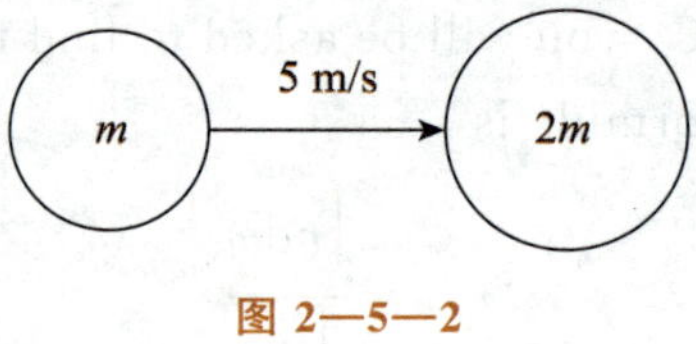

图 2—5—2

5. Two same identical balls have initial velocities $v_1=4$m/s to the right and $v_2=3$m/s to the left, respectively. The balls collide head on and stick together. What is the velocity of the combined balls after the collision?

(A) $\frac{1}{7}$ m/s to the right　　　　(B) $\frac{3}{7}$ m/s to the right

(C) $\frac{1}{2}$ m/s to the right　　　　(D) $\frac{4}{7}$ m/s to the right

Multi-Select（多选题）

6. A 2 kg object initially moving with a constant velocity is subjected to a force of magnitude F in the direction of motion. A graph of F as a function of time t is shown in the figure（图 2—5—3）. What are the increase of the momentum of the object and the increase of the velocity of the object during the time the force is applied?

(A) The momentum of the object increases 3.0kg m/s

(B) The velocity of the object increases 2.0m/s

(C) The velocity of the object increases 3.0m/s

(D) The momentum of the object increases 6.0kg m/s

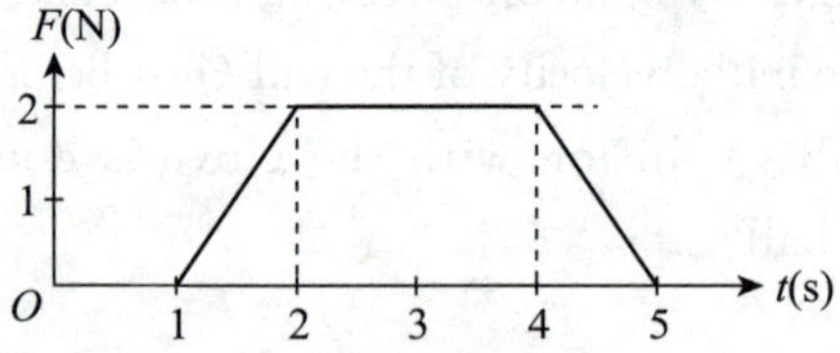

图 2—5—3

Free-Response（计算题）

7. A 75kg skier skis down a hill. The skier collides with a 40kg child who is at rest on the flat surface near the base of the hill, 100 m from the skier's starting point, as shown in the figure（图 2—5—4）. The skier and the child become entangled. Assume all surfaces are frictionless.

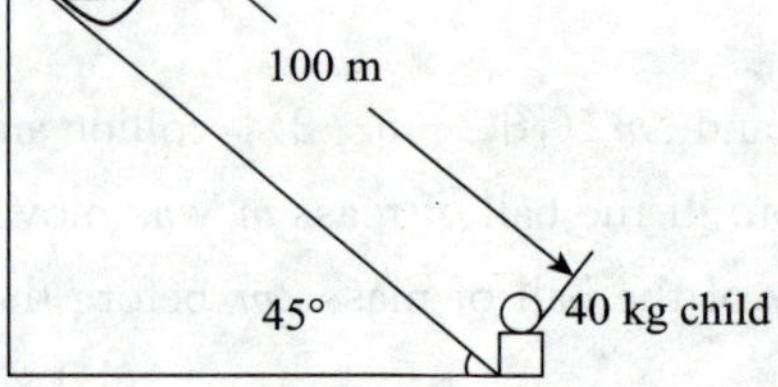

图 2—5—4

(a) How fast will the skier be moving when he reaches the bottom of the hill? Assume the skier is at rest when he begins his descent.

(b) What will be the speed of the skier and child just after they collide?

(c) If the collision occurs in half a second, how much force will be experienced by each person?

8. In the figure（图 2—5—5）, mass 1 does not slide with respect to the surface when the horizontal force F_a is applied. Determine the magnitude of the force F_a. Assume there is no friction.

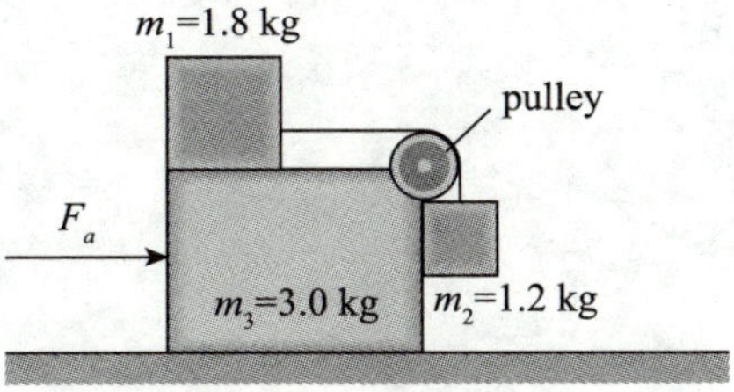

图 2—5—5

题解

单选题

1. 选（D)。因为碰撞后，两物体粘在一起，所以是完全非弹性碰撞。$mv=(M+m)v'$，所以 $v'<v$。

2. 选（B)。根据动量定理 $F=\Delta p/\Delta t$，球的动量变化相同，接球时间长一倍，作用力小一半。

3. 选（D)。用 $F=\Delta p/\Delta t$，由于球的速度大了一倍，即动量变化大了一倍，所以接球的力要大一倍。

4. 选（B)。由于两球碰撞后的共同速度为零，总动量为零，因此，碰撞前总动量也为零。质量为 m 的球碰撞前向右以 5m/s 速度运动，动量为 $5m$ m/s，所以右边的大球必须有同样大小向左的动量，由于质量大一倍，所以其向左的速度为 2.5m/s。

5. 选（C)。两球质量相同，假设为 1kg。由于碰撞是完全非弹性的，所以，根据动量守恒：(1kg)（+4m/s）+（1kg)（−3m/s）=（2kg)（v'）。

式中 v' 代表碰撞后两物体共同的速度，解出为，$v'=+0.5$m/s，+号代表向右。

多选题

6. 选（C）和（D)。在力作用的 5 秒内，动量的增加等于力的冲量，即曲线下的图形的面积，为 6.0kg m/s。速度增加了 3.0m/s。

计算题

7. (a) 该问为动力学问题，可以用牛顿第二定律。选滑雪者为研究对象，沿斜面的分力为 $mg\sin 45°$。根据牛顿第二定律，沿斜面的加速度为 $g\sin 45°=7.1\ \text{m/s}^2$。用运动学公式可以解出经 100 m 滑行后，其末速度为 38 m/s。

(b) 滑雪者和儿童碰撞后在一起，因此是完全非弹性碰撞。动量守恒，碰前动量为 (75kg)（38m/s）= 2 850 kg m/s。碰撞后滑雪者和儿童总质量为 115 kg。因此，碰撞后的速度=2 850 kg m/s ÷115 kg=25 m/s。

（c）根据动量定理，儿童的动量变化为（40kg）（25m/s）＝1 000 kg m/s，除以时间 0.5 秒，为 2 000 N。可以看出，有很大的力作用在滑雪者和儿童之间。

8. 由于 m_1，m_2，m_3都没有相对滑动，因此，不能将 m_1，m_2视为连体运动处理。这时，绳上张力为 m_2g，其大小为 1.2kg×9.8 m/s^2＝11.76 N，使 m_1产生的加速度为 6.53 m/s^2。F_a使整体产生同样的加速度，因此 F_a＝（m_1＋m_2＋m_3）×a＝(1.2＋1.8＋3.0) kg ×6.53m/s^2＝ 39.2 N。

Chapter 6

Work and energy, conservation of energy

第六章 功和能、能量守恒

该知识点复习功、能、动能、势能等概念，介绍动能定理、功能原理、机械能守恒定律。应用这些定律解决一些实际问题。重点放在机械能守恒定律及其应用上。

Main points（主要知识点）

- 动能定理，把做功和动能联系起来的定理
- 重力势能和弹性势能
- 机械能守恒定律
- 其他形式的能量守恒

Relevant equations（相关公式）

The definition of work（功的定义）: $W = F \cdot d_{Parallel\ to\ F}$ （For Physics 1&2）

$W = \vec{F} \cdot \vec{d}$ （For Physics C）

The work-energy theorem（功能定理）: $W_{net} = \Delta E_k$

The force of a spring（弹性恢复力）: $F = -kx$

Two different types of potential energy（两种不同的势能）:

$PE = mgh$（重力势能）

$PE = \frac{1}{2}kx^2$（弹性势能）

1 Kinetic energy and the work-energy theorem 动能和功能原理

(1) Energy is the ability to do work by an object.

(2) Work is the product of force and the displacement which is parallel to the force.

- Work is a scalar. So is energy.
- The units of work and energy are joules. One joule is the product of one newton

and one meter, abbreviated as 1 J.

- Work can be negative, which just means that the force is applied in the direction opposite displacement.

Kinetic energy (动能): Energy of motion. Abbreviated as KE or E_k.

$$KE = E_k = \frac{1}{2}mv^2$$

This means that the kinetic energy of an object equals one half the object's mass times its speed squared.

Work-energy theorem: $W_{net} = \Delta KE$

The net work done on an object is equal to that object's change in kinetic energy.

2 Potential energy 势能

Potential energy is the energy which depends on the position. Abbreviated as PE or E_p.

Potential energy comes in many forms: gravitational potential energy, spring potential energy, electrical potential energy and so on. For starters, we will concern ourselves with gravitational potential energy.

Gravitational PE is described by the following equation:

$$PE = mgh$$

In this equation, m is the mass of an object, g is the gravitational field strength (10 N/kg on the Earth), and h is the height of an object above certain point (called the zero potential). That point can be wherever you want it to be, depending on the problem.

REMINDER: h in the potential energy equation stands for vertical height above the zero of potential.

3 Conservation of energy 机械能守恒

The general approach of solving energy conservation problem is that write out all the terms for the initial energy of the system, and set the sum of those terms equal to the sum of all the terms for the final energy of the system.

Example:

A block of mass m is placed on a frictionless plane inclined at a 30° angle above the horizontal. It is released from the rest and allowed to slide 5 m down the plane, as shown in the figure (图 2—6—1). What is the final velocity?

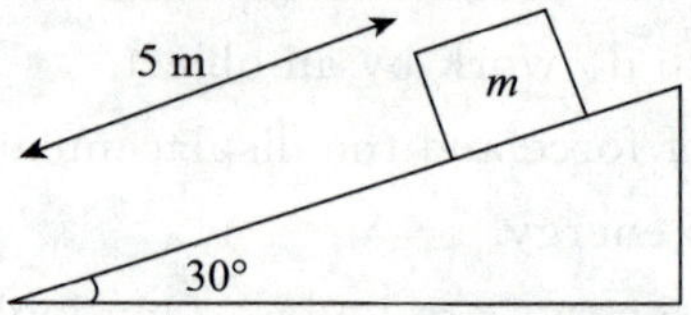

图 2—6—1

Solution:

If we were to approach this problem using kinetic equations, it would take about a page of work to solve. Instead, observe how quickly it can be solved using conservation of energy.

$$KE_i + PE_i = KE_f + PE_f$$

$$\frac{1}{2}mv_i^2 + mgh_i = \frac{1}{2}mv_f^2 + mgh_f$$

We will define our zero of potential to be the height of the box after it has slid the 5 m down the plane. By defining it in this way, the PE term on the right side of the equation will cancel out. Furthermore, because the box starts from rest, its initial KE also equals zero.

$$0 + mgh_i = \frac{1}{2}mv_f^2 + 0$$

The initial height can be found using trigonometry: $h_i = 5\text{m} \times \sin 30° = 2.5\ \text{m}$.

$$mg\,(2.5\text{m}) = \frac{1}{2}mv_f^2$$

$$50\ (\text{m/s})^2 = v_f^2$$

$$v_f \approx 7\ \text{m/s}$$

总体来说，能量守恒可以用下式表示：

$$KE_i + PE_i + W = KE_f + PE_f$$

其中 W 代表外力对物体做的功。

Example:

If there had been friction between the box and the plane in the previous example, the work done by friction would be the W term. When it comes to the AP exam, you will include this W term only when there is friction involved. When friction is involved, $W = F_f d$, where F_f is the force of friction on the object, and d is the distance the object travels.

Let's say that there was friction between the box and the inclined plane.

A box of mass m is placed on a plane inclined at a 30° angle above the horizontal. The coefficient of friction between the box and the plane is 0.20. The box is released from rest and allowed to slide 5.0 m down the plane. What is its final velocity?

Solution:

We start by writing the general equation for energy conservation:

$$KE_i + PE_i + W = KE_f + PE_f$$

W equals $F_f d$, where F_f is the force of friction, and d is 5 m,

$$W = -F_f d = -\mu F_N d$$

$$F_N = mg\cos\theta = mg\cos 30°$$

The value for W is negative because friction acts opposite displacement. You may want to draw a free-body diagram to understand how we derived this value for F_N.

Now, plugging in values, we have:

$$\frac{1}{2}mv_i^2 + mgh_i - 0.2 \times mg\cos 30° \times 5\text{m} = \frac{1}{2}mv_f^2 + mgh_f$$

$$0+mg\times 2.5\text{m}-0.2\times mg\times 0.87\times 5\text{m}=\frac{1}{2}mv_f^2$$

We rearrange some terms and cancel out m from each side to get:

$$v_f=5.7\text{m/s}$$

结果说明，由于下滑力要克服摩擦力做功，因此末速度比没有摩擦时要小。

4 Spring 弹簧

1. Restored force（弹性恢复力）

The force exerted by a spring is directly proportional to the amount that the spring compressed. That is, $F_{spring}=-kx$.

In this equation, k is a constant (called the spring constant), and x is the distance that the spring has been compressed or extended from its equilibrium position.

注意：公式中的负号代表弹性恢复力总是和位移相反，弹簧伸长时，x 为正，力为负。弹簧压缩时，x 为负，力为正，力总是指向平衡位置。

2. Elastic potential energy（弹性势能）

The amount of energy stored in a spring is given as follows:

$$PE_{spring}=\frac{1}{2}kx^2$$

弹性力做功大小和弹性势能相同：$W_{spring}=\frac{1}{2}kx^2$

5 Power 功率

Power is defined as the work done in unit time:

$$\text{power}=\frac{\text{work}}{\text{time}}$$

The units of power is joules/second（焦耳/秒）, also known as watts（瓦特）.

6 Practice questions 练习题

Multiple-Choice（选择题）

Single-Select（单选题）

1. The figure（图 2—6—2）shows a rough semicircular track whose ends are at a vertical height h. A block placed at point P at one end of the track is released from rest and slides past the bottom of the track. Which of the following is true of the height to which the block rises on the other side of the track?

(A) It is equal to $h/2$

(B) It is equal to $h/4$

(C) It is equal to h

(D) It is between zero and h; the exact height depends on how much energy is lost to friction

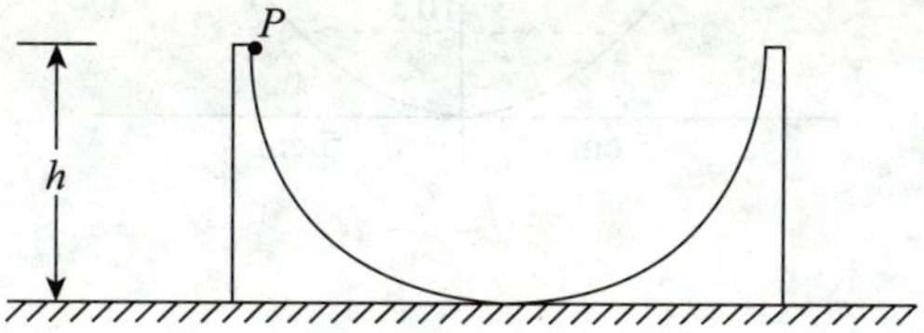

图 2—6—2

Questions 2 and 3:

A block released from rest at position A, slide with negligible friction down an inclined track, around a vertical loop, and then along a horizontal track as shown in the figure (图 2—6—3). The block never leaves the track.

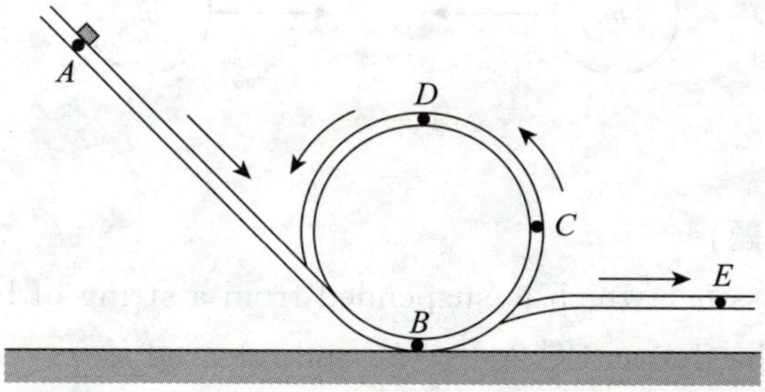

图 2—6—3

2. After the block is released, in which of the following sequences of positions is the speed of the block ordered from fastest to slowest?

(A) $B\ C\ D\ E$ (B) $B\ E\ C\ D$

(C) $D\ C\ E\ B$ (D) $E\ B\ C\ D$

3. The gravitational potential energy and kinetic energy of the block are most nearly equal at which position? (Consider the potential energy to be zero at position B)

(A) A (B) B

(C) C (C) D

4. A mass experiences a potential energy U that varies with distance x as shown in the figure (图 2—6—4). The mass is released from position $x=0$ with 10 J of kinetic energy. Which of the following describes the long-term motion of the mass?

(A) The mass eventually comes to rest at $x=0$

(B) The mass slows down with constant acceleration, stopping at $x = 5$ cm

(C) The mass speeds up with constant acceleration

(D) The mass oscillates, never getting farther than 5 cm from $x = 0$

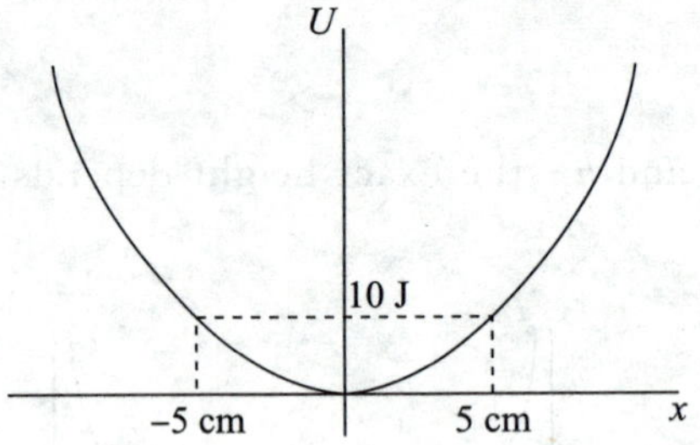

图 2—6—4

5. Two identical balls of mass $m = 1.0$ kg are moving towards each other, as shown in the figure (图 2—6—5). What is the initial kinetic energy of the system consisting of the two balls?

(A) 0 joules (B) 1 joules

(C) 12 joules (D) 36 joules

图 2—6—5

Multi-Select (多选题)

6. A sphere of mass m, which is suspended from a string of length l, is displaced to the right as shown in the figure (图 2—6—6) and released from rest so that is swings as a simple pendulum with small amplitude. Which of the following statements are true?

(A) The maximum kinetic energy is attained as the sphere passes through its equilibrium position

(B) The maximum kinetic energy is attained as the sphere reaches its point of release

(C) The minimum gravitational potential energy is attained as the sphere passes through its equilibrium position

(D) The minimum gravitational potential energy is attained as the sphere reaches its point of release

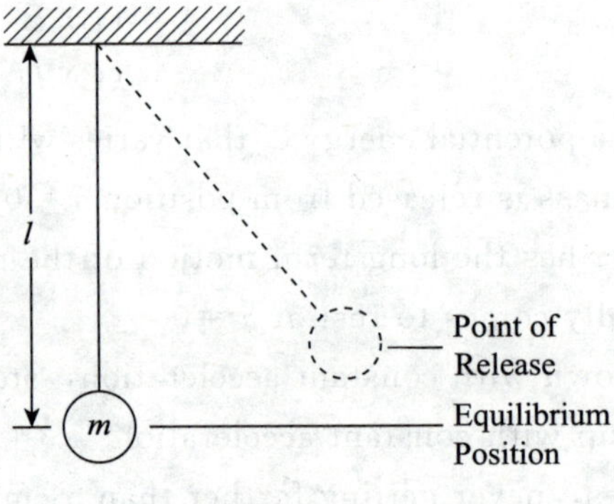

图 2—6—6

Free-Response（计算题）

7. A 1 500 kg car moves north according to the velocity-time graph shown in the figure（图 2—6—7）.

(a) Determine the change in the car's kinetic energy during the first 7s.

(b) To determine how far the car traveled in these 7s, the three basic kinematics equations can not be used. Explain why not.

(c) Use the velocity-time graph to estimate the distance the car traveled in 7s.

(d) What was the net work done on the car in these 7s?

(e) Determine the average power necessary for the car to perform this motion.

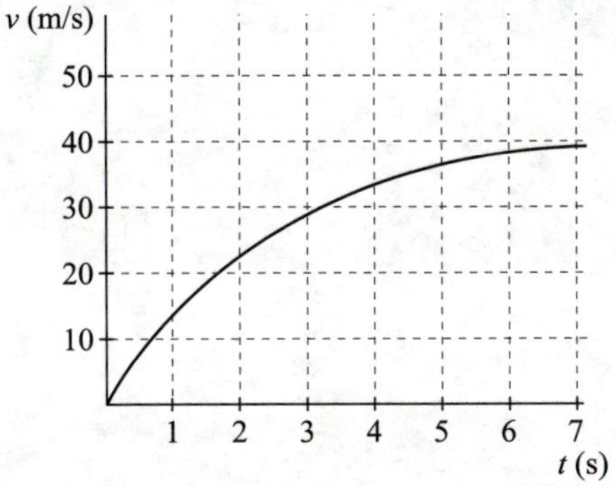

图 2—6—7

题解

单选题

1. 选（D）。由于部分能量用于克服摩擦力做功，因此，物块在通过最低点后，上升的高度界于 0 和 h 之间。

2. 选（B）。部分势能将转换为动能，对于 B、C、D、E 点，位置越低转换的动能越多，因此，根据动能的大小，速度从大到小的排序为 B、E、C、D。

3. 选（C）。C 点的高度接近 A 点的一半，因此，一半势能转换成动能，即势能和动能近似相同。

4. 选（D）。该物体将围绕平衡点 $x=0$ 处做往复振动。由于开始时动能（KE）为 10 J，它的最大势能也为 10 J，对应图上的最大位移为 5 cm。

5. 选（D）。动能是标量，每个球的动能均为 $\frac{1}{2}\times 1\ \text{kg}\times(6\ \text{m/s})^2$，两个球的动能的和为 36 J。

多选题

6. 选（A）和（C）。小球释放时势能最大，动能为零，根据机械能守恒，到最低点时，势能最小，动能最大。

计算题

7. (a) 汽车从静止出发，动能为零，最后加速到 40m/s，动能为 $\frac{1}{2}\times 1\ 500\text{kg}\times(40\text{m/s})^2=1.2\times 10^6\ \text{J}$。即动能变化为 $1.2\times 10^6\text{J}$。

(b) 由于速度—时间曲线不是直线，所以，加速度不是常数，不是匀变速直线运动，所以不能用相应的公式。

(c) 汽车行驶距离可以由速度曲线下的图形的面积估算出来。一个方格代表 10m，曲线下约 19 个方格，因此距离为 190m。

(d) 我们不能用 $W=f\times d$ 来计算功，因为力是变力。但功能定理 $W_{net}=\Delta KE$ 总是成立的。由 (a) 可知，动能变化为 1.2×10^6J，因此，合力做功为 1.2×10^6J。

(e) 功率为做的功除以所用的时间：$1.2\times10^6\text{J}\div7\text{s}=170\text{kW}$。

Chapter 7

Circular motion and gravitation

第七章 圆周运动和万有引力

Main points（主要知识点）

- 做匀速率圆周运动物体的向心加速度为：$\frac{v^2}{r}$，或 $r\omega^2$
- 圆周运动的关键是什么力提供了向心力，因此进行隔离体受力分析很重要
- 万有引力定律描绘了宇宙中任意两个可以看成质点的物体之间的相互吸引力。该力和两物体的质量乘积成正比，和两物体的距离平方成反比
- 了解关于行星运动的开普勒定律

Relevant equations（相关公式）

Centripetal acceleration（向心加速度）：$a_c = \frac{v^2}{r}$

Gravitational force（万有引力）：$F_G = G\frac{m_1 m_2}{r^2}$

Gravitational potential energy（引力势能）：$PE = \frac{Gm_1 m_2}{r}$

1 Velocity and acceleration in uniform circular motion 匀速率圆周运动中的速度和加速度

When an object maintains the same speed but turns in a circle, the magnitude of its acceleration is constant and directed toward the center of the circle. This means that the acceleration vector is perpendicular to the velocity vector at any given moment, as shown in the figure（图 2—7—1）.

The velocity vector is always directed tangent to the circle, and the acceleration vector is always directed toward the center of the circle.

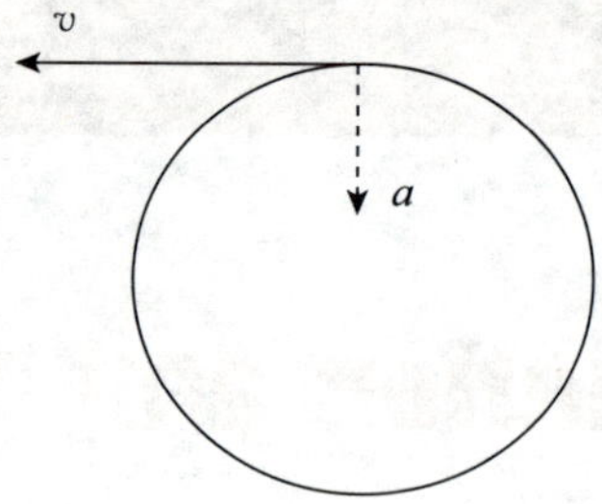

图 2—7—1 velocity and acceleration of an object traveling in uniform circular motion

2 Centripetal acceleration 向心加速度

Centripetal acceleration: The acceleration keeping an object in uniform circular motion, abbreviated as a_c. It can be proved mathematically that the formula for centripetal acceleration is $a_c = \frac{v^2}{r} = r\omega^2$.

In this equation, v is the object's linear velocity, ω is the angular velocity, and r is the radius of the circle in which the object is traveling.

注意：在解圆周运动问题时，首先用隔离体受力分析法找出提供向心力的合力，再用牛顿第二定律解出向心加速度。下面两节，我们举两个在 AP 物理中常见的例子。

3 Mass on a string 绳牵旋转体

Example:

A block of mass $M=2$ kg is swung on a rope in a vertical circle of radius r at constant speed v (图 2—7—2). When the block is at the top of the circle, the tension in the rope is measured to be 10 N. What is the tension in the rope when the block is at the bottom of the circle?

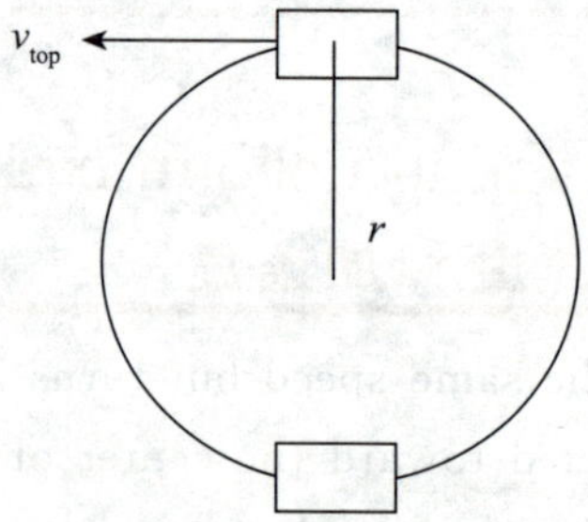

图 2—7—2 a block is swung in a vertical circle

Solution:

Let's begin by drawing a free-body diagram of the block at the top of the circle and another of the block at the bottom of the circle as shown in the figure (图 2—7—3).

(a) a free-body diagram of the block at the top of the circle

(b) a free-body diagram of the block at the bottom of the circle

图 2—7—3

Next, we write Newton's second law for each diagram. Acceleration is always toward the center of the circle.

$$Mg + T_{top} = Ma \ ; \qquad T_{bottom} - Mg = Ma$$

The acceleration is centripetal, so we can plug in $\frac{v^2}{r}$ for both accelerations.

$$Mg + T_{top} = M\frac{v^2}{r} \ ; \ T_{bottom} - Mg = M\frac{v^2}{r}$$

At both top and bottom, the speed v and the radius r are the same. So $M\frac{v^2}{r}$ has to be the same at both the top and bottom, allowing us to set the left side of each equation equal to one another.

$$Mg + T_{top} = T_{bottom} - Mg$$

With $M = 2$ kg and $T_{top} = 10$ N, we solve to get $T_{bottom} = 50$ N.

这个结果说明当物体在最低处时，绳子的拉力是最高处的 5 倍。

4 Car on a curve
弯曲路径上运行的车

Example:

A car of mass m travels around a curve that has a radius of curvature r. What is the necessary coefficient of friction such that the car can round the curve with a velocity v?

Solution:

Friction provides the centripetal force necessary to round the corner. The force of friction must point in toward the center of the curve (图 2—7—4).

We can now write some equations and solve for μ (the coefficient of friction).

The net force in the horizontal direction is F_f, which can be set equal to mass times (centripetal) acceleration. $F_f = ma = m\frac{v^2}{r}$

We also know that: $F_f = \mu F_N$

So, $\mu F_N = m\frac{v^2}{r}$

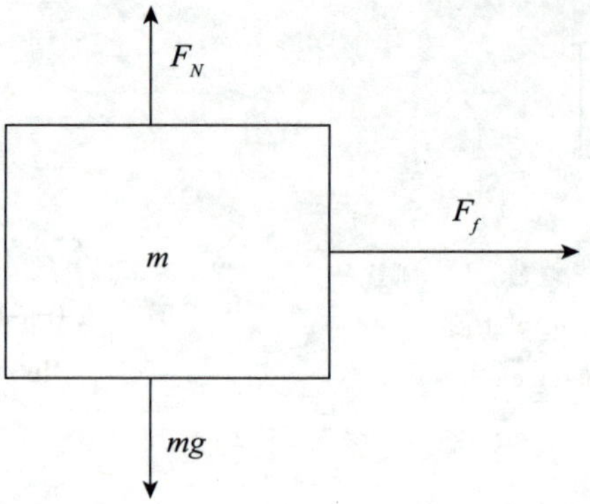

图 2—7—4 a car of mass m travels around a curve

Solving for μ, we have: $\mu=\frac{v^2}{gr}$

结果说明：车速越大，弯路的曲率半径越小，需要地面的摩擦系数越大。

5 Newton's law of gravitation 牛顿万有引力定律

Newton's law of gravitation can be written as:

$$F_G=G\frac{M_1M_2}{r^2}$$

Where M_1 is the mass of one of the objects, M_2 is the mass of the other object, r is the distance between the center of mass of each object, and G is called the "Universal Gravitational Constant" and is equal to $6.67\times10^{-11}\ \text{N}\cdot\text{m}^2/\text{kg}^2$.

Example:

The mass of the Earth, M_E, is 5.97×10^{24} kg. The mass of the Sun, M_S, is 1.99×10^{30} kg. The two objects are about 154 000 000 km away from each other. How much gravitational force does the Earth exert on the Sun?

Solution:

$$F_G=G\frac{M_EM_S}{r^2}$$

$$F_G=\frac{6.67\times10^{-11}\times5.97\times10^{24}\times1.99\times10^{30}}{(1.54\times10^{11})^2}=3.3\times10^{22}\ (\text{N})$$

注意：计算时采用国际单位制，太阳作用于地球的力和地球作用于太阳的力是作用力和反作用力，其大小相等，方向相反。

万有引力问题常见题型：

Example:

What is the speed of the Earth as it revolves on orbit around the Sun?

Solution:

The force of gravity exerted by the Sun on the Earth is what keeps the Earth in motion; it is the centripetal force acting on the Earth.

$$F_G = M_E a = M_E\left(\frac{v^2}{r}\right)$$

$$G\frac{M_E M_S}{r^2} = M_E\frac{v^2}{r}$$

$$v = \sqrt{\frac{GM_S}{r}} = 29\ 000\ \text{m/s}$$

Example:

What is the minimum velocity a satellite turns around the Earth?

Solution:

$$v = \sqrt{\frac{GM_E}{r_E}} = 7\ 900\ \text{m/s}$$

It is called as the first universal velocity at which the satellite can be launched into the space and travels around the Earth.

6 Kepler's laws
开普勒行星运动定律

In Physics C, we have to deal with planetary and satellite orbits in a bit more detail. Johannes Kepler, the late 1500s theorist, developed three laws of planetary motion based on the detailed observations of Tycho Brahe. You need to understand each law and its consequences.

1. Planetary orbits are ellipses, with the Sun as one focus. Of course, we can apply this law to a satellite orbiting the Earth, in which case the orbit is an ellipse, with the Earth at one focus. (When we mean the center of the Earth—for the sake of Kepler's laws, we consider the orbiting bodies to be point particles.) In the simple case of a circular orbit, this law still applies because a circle is just an ellipse with both foci at the center.

2. An orbit sweeps out equal areas in equal times. If you draw a line from a planet to the Sun, this line crosses an equal amount of area every minute (or hour, or month or whatever). See the following figure (图 2—7—5):

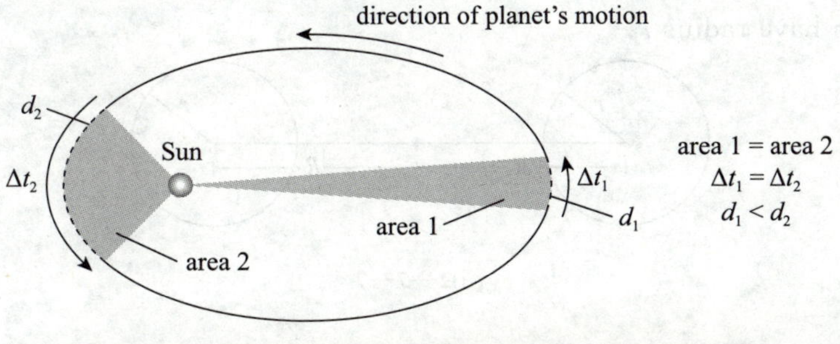

图 2—7—5 an orbit sweeps out equal areas in equal times

3. A planet's orbital period squared is proportional to its orbital radius cubed. In mathematics, we write this as $T^2 = cR^3$.

注意：在太阳系里，如果一个行星的总能量为负值，它将围绕太阳作轨道运动。反之，若总能量为正，它可能从太阳的引力范围逃逸。同样的结论可应用于地球和月球系统以及地球和卫星系统。

7 Practice questions
练习题

Multiple-Choice（选择题）

Single-Select（单选题）

1. The horizontal turntable shown in the figure（图 2—7—6）rotates at a constant rate. As viewed from above, a coin on the turntable moves counterclockwise in a circle as shown. Which of the following vectors best represents the direction of the frictional force exerted on the coin by the turntable when the coin is in the position shown?

(A) ←　　　(B) →

(C) ↙　　　(D) ↓

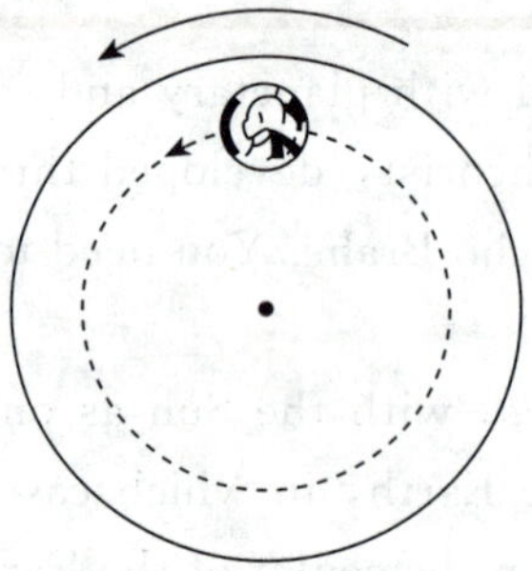

图 2—7—6

Questions 2 and 3:

Two stars, each of mass M, are in a binary system. The stars orbit about a point a distance R from the center of each star as shown in the figure（图 2—7—7）. The stars themselves each have radius r.

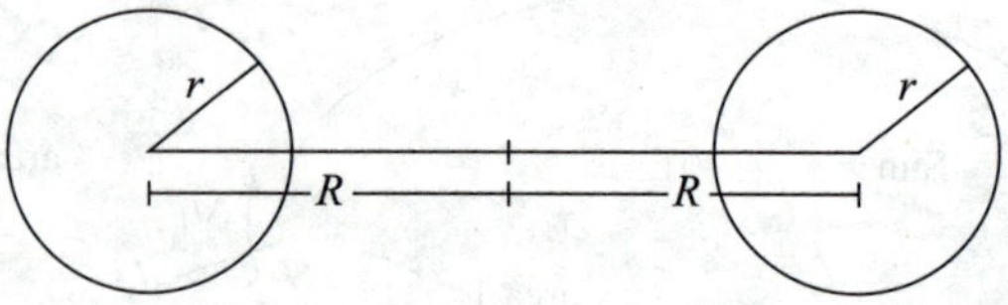

图 2—7—7

2. What is the force each star exerts on the other?

(A) $G\dfrac{M^2}{(2r+2R)^2}$ (B) $G\dfrac{M^2}{(r+R)^2}$

(C) $G\dfrac{M^2}{R^2}$ (D) $G\dfrac{M^2}{4R^2}$

3. In terms of each star's tangential speed v, what is the centripetal acceleration of each star ?

(A) $\dfrac{v^2}{2R}$ (B) $\dfrac{v^2}{(r+R)}$

(C) $\dfrac{v^2}{2(r+R)}$ (D) $\dfrac{v^2}{R}$

Questions 4 and 5:

In the movie *Return of the Jedi*, the Ewoks throw rocks using a circular motion device. A rock is attached to a string. An Ewok whirls the rock in a horizontal circle above his head, then lets go, sending the rock careening into the head of an unsuspecting stormtrooper.

4. What force provides the rock's centripetal acceleration?

(A) The vertical component of the string's tension

(B) The horizontal component of the string's tension

(C) The entire tension of the string

(D) The gravitational force on the rock

5. The Ewok whirls the rock and releases it from a point above his head and to his right. The rock initially goes straight forward. Which of the following describes the subsequent motion of the rock?

(A) It will continue in a straight line forward, while falling due to gravity

(B) It will continue forward but curve to the right, while falling due to gravity

(C) It will continue forward but curve to the left, while falling due to gravity

(D) It will fall straight down to the ground

6. A Space Shuttle orbits the Earth 300 km above the surface. Why can not the Shuttle orbit 10 km above the Earth?

(A) The Space Shuttle cannot go fast enough to maintain such an orbit

(B) Kepler's laws forbid an orbit so close to the surface of the Earth

(C) Because r appears in the denominator of Newton's law of gravitation, the force of gravity is much larger closer to the Earth; this force is too strong to allow such an orbit

(D) Much of the Shuttle's kinetic energy would be dissipated as heat in the atmosphere, degrading the orbit

Multi-Select (多选题)

7. A satellite of mass m and speed v moves in a stable, circular orbit around a planet of mass M. The radius of the satellite's orbit

(A) is directly proportional to the mass of planet M

(B) is directly proportional to the mass of satellite m

(C) is inversely proportional to v^2

(C) is inversely proportional to v

Free-Response（计算题）

8. Consider two points on a rotating turntable: Point A is very close to the center of rotation, while point B is on the outer rim of the turntable. Both points are shown in the figure (图 2—7—8). A penny could be placed on the turntable at either point A or point B.

(a) In which case would the speed of the penny be greater, if it were placed at point A, or if it were placed at point B? Explain.

(b) At which point would the penny require the larger centripetal force to remain in place? Justify your answer.

(c) Point B is 0.25 m from the center of rotation. If the coefficient of friction between the penny and the turntable is $\mu = 0.30$, calculate the maximum linear speed the penny can have there and still remain in circular motion.

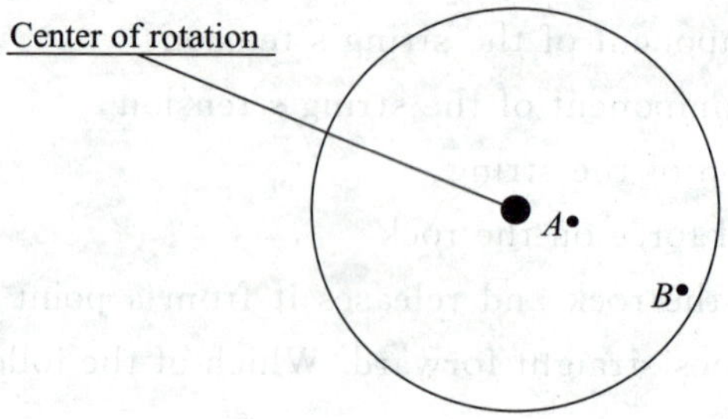

图 2—7—8

题解

单选题

1. 选 (D)。转盘和硬币之间有摩擦力，转盘作用在硬币上的摩擦力提供了该硬币做圆周运动的向心力，所以摩擦力的方向向下。

2. 选 (D)。根据牛顿万有引力定律，

$$F = G\frac{m_1 m_2}{r^2}$$

距离为两个行星间的距离 $2R$，而 $(2R)^2 = 4R^2$，所以选 (D)。

3. 选 (D)。向心加速度为：

$$a_c = \frac{v^2}{r}$$

因为两行星围绕其连线的中点运动，所以，半径为 R。

4. 选 (B)。考虑有竖直方向的力作用在石头上，石头的重力 mg 方向垂直向下。但石头没有下落，必然有向上的垂直分力与之抵消，这就是绳子张力的垂直分量。绳子张力

的水平分量提供了向心力。

5. 选（A）。当 Ewok 松手后，除重力以外，没有其他力作用在石块上，因此，石块在水平方向做匀速直线运动，竖直方向做自由落体运动。

6. 选（D）。根据圆周运动规律，飞船可以在离地球表面任意距离飞行，但在距离地表面 10 公里处，大气仍然密集，飞船会因为和空气摩擦产生高热而烧毁。在 300 公里以外，飞船已经脱离大气层，飞船可以自由飞行。

多选题

7. 选（A）和（C）。根据牛顿第二定律，可以列出卫星做匀速率圆周运动方程。

$\frac{GMm}{r^2}=m\frac{v^2}{r}$，可以解出：$r=\frac{GM}{v^2}$，因此，卫星轨道半径 r 和 M 成正比，和 v^2 成反比。

计算题

8. (a) A，B 两处，角速度相同，但因为 B 处远离转动中心，所以线速度大。

(b) 由于 B 点距离中心远，所以硬币要保持圆周运动需要更大的向心力。向心力等于 mv^2/r，线速度等于 $v=2\pi r/T$，因此，向心力为：$F_{net}=4m\pi^2 r/T^2$，所以，硬币在距离中心较远的 B 处需要更大的向心力。

(c) 由于摩擦力提供向心力，摩擦力等于 μF_N，正压力为 $F_N=mg$，所以 $\mu mg=mv^2/r$，解出 v：

$$v=\sqrt{\mu rg}$$

代入数据：$r=0.25$ m，$\mu=0.30$，可得：$v=0.87$ m/s。如果速度更大，则需要更大的向心力。

Chaptr 8

Rotational motion of a rigid body

第八章 刚体的转动（AP 物理 1 新增加的内容）

该知识点为改革后 AP 物理 1 新增加的内容。相对质点的平动，在讨论物体转动时，描述转动的物理量为：角位移、角速度、角加速度。对于平动，物体受到的合外力产生加速度。对于转动，物体受的合外力矩产生角加速度。对于转动，物体的质量和质量分布决定了转动惯量。同样，对应平动物体的动量，转动物体有角动量的概念，平动动能对应转动动能的概念。我们将介绍转动的动量定理和动能定理，并用它们解决实际问题。

Main points（主要知识点）

- 转动的角量表示：角位移、角速度、角加速度
- 转动惯量、力矩和转动的牛顿第二定律
- 物体的转动动能和平动动能
- 角动量和角动量守恒定律

Relevant equations（相关公式）

Rotational kinematics equations（匀加速转动的运动学公式）：When the acceleration is constant，we have a similar expression as translational motion：

$$\omega_f = \omega_0 + \alpha t$$

$$\Delta\theta = \omega_0 t + \frac{1}{2}\alpha t^2$$

$$\omega_f^2 = \omega_0^2 + 2\alpha\Delta\theta$$

Conversion between linear and rotational variables（线角关系）：

$$x = r\theta$$

$$v = r\omega$$

$$a = r\alpha$$

Moment of inertia for point particle（质点对轴的转动惯量）：$I = mr^2$

Moment of inertia for a continuous body（质量连续分布物体的转动惯量）：$I = \int r^2 \mathrm{d}m$

Newton's second law for rotation（转动的牛顿第二定律）: $\tau_{net} = I\alpha$

Rotational kinetic energy（转动动能）: $K_{rot} = \frac{1}{2}I\omega^2$

Angular momentum（角动量）: $L = I\omega = mvr$

1 Rotational kinematics 转动运动学

（1）物体运动的线量描述：

v_0 initial velocity

v_f final velocity

Δx displacement

a acceleration

t time interval

（2）物体运动的角量描述：

ω_0 initial angular velocity, measured in radians per second

ω_f final angular velocity, measured in radians per second

$\Delta\theta$ the total angle through which the spinning object rotates, measured in radians

α angular acceleration, telling how angular velocity changes with time

t time interval

（3）匀变速转动的运动方程

$$\omega_f = \omega_0 + \alpha t$$

$$\Delta\theta = \omega_0 t + \frac{1}{2}\alpha t^2$$

$$\omega_f^2 = \omega_0^2 + 2\alpha\Delta\theta$$

注意：我们可以用对比的方法将描述匀变速直线运动的方程和转动方程进行比较，得到事半功倍的效果。

（4）线角关系：

$$x = r\theta$$

$$v = r\omega$$

$$a = r\alpha$$

where r represents the distance from the spot you are interested in to the center of the object.

注意：以上描述的线角关系，是指没有滑动的转动。有滑动时，情况会更复杂。

2 Moment of inertia 转动惯量

1. Inertia for a single particle of mass m（质点对轴的转动惯量）

The moment of inertia, I, is the rotational equivalent of mass. It tells how difficult it is for an object to speed up or slow its rotation. For a single particle of mass m, a distance r from the axis of rotation, the moment of inertia is

$$I = mr^2$$

To find the moment of inertia of several masses—for example, two weights connected by a thin, light rod—just add the I due to each mass.

2. Inertia for a body with continuous mass distribution（质量连续分布物体的转动惯量）

For a complicated, continuous body, like a sphere or a disk, I can be calculated through integration:

$$I = \int r^2 \mathrm{d}m$$

注意：这已经超出物理 1 的范围。对于参加物理 C 的考生，应该掌握用积分的方法计算转动惯量。在上述积分中，关键是能正确掌握在不同坐标系中质量微元 dm 的写法，掌握变量 r 的选择。同时要掌握平行轴定理。在转轴平行移动时，转动惯量相应改变。

3. The theorem of inertia for a body about parallel axis（平行轴定理）

The rotational inertia of an object about an axis, a distance D from its center of mass ($I_{parallel\ axis}$) is related to the rotational inertia of the object about the parallel axis running through its center of mass (I_{CM}) according to the equation:

$$I_{parallel\ axis} = I_{CM} + MD^2$$

注意：平行轴定理中，I_{CM}代表通过质心轴的转动惯量，$I_{parallel\ axis}$ 是距离通过质心的轴为 D 的一个平行轴的转动惯量。知道二者任意一个，可求另外一个。

3 Newton's second law for rotation 转动牛顿第二定律

The Newton's second law for rotation is

$$\tau_{net} = I\alpha$$

Where τ_{net} is the net torque on an object. I is the rotational inertia and α is the angular acceleration.

Example:

A 3.0kg block on a smooth table is connected to a hanging 4.0kg block with a light string. This string passes over a pulley of mass 1.0 kg, as shown in the figure（图 2—8—1). Determine the acceleration of the masses. (The moment of inertia for a solid disc of mass m and radius r is (1/2) mr^2.)

Solution:

Now let us draw a free-body diagram（图 2—8—2) for each block, and use $F_{net} = ma$. So we start that way.

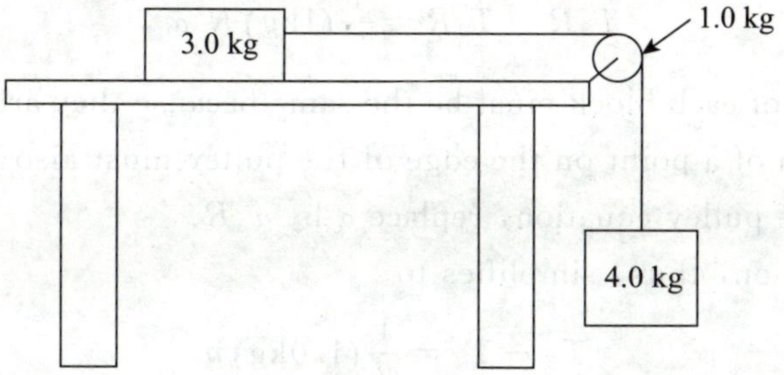

图 2—8—1

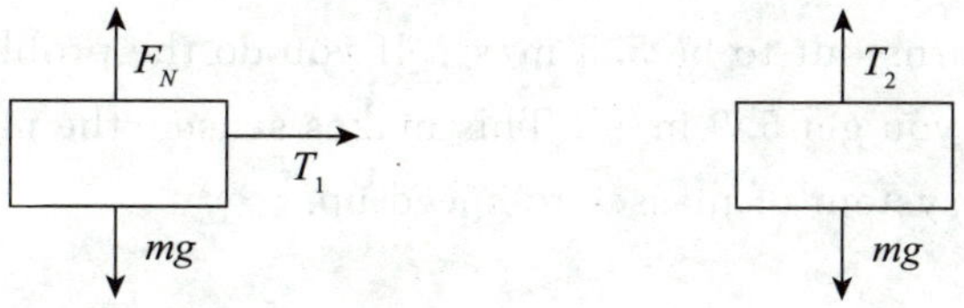

图 2—8—2

The twist in this problem is the massive pulley, which causes two changes in the problem solving approach:

(1) Draw a free-body diagram (图 2—8—3) for the pulley as well as the blocks. Even though it does not move, it still requires torque to accelerate its spinning speed.

(2) The tension in the rope may be different on each side of the pulley, because the pully has mass.

The Physics 1 corollary means that the free-body diagrams indicate T_1 and T_2 which must be treated as different variables. The free-body diagram for the pulley includes only these two tensions:

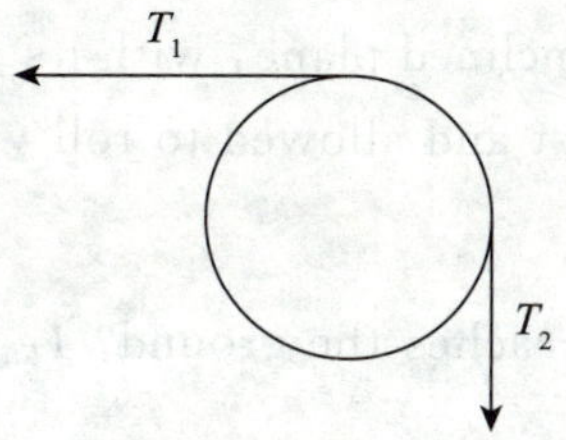

图 2—8—3

Now, we write Newton's second law for each block:

$$T_1 - 0 = (3\text{kg})a$$

$$(4\text{kg})\,g - T_2 = (4\text{kg})\,a$$

For the pulley, because it is rotating, we write Newton's second law for rotation. The torque provided by each rope is equal to the tension in the rope times the distance to the center of rotation:

$$T_2R - T_1R = \frac{1}{2}(1\text{kg})R^2\alpha$$

The acceleration of each block must be the same because they are connected by a rope; the linear acceleration of a point on the edge of the pulley must also be the same as that of the blocks. So, in the pulley equation, replace α by a/R.

The pulley equation, thus, simplifies to

$$T_2 - T_1 = \frac{1}{2}(1.0\text{kg})\,a$$

Now we have three equations and three variables (T_1, T_2 and a), and then we can get the solution:

The acceleration turns out to be 5.3 m/s^2. If you do the problem neglecting the mass of the pulley (try it!) you get 5.9 m/s^2. This makes sense—the more massive the pulley, the harder it is for the system of masses to speed up.

4 Rotational kinetic energy 转动动能

When an object is rotating, its rotational kinetic energy is found by the following equation:

$$KE_{rotational} = \frac{1}{2}I\omega^2$$

Notice that this equation is very similar to the equation for linear kinetic energy. But, because we're dealing with rotation here, we use moment of inertia in place of mass and if an object is moving linearly at the same time that it's rotating, its total kinetic energy equals the sum of the linear KE and the rotational KE.

Example:

A ball of mass m sits on an inclined plane, with its center of mass at a height h above the ground. It is released from rest and allowed to roll without slipping down the plane as shown in the figure (图 2—8—4).

What is its velocity when it reaches the ground? $I_{ball} = \frac{2}{5}mr^2$.

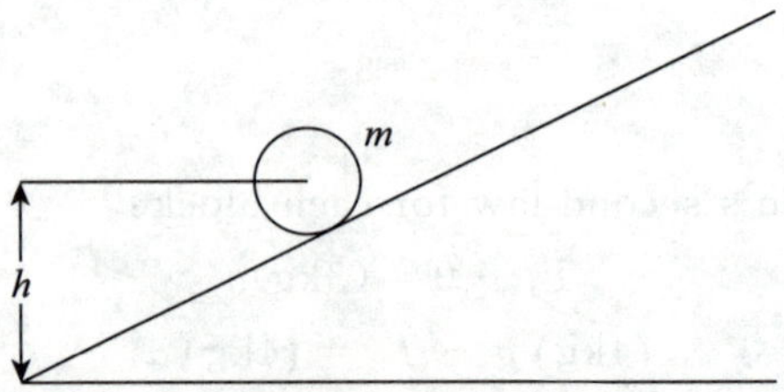

图 2—8—4

Solution:

This is a situation you've seen before, except that there's a twist: this time, when the object moves down the inclined plane, it gains both linear and rotational kinetic energy. However, it's still just a conservation of energy problem at heart. Initially, the ball just has gravitational potential energy, and when it reaches the ground, it has both linear kinetic and rotational kinetic energy.

$$mgh = \frac{1}{2}mv^2 + \frac{1}{2}I\omega^2$$

$$mgh = \frac{1}{2}mv^2 + \frac{1}{2}\left(\frac{2}{5}mr^2\right)\left(\frac{v}{r}\right)^2$$

We can find that: $v = \sqrt{\frac{10}{7}gh}$

If the ball in this problem hadn't rolled down the plane—if it had just slid—its final velocity would have been $\sqrt{2gh}$. So, it makes sense that if the final velocity of the ball when it does roll down the plane is less than $\sqrt{2gh}$, only a fraction of the initial potential energy is converted to linear kinetic energy.

5 Angular momentum and its conservation 角动量守恒

1. Angular momentum for a rigid body (刚体的角动量): In comparision with the linear momentum, the angular momentum should be in the following formula:

$$L = I\omega$$

2. Angular momentum for a point particle (质点对轴的角动量): Angular momentum for a point particle can be rewritten as

$$L = mvr$$

Where v is linear velocity, and r is either (1) the radius of rotation, if the particle is moving in a circle, or (2) distance of closest approach if the particle is moving in a straight line (see the figure (图 2—8—5)).

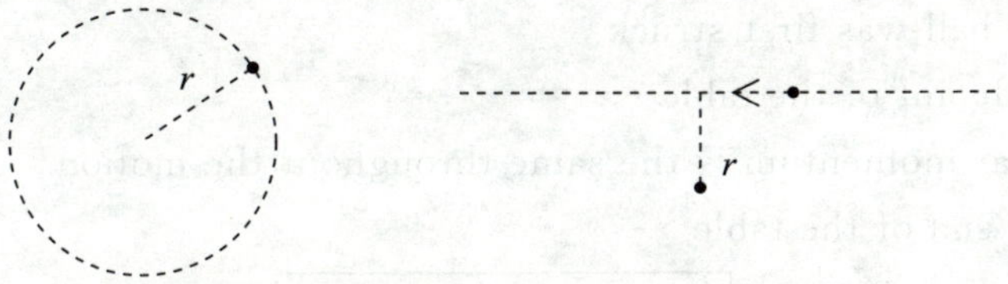

图 2—8—5 the angular momentum for a particle

Angular momentum is conserved (角动量守恒) in a closed system; that is, when no external torques act on the objects in the system, the angular momentum is conserved. Among the most famous examples of conservation of angular momentum is a satellite's orbit around a planet. As shown as follows (图 2—8—6): a satellite is traveling with an elliptical orbit around a planet. This means that the satellite is closer to the planet at some times than at others.

Obviously, at point A, the satellite is farther from the center of rotation than at point B.

Conservation of angular momentum tells us that, correspondingly, the angular speed at point A must be less than at point B.

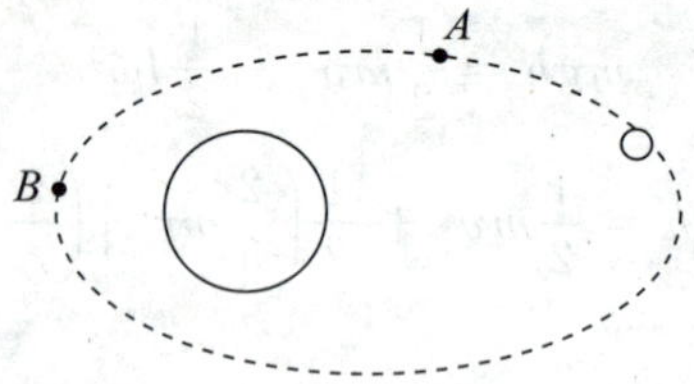

图 2—8—6 a satellite is traveling with an elliptical orbit around a planet

6 Practice questions 练习题

Multiple-Choice（选择题）

Single-Select（单选题）

1. All of the objects mentioned in the choices below have the same total mass and length. Which has the greatest moment of inertia about its midpoint?

(A) a very light rod with heavy balls attached at either end

(B) a uniform rod

(C) a nonuniform rod, with the linear density increasing from one end to the other

(D) a nonuniform rod, with the linear density increasing from the middle to the ends

2. A pool ball is struck at one end of the table; it moves at constant speed to the far end of the table. A camera is mounted at the side pocket at the table's midpoint, as shown in the figure（图 2—8—7）. From the camera's point of view, the pool ball travels from right to left. At which point in its motion does the ball have the greatest angular momentum about the camera's position?

(A) when the ball was first struck

(B) at the midpoint of the table

(C) the angular momentum is the same throughout the motion

(D) at the far end of the table

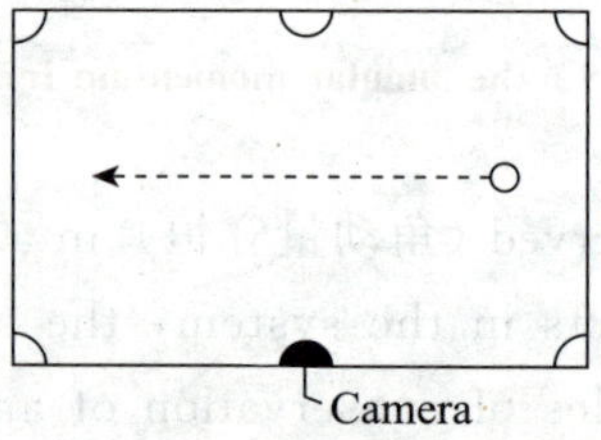

图 2—8—7

3. A ladder of length L leans against a wall at an angle of θ from the horizontal, as

shown in the figure（图 2—8—8）. The normal force F_N applied from the ground on the ladder applies what torque about the ladder's center of mass?

(A) F_N $(L/2)$ (B) $F_N L \cos\theta$

(C) $F_N L \sin\theta$ (D) F_N $(L/2)$ $\cos\theta$

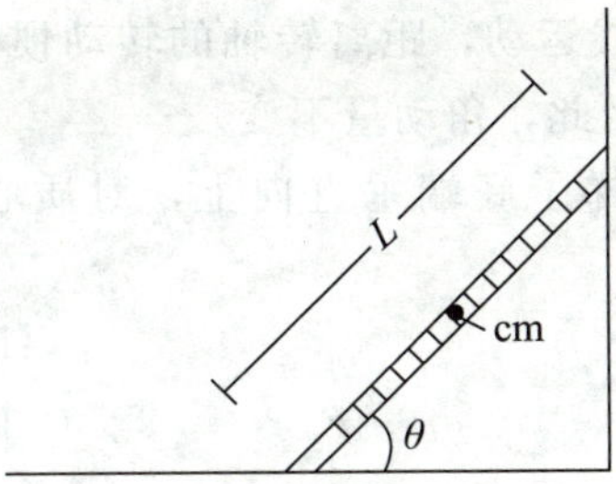

图 2—8—8

4. The front wheel on an ancient bicycle has radius 0. 5 m. It moves with angular velocity given by the function $\omega(t)=2+4t^2$, where t is in seconds.

About how far does the bicycle move between $t=2$ and $t=3$ seconds?

(A) 36m (B) 27m

(C) 21m (D) 14m

Multi-Select（多选题）

5. In rotational motion, the rotational inertia depends on

(A) the mass of object only

(B) the position of axis which the object turns around

(C) the mass distribution of the object

(D) the volume of the object only

Free-Response（计算题）

6. A stick of mass M and length L is pivoted at one end. A small mass $m \ll M$ is attached to the right-hand end of the stick. The stick is held horizontally and released from rest（图 2—8—9）.

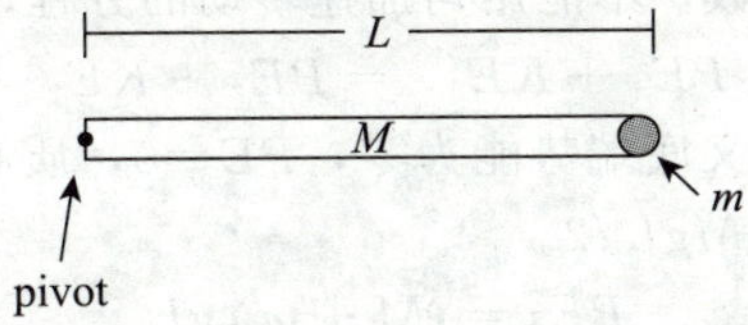

图 2—8—9

(a) Given that the moment of inertia of a uniform rod pivoted around one end is $\frac{1}{3}ML^2$, determine the moment of inertia of the described contraption.

(b) Calculate the angular velocity of the contraption when it reaches a vertical position.

(c) Calculate the linear velocity of the small mass m when it is at its lowest position.

题解

单选题

1. 选（A）。质量分布离中点越远，对转动惯量的贡献就越大。

2. 选（C）。一个质点沿直线运动，距离转轴的转动惯量用 mvr 表示。因为 r 在整个运动过程中不变，速度不变，因此，角动量不变。

3. 选（D）。力 F_N 作用在梯子底端垂直向上，对质心的距离为（$L/2$）$\cos\theta$，如图 2—8—10 所示。

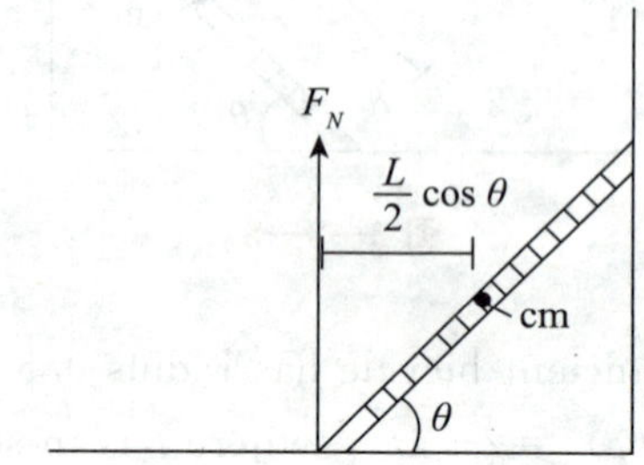

图 2—8—10

4. 选（D）。角位移等于角速度对时间的积分，积分限为 2～3 秒：

$$\int_2^3 (2+4t^2)\,dt = \left[2t+\frac{4t^3}{3}\right]_2^3 \approx 27 \text{ 弧度}$$

应用 $x = r\theta$，计算出距离约为 14 m。

（此题对于参加物理 1 的同学，有点超纲。）

多选题

5. 选（B）和（C）。转动惯量和质量以及质量在转动轴的分布有关。

计算题

6. (a) 总转动惯量为质量为 M 的杆对轴的转动惯量加上质点 m 对轴的转动惯量：

$$I_{total} = \frac{1}{3}ML^2 + mL^2$$

(b) 由于角加速度不是常数，不能用匀加速转动的方程，可以用能量守恒定律解：

$$PE_1 + KE_1 = PE_2 + KE_2$$

当杆转到垂直位置时，定义底端势能为零，$PE=0$，质心对势能的贡献为 PE_2，其位置在 0 点以上 $L/2$ 处，$PE_2 = MgL/2$。

$$PE_1 = (M+m)gL$$

$KE_1 = 0$，KE_2 未知。于是：

$$(M+m)gL + 0 = Mg\frac{L}{2} + \frac{1}{2}I\omega^2$$

代入 I 的表达式，可以解出 ω：

$$\omega = \sqrt{\frac{g}{L}\left(\frac{\frac{1}{2}M+m}{\frac{1}{3}M+m}\right)}$$

（c）用线角关系 $v = r\omega$，这里 $r = L$，

可得：$v = \sqrt{gL\left(\dfrac{\dfrac{1}{2}M + m}{\dfrac{1}{3}M + m}\right)}$

Chapter 9

Simple harmonic motion

第九章 简谐运动

该知识点讨论了简谐运动的条件、运动方程和位移。给出了运动振幅、周期、频率的表达式，位移、速度、加速度与时间的关系，以及运动过程中的能量关系。

Main Points（主要知识点）

- 判定物体是否做简谐运动有以下三个判据，参加物理 1 和 2 考试的同学，只要掌握前两个，参加物理 C 考试的同学还应掌握第三个

1. 做简谐运动的物体的运动方程应为正弦或余弦函数方程
2. 做简谐运动的物体受弹性恢复力 $F=-kx$ 的作用
3. 做简谐运动的物体运动的微分方程为 $\frac{d^2x}{dt^2}+\omega^2x=0$，其中$\omega$ 为圆频率

- 在无摩擦的平面上，质量为 m 的物块固定在弹簧一端的运动是简谐运动的基本范例
- 小摆幅单摆是简谐运动的另一个范例

Relevant equations（相关公式）

The equation of SHM（简谐振动的运动方程）：

$x=A\sin(\omega t+\varphi)$ ，其中 A 为振幅，φ 为初相位

Period of a mass on a spring（弹簧振子的周期）：

$$T=2\pi\sqrt{\frac{m}{k}}$$

Period of a pendulum（单摆周期）：

$$T=2\pi\sqrt{\frac{l}{g}}$$

Relationship between period and frequency（周期和频率的关系）：

$$T=\frac{1}{f}$$

1 Amptitude, period and frequency
振幅、周期和频率

Objects undergo oscillation when they experience a restoring force. This is a force that restores an object to the equilibrium position. For example, if the spring is stretched out, the restoring force tries to squish it back in, and if the spring is compressed, the restoring force tries to stretch it back out. Some books present this equation as $F=-kx$. The negative sign simply signifies that this is a restoring force.

One repetition of periodic motion is called a cycle. The maximum displacement from the equilibrium position during a cycle is the amplitude.

In the figure (图 2—9—1), the equilibrium position is denoted by "0" and the maximum displacement of the object on the end of the spring is denoted by "A".

The equation of the SHM can be expressed as: $x=A\sin(\omega t+\varphi)$.

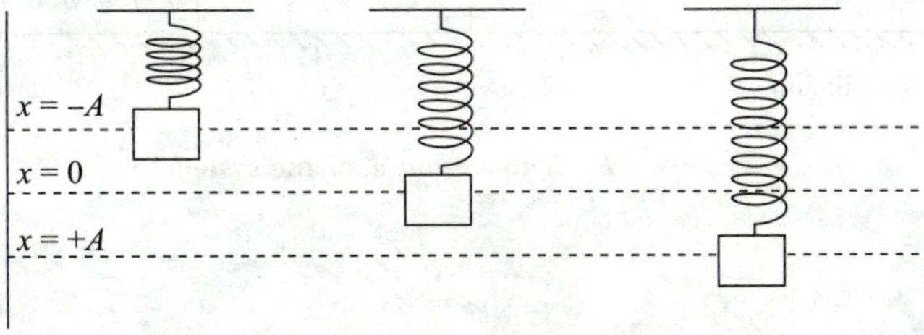

图 2—9—1 a spring block in SHM

The time it takes for an object to pass through one cycle is the period, abbreviated as "T". The number of cycles per second is the frequency. "f" is the standard abbreviation for frequency; the unit of frequency, the cycle per second, is called a hertz, abbreviated as Hz. Period and frequency are related by this equation:

$$T=\frac{1}{f}$$

2 Vibrating mass on a spring
弹簧振子

A mass attached to the end of a spring will oscillate in simple harmonic motion. The period of the oscillation is found by this equation:

$$T=2\pi\sqrt{\frac{m}{k}}$$

In this equation, m is the mass of the object on the spring, and k is the "spring constant". As far as equations go, this is one of the most difficult ones to memorize, but once you have committed it to memory, it becomes very simple to use.

Example:

A block of mass M is resting on a horizontal, frictionless table and is attached as shown in the figure (图 2—9—2) to a relaxed spring of spring constant k. A second block of mass $2M$ and initial speed v_0 collides with and sticks to the first block. Develop expressions for the following quantities in terms of M, k, and v_0.

(a) v, the speed of the blocks immediately after impact.

(b) x, the maximum distance the spring is compressed.

(c) T, the period of the subsequent simple harmonic motion.

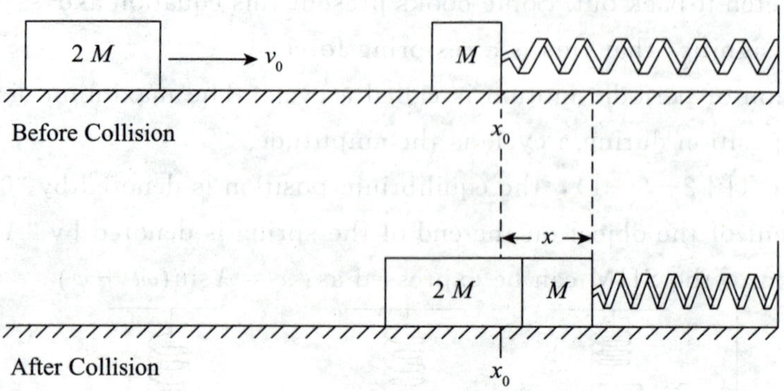

图 2—9—2 A mass and a spring system

Solution:

(a) In the collision of blocks, the linear momentum is conserved.

Before collision: $p = 2mv_0$

After collision: $p = 3mv$

So, $3mv = 2mv_0$

And $v = \frac{2}{3}v_0$

(b) After the collision, all the kinetic energy of the blocks is converted to the potential energy of spring:

$$KE = \frac{1}{2}mv^2 = \frac{1}{2}(3M)\left(\frac{2}{3}v_0\right)^2 = \frac{2}{3}Mv_0^2$$

So, $\frac{1}{2}kx^2 = \frac{2}{3}Mv_0^2$

$$x = \sqrt{\frac{4Mv_0^2}{3k}}$$

(c) Because the period of oscillation for a mass m attached to a spring of a spring constant k is

$$T = 2\pi\sqrt{\frac{m}{k}}$$

Here, m is $3M$, $T = 2\pi\sqrt{\frac{3M}{k}}$

3 Simple pendulums
单摆

The formula for the period of a simple pendulum is this:

$$T = 2\pi\sqrt{\frac{L}{g}}$$

In this equation, L is the length of the pendulum, and g is the acceleration attributable to gravity (about 10 m/s^2). Of course, if your pendulum happens to be swinging on another planet, g will have a different value.

One interesting thing about this equation: The period of a pendulum does not depend on the mass which is hanging on the end of the pendulum.

4 Physical pendulums
物理摆

The angular velocity: $\omega = \sqrt{\frac{mgd}{I}}$

The period: $T = 2\pi\sqrt{\frac{I}{mgd}}$

Example:

A string with a bowling ball tied to its end is attached to the ceiling (图 2—9—3). The string is pulled back such that it makes a 10°angle with the vertical, and it is then released. When the bowling ball reaches its lowest point, it has a speed of 2 m/s. What is the frequency of this bowling ball-pendulum?

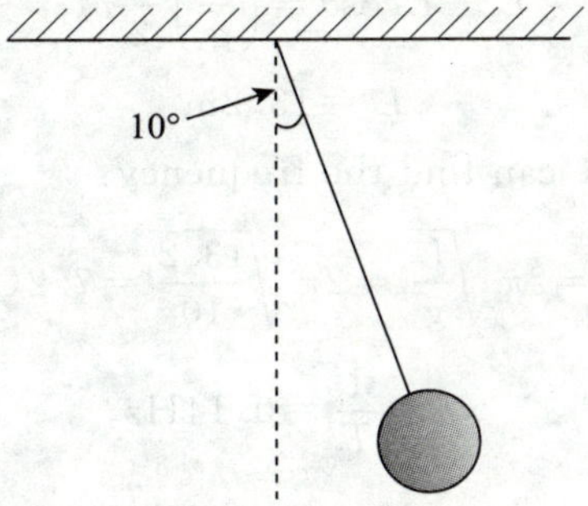

图 2—9—3 a simple pendulum

Solution:

To calculate the period of this pendulum, we must know the length of the string. We can calculate this using conservation of energy. Then, we will convert the period to a frequency.

Before the string is released, all of the bowling ball's energy is in the form of gravitational PE. If we define the zero of potential to be at the ball's lowest point, then at that point all the bowling ball's energy is in the form of KE. We will use a subscript "a" to represent values before the bowling ball is released and "*b*" to represent values when the bowling ball is at its lowest point.

$$KE_a + PE_a = KE_b + PE_b$$

$$\frac{1}{2}mv_a^2 + mgh_a = \frac{1}{2}mv_b^2 + mgh_b$$

$$0 + mgh_a = \frac{1}{2}mv_b^2 + 0$$

The height of the bowling ball before it is released, h_a, can be calculated using trigonometry (图 2—9—4).

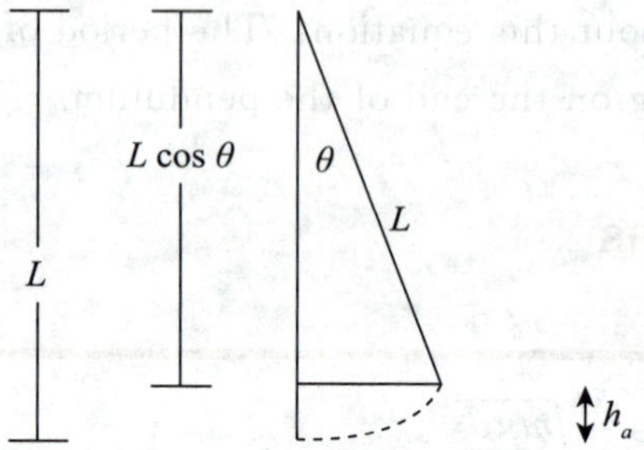

图 2—9—4 the relation between *L* and *θ*

$$h_a = L - L\cos\theta$$

So, getting back to our previous equation, we have

$$0 + mg(L - L\cos\theta) = \frac{1}{2}mv_b^2 + 0$$

We know θ and v_b, So we can solve for L.

$$L - L\cos\theta = \left(\frac{1}{g}\right)\frac{1}{2}v_b^2$$

$$L = 13.2\text{m}$$

Now that we know L, we can find the frequency:

$$T = 2\pi\sqrt{\frac{L}{g}} = 2\pi\sqrt{\frac{13.2}{10}} = 7.2(\text{s})$$

$$f = \frac{1}{T} = 0.14\text{Hz}$$

5 Practice questions 练习题

Multiple-Choice (选择题)

Single-Select (单选题)

1. An object swings on the end of a cord as a simple pendulum with period T. Another object oscillates up and down on the end of a vertical spring also with period T. If the masses of both objects are doubled, what are the new values for the periods?

	pendulum	mass on spring
(A)	$\frac{T}{\sqrt{2}}$	$\sqrt{2}T$
(B)	T	$\sqrt{2}T$
(C)	T	T
(D)	$\sqrt{2}T$	T

2. A pendulum has a period of five seconds on the Earth. On the Jupiter, where $g = 30\text{m/s}^2$, the period of this pendulum would be closest to

(A) 1s (B) 3s

(C) 5s (D) 8s

3. A pendulum and a mass on a spring are designed to vibrate with the same period T. These devices are taken onto the Space Shuttle in orbit. What is the period of each on the Space Shuttle ?

	pendulum	mass on a spring
(A)	will not oscillate	T
(B)	T	will not oscillate
(C)	will not oscillate	will not oscillate
(D)	$\sqrt{2}T$	T

4. A mass on a spring has a frequency of 2.5 Hz and an amplitude of 0.05 m. In one complete period, what distance does the mass traverse? (This question asks for the actual distance, not the displacement)

(A) 0.05 cm (B) 0.01 cm

(C) 20 cm (D) 10 cm

5. Increasing which of the following will increase the period of a simple pendulum?

Ⅰ. the length of the string

Ⅱ. the local gravitational field

Ⅲ. the mass attached to the string

(A) Ⅰ only

(B) Ⅱ only

(C) Ⅲ only

(D) Ⅰ and Ⅱ only

Multi-Select (多选题)

6. A simple pendulum and a mass hanging on a spring both have a period of 1s when set into small oscillatory motion on the Earth. They are taken to Planet X, which has the same diameter as the Earth but twice the mass. Which of the following statements are true about the periods of the two objects on Planet X compared to their periods on the Earth?

(A) Both are shorter

(B) Both are longer

(C) The period of the pendulum on planet X is shorter

(D) The period of the pendulum on the Earth is longer

Free-Response（计算题）

7. A mass m is attached to a horizontal spring of spring constant k. The spring oscillates in simple harmonic motion with amplitude A. Answer the following in terms of A.

(a) At what displacement from equilibrium is the speed half of the maximum value ?

(b) At what displacement from equilibrium is the potential energy half of the maximum value?

(c) When is the mass farther from its equilibrium position, when its speed is half maximum, or when its potential energy is half maximum?

题解

单选题

1. 选（B）。由于单摆运动周期和质量无关，因此，考虑选（B）或（C）。弹簧振子周期为：

$$T=2\pi\sqrt{\frac{m}{k}}$$

当质量增大一倍时，振动周期变为：

$$\sqrt{2}\,T$$

2. 选（B）。根据单摆周期公式

$$T=2\pi\sqrt{\frac{L}{g}}$$

木星上的重力加速度是地球的三倍，即 $3g$，所以单摆在木星上的周期为在地球上周期的$\frac{1}{\sqrt{3}}$，于是，在地球上周期为 5 秒，则在木星上变为约 3 秒。

3. 选（A）。因为引起单摆振动的弹性恢复力是重力，宇宙飞船上是失重的，所以单摆不会振动。然而，引起弹簧振子振动的力是弹性恢复力，在飞船上不受影响，周期不变。

4. 选（C）。简谐运动的振幅为振子距离平衡位置的最大距离。在一个周期内，振子移动四倍振幅距离，为 0.20m，即 20cm。

5. 选（A）。根据单摆周期公式：

$$T=2\pi\sqrt{\frac{L}{g}}$$

单摆振动周期随摆长增加而增加，随重力加速度增大而变小，与质量无关。

多选题

6. 选（C）和（D）。由于 X 行星质量比地球大，所以在该星球表面重力加速度比地球大。

根据单摆周期公式可知，在 X 行星上单摆的周期比在地球上短。

计算题

7. (a) 简谐运动中，平衡位置势能为零，速度最大，即动能最大。根据机械能守恒：

$$\frac{1}{2}kA^2=\frac{1}{2}mv_{\max}^2$$

可以解出最大速度为：

$$v_{\max}=A\sqrt{\frac{k}{m}}$$

在$\frac{1}{2}A$处，速度为最大值的一半。于是，根据能量守恒：

$$\frac{1}{2}kA^2=\frac{1}{2}mv_2^2+\frac{1}{2}kx_2^2$$

速度 v_2是最大速度的一半，即 $\frac{1}{2}\left(A\sqrt{\frac{k}{m}}\right)$，代入方程，解出 x_2：

$$\frac{1}{2}kA^2=\frac{1}{2}m\left(\frac{1}{2}A\sqrt{\frac{k}{m}}\right)^2+\frac{1}{2}kx_2^2$$

$x_2=A\sqrt{\frac{3}{4}}$，约为振幅最大值的 86%。

(b) 由于总能量为 $\frac{1}{2}kA^2$，某一位置 x，势能为最大值的$\frac{1}{2}$，因此，在该点有：

$$\frac{1}{2}kx^2=\frac{1}{2}\left(\frac{1}{2}kA^2\right)$$

解出

$$x=\frac{1}{\sqrt{2}}A$$

该值大约为振幅最大值的 70%。

(c) 根据 (a) 和 (b) 的结果，当位置在 $\sqrt{\frac{3}{4}}A$，或 86% A 时，速度为最大值的一半；当位置在 $\frac{1}{\sqrt{2}}A$，或 71%A 时，势能为最大值的一半。因此，速度为最大值一半时，物块离平衡位置更远。

Chapter 10

Electrostatics

第十章　静电学*

本知识点主要讲述了电荷、电场力、库仑定律和电场强度的概念。注意电场强度是矢量。另外，从电场力做功的角度介绍了电势和电势差的概念。注意电势是标量。要注意正电荷所受电场力指向电场的方向，即电势降落的方向；负电荷与之相反。点电荷激发的电场是非匀强电场，带电平行板之间的电场可视为匀强电场。这些知识是物理 1 和 2 考试所要求的，同时也是物理 C 的基础。

Main points（主要知识点）

- 电荷和电荷守恒定律
- 电场强度和电场力
- 电场力做功、电场能量、电势的概念
- 匀强电场和非匀强电场

Relative equation（相关公式）

Electric force between two point charges—Coulomb's law（库仑定律）：$F=\frac{kQq}{r^2}=\frac{Qq}{4\pi\varepsilon_0 r^2}$

Electric force on a charge in all electric field（电场力）：$F=qE$

Electric field produced by a point charge（点电荷激发的电场）：$E=\frac{kQ}{r^2}=\frac{Q}{4\pi\varepsilon_0 r^2}$

Electric field produced by parallel plates（平行板电容器的电场）：$E=\frac{V}{d}$

Electric potential energy（电势能）：$PE=qV$　or　$PE=\frac{1}{4\pi\varepsilon_0}\frac{q_1q_2}{r}$

---

* AP 物理 1 和 2 均涉及静电学内容，物理 1 重点介绍电荷、电荷守恒、库仑定律等内容。物理 2 重点介绍电场、电势、静电平衡等内容。在知识体系上，这部分内容互相联系，不可分割。因此，本知识点内容兼顾 AP 物理 1 和 2。

Voltage produced by a point charge (点电荷激发的电势): $V=\frac{kQ}{r}=\frac{Q}{4\pi\varepsilon_0 r}$

Capacitance of a capacitor (电容器的电容): $C=\frac{Q}{V}$

Capacitance of a parallel plate capacitor (平行板电容器的电容): $C=\frac{\varepsilon_0 A}{d}$

1 Electric charge & the conservation of electric charge 电荷和电荷守恒

All matter is made up of three types of particles: protons, neutrons, and electrons. Protons have an intrinsic property called "positive charge." Neutrons don't contain any charge, and electrons have a property called "negative charge."

The unit of charge is the coulomb, abbreviated as C. One proton has a charge of 1.6×10^{-19} C. One electron has a charge of -1.6×10^{-19} C. This is called elementary charge in the nature.

In an isolated system, the charges are conserved.

2 Electric fields 电场

Field (场): A property of a region of space that can apply a force to an object with mass found in that region of space.

Electric field (电场): A property of a region of space that applies a force to charged objects in that region of space. A charged particle in an electric field will experience an electric force.

1. The units of electric field (电场的单位): Newtons/Coulomb, abbreviated as N/C, or Voltage/meter, V/m.

注意：电场和引力场的区别：引力场非常弱，和电场比较，可以忽略。电场可以由正、负两种电荷激发，而引力场只和负电荷激发的电场类似。

2. Electric force and the strength of electric field (电场力和电场强度):

Electric field strength (电场强度): The ratio between the force exerted on the test charge and the quantity of the test charge: $E=\frac{F}{q}$

特别注意：电场强度只和激发电场的场源电荷有关，和检验电荷 q 无关！

Electric force (电场力): The force felt by a charged particle in an electric field is described by a simple equation:

$$F=qE.$$

3 Electric potential
电势

1. Electric potential energy (电势能): The electric potential energy is the work done by the external force against the electric force exerted on a positive charge.

2. Electric potential (电势): Potential energy provided by an unit charge.

Electric potential is a scalar quantity. The units of electric potential are volts. 1 volt = 1 J/C.

For a point charge Q, the potential created by Q in distance of r can be expressed by:

$$V=\frac{kQ}{r}=\frac{Q}{4\pi\varepsilon_0 r}$$

3. Zero electric potential (零电势位置的选择):

(1) For point charge or a limited size charged body: The zero potential position is chosen in infinity.

(2) For an infinity large charged body, the zero potential position can NOT be chosen in infinity.

4. The calculation of electric potential energy (计算电势能的公式):

The electric potential energy of a charged particle is given by this equation:

$$PE=qV$$

Here, q is the charge on the particle, and V is the voltage.

Example:

Object Ⅰ, shown in figure 2—10—1 (a), has a charge of $+3\times10^{-6}$ C and a mass of 0.002 5 kg.

Object Ⅱ, of the same mass as object Ⅰ, but having a charge of $+1\times10^{-6}$ C, is brought from infinity to point P, as shown in the figure (b).

(a) What is the electric potential at point P, 0.30 meter from object Ⅰ?

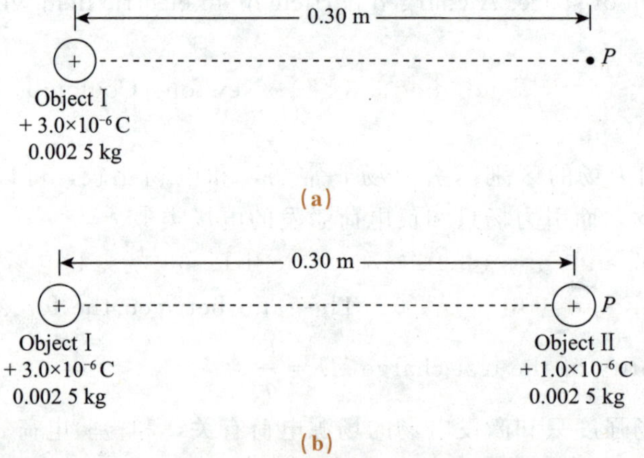

图 2—10—1

(b) How much work must be done to bring object Ⅱ from infinity to point P?

(c) What is the magnitude of the electric force between the two objects when they are

0.30 meter apart?

(d) What are the magnitude and direction of the electric field at the point midway between the two objects?

Solution:

(a) According to the equation: $V=\frac{kQ}{r}=(9\times10^{9})\frac{(3\times10^{-6})}{0.3}=9.0\times10^{4}(\mathrm{V})$

(b) Based on the equation: $W=qV=(1\times10^{-6})(9\times10^{4})=0.09(\mathrm{J})$

(c) According to Coulomb's law:

$$F=\frac{kQq}{r^{2}}=(9\times10^{9})\frac{(3\times10^{-6})(1\times10^{-6})}{(0.3)^{2}}=0.3(\mathrm{N})$$

(d) The electric field at the point midway between the two charges is the sum of vector:

The electric field created by object Ⅰ:

$$E_{1}=\frac{kQ}{r^{2}}=(9\times10^{9})\frac{(3\times10^{-6})}{(0.15)^{2}}=1.2\times10^{6}(\mathrm{N/C})\quad \text{to right}$$

The electric field created by object Ⅱ:

$$E_{2}=\frac{kQ}{r^{2}}=(9\times10^{9})\frac{(1\times10^{-6})}{(0.15)^{2}}=0.4\times10^{6}(\mathrm{N/C})\quad \text{to left}$$

When we choose right as the positive direction, the magnitude of the sum of electric field is:

$$E=E_{1}-E_{2}=8.0\times10^{5}(\mathrm{N/C})$$

The direction of the field is to the right.

5. Equipotential lines (等势线):

Figure 2—10—2 shows a few examples of equipotential lines (shown with solid lines) and their relationship to electric field lines (shown with dotted lines): the electric field points away from the positive charge.

图 2—10—2 equipotential lines

The equipotential lines are always drawn perpendicular to electric field lines, but when the field lines are not parallel (as on the left of the figure), this fact is harder to see, but it is clear at every point.

4 Special geometries for electrostatics
两个特殊的静电场

1. Parallel plates (带电平行板间的场)

If you take two metal plates, charge one positive and one negative, and then put them parallel to each other, you create a uniform electric field in the middle, as shown in figure 2—10—3:

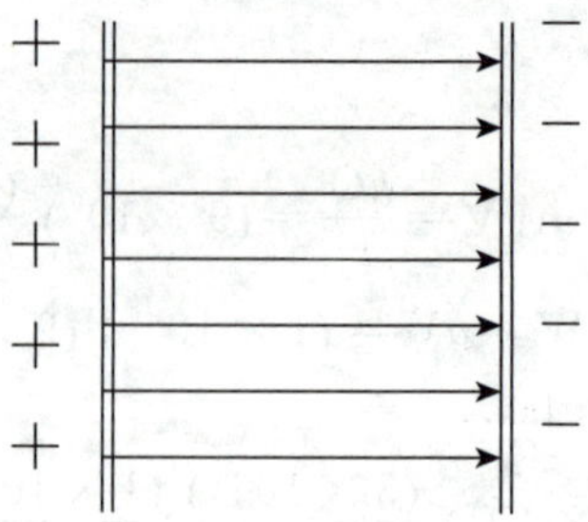

图 2—10—3

The electric field between the plates has a magnitude of

$$E=\frac{V}{d}$$

注意：这个公式只适用于匀强电场，对于非匀强电场，AP 物理 C 中要进一步讨论。

Capacitance of parallel plates（带电平行板的电容）:

If each plate has charge of Q, the voltage across the plates is V, and C is the ratio between Q and V, it can be written as:

$$C=\frac{Q}{V}$$

It is determined primarily by the size of the plates and the distance between the plates, as well as by the material that fills the space between the plates. The units of capacitance are farads, abbreviated as F; 1 coulomb/volt = 1 farad.

The capacitance of a parallel plate capacitors can be expressed by:

$$C=\frac{\varepsilon_0 A}{d}$$

In this equation, A is the area of each plate (in m^2), and d is the distance between the plates (in m). The term ε_0 is called the "permittivity of free space". The value of ε_0 is 8.84×10^{-12} C/V · m.

2. Point charges（点电荷的场）

(1) Coulomb's law（库仑定律）——两个点电荷之间的相互作用：

$$F=\frac{kQ_1Q_2}{r^2}=\frac{Q_1Q_2}{4\pi\varepsilon_0 r^2}$$

In this equation, Q_1 is the charge of one of the point charges, and Q_2 is the charge of the other one. This equation is known as Coulomb's law. k is called the Coulomb's law's constant ($k=9\times10^9$ N m^2/C^2) and r is the distance between the two point charges. Writing it using ε_0 in this way is very helpful for students in Physics C.

(2) Electric field strength by point charge（点电荷激发的场强）

Based on definition of electric field strength, $E=\frac{F}{q}$

and Coulomb's law: $F=\frac{kQq}{r^2}=\frac{Qq}{4\pi\varepsilon_0 r^2}$

The electric field: $E=\frac{kQ}{r^2}=\frac{Q}{4\pi\varepsilon_0 r^2}$

Q is the charge which created the electric field, and r is the distance away from the point charge. The electric field created by point charge is shown in figure 2—10—4.

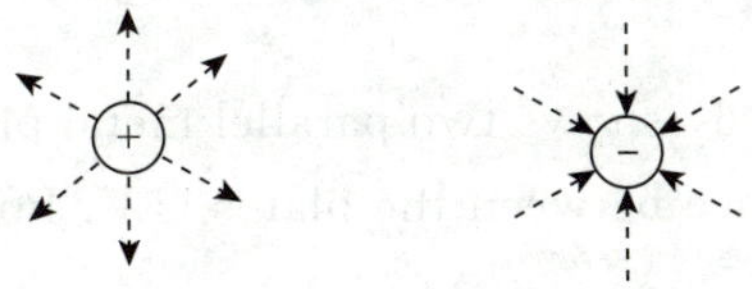

图 2—10—4 electric field

注意：电场强度是矢量。

(3) Electric potential by point charge（点电荷周围的电势）：

$$V=\frac{kQ}{r}=\frac{Q}{4\pi\varepsilon_0 r}$$

When using this equation, you must include a + or − sign on the charge creating the potential.

要特别注意，平行板间的电场可以视为匀强电场，而点电荷周围的电场不是匀强电场。

5 Practice questions
练习题

Multiple-Choice（选择题）

Single-Select（单选题）

Questions 1 and 2:

Two positive charges of magnitude q, each has a distance d from the origin A of a coordinate system, are shown in figure 2—10—5.

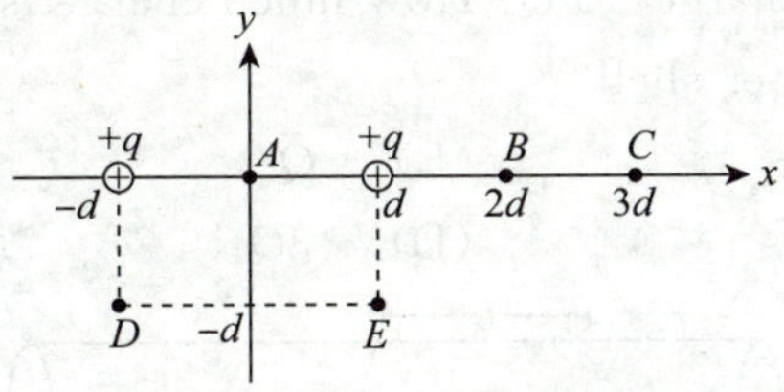

图 2—10—5

1. At which of the following points is the electric field least in magnitude?

(A) A (B) B

(C) C (D) D

2. At which of the following points is the electric potential greatest in magnitude?

(A) E (B) D

(C) C (D) A

Questions 3 and 4:

The diagram (图 2—10—6) shows two parallel metal plates that are separated by distance d. The potential difference between the plates is V. Point A is twice as far from the negative plate as point B.

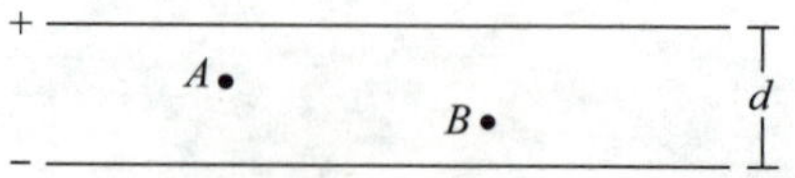

图 2—10—6

3. Which of the following statements about the electric potential between the plates is correct?

(A) The electric potential is the same at points A and B

(B) The electric potential is two times larger at A than at B

(C) The electric potential is two times larger at B than at A

(D) The electric potential is four times larger at A than at B

4. Which of the following statement about the electric field between the plates is correct?

(A) The electric field is the same at points A and B

(B) The electric field is two times larger at A than at B

(C) The electric field is two times larger at B than at A

(D) The electric field is four times larger at A than at B

5. (This problem is for AP Physics C students only) A very long cylindrical conductor is surrounded by a very long cylindrical conducting shell, as shown in figure 2—10—7. A length L of the inner conductor carries positive charge Q. The same length L of the outer shell carries total charge $-3Q$. How much charge is distributed on a length L of the outside surface of the outer shell?

(A) none (B) $-Q$

(C) $-2Q$ (D) $-3Q$

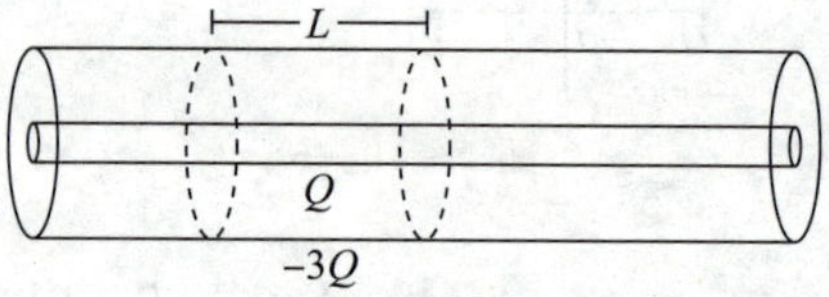

图 2—10—7

Multi-Select（多选题）

6. The two charged metal spheres X and Y shown in figure 2—10—8 are far apart, and each is isolated from all other charges. The radius of sphere X is greater than that of sphere Y, and the magnitudes of the charge on the surface of X is Q, and the magnitude of the charge on the surface of Y is q. If a conductive wire connects the two spheres, which statements are correct?

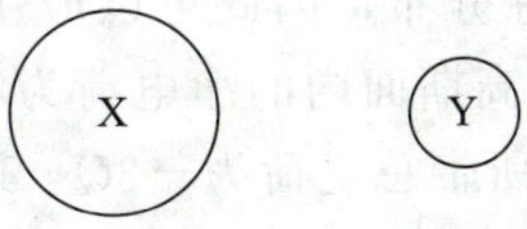

图 2—10—8

(A) The electric potential of two spheres is the same

(B) The charge on both spheres is equal to $(Q+q)/2$

(C) The electric potential at X is larger than that at Y

(D) The charge on sphere X is more than that on sphere Y

Free-Response（计算题）

7. Two conducting metal spheres of different radius are as shown in figure 2—10—9, each has charge $-Q$.

(a) Consider one of the spheres. Is the charge on that sphere likely to clump together or to spread out? Explain briefly.

(b) Is charge more likely to stay inside the metal spheres or on the surface of the metal spheres? Explain briefly.

(c) If the two spheres are connected by a metal wire, will charge flow from the big sphere to the little sphere, or from the little sphere to the big sphere? Explain briefly.

(d) Which one of the following two statements is correct? Explain briefly.

i. If the two spheres are connected by a metal wire, charge will stop flowing when the electric field at the surface of each sphere is the same.

ii. If the two spheres are connected by a metal wire, charge will stop flowing when the electric potential at the surface of each sphere is the same.

(e) Explain how the correct statement you chose from part (d) is consistent with your answer to (c).

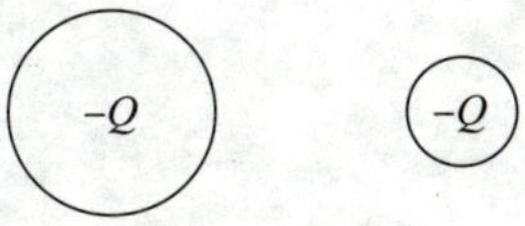

图 2—10—9

题解

单选题

1. 选（A）。电场强度是矢量，A 点场强为零，其他各点场强均不为零。

2. 选（D）。电势是标量，由于 A 点距离两个正电荷最近，均为 d，因此，该点电势最大。

3. 选（B）。如果两板间的电势差为 100V，可以说一板电势为 +100 V，另一板电势为零。两板间电势在变化。由于是匀强电场，场强为 V/d，电势变化为线性的，所以 A 点电势为 B 点 2 倍。

4. 选（A）。板间电场为匀强电场，各点电场强度相同。

5. 选（C）。由于电场是柱对称分布，因此可以应用高斯定理。在外柱壳中建立一个同轴高斯面，壳内电场为零，因此高斯面内的净电荷为零，因为内柱电荷为 $+Q$，外壳的内表面必然有电荷 $-Q$，由于外壳所带总电荷为 $-3Q$，其他 $-2Q$ 电荷必然在外柱壳表面。

多选题

6. 选（A）和（D）。导体球 X 和 Y 用金属线连接成为一个等势体。由于 X 体积比 Y 大，所以 X 电容比 Y 大，储存电荷比 Y 多。

计算题

7. (a) 由于电荷同性相斥，电荷必将尽可能向两侧分布。

(b) 电荷都将分布在球表面。

(c) 因为大球的电荷分布比小球稀疏，电荷将从小球流向大球。

(d) ii 是正确的。由于两球用导线连接，它们是等势体，电荷不流动。

(e) 球面电势为 $-kQ/r$，式中 r 是球半径，小球电势比大球更负，负电荷将从低电势处流向高电势处。

Chapter 11

Electric circuits

第十一章 直流电路*

本知识点介绍基本的直流电路、欧姆定律和电功率的表达式。讨论电阻串联、并联的特点和公式。同时，介绍电容的概念和电容器。

Main points（主要知识点）

- 串联电路中电流处处相同，总电压等于分电压的和
- 并联电路中分支路电压相同，总电流等于分电流的和
- 部分电路的欧姆定律和全电路欧姆定律
- 电容有储存电荷的能力，但在直流电路中电容器会阻隔电流

Relevant equations（相关公式）

Definition of current（电流强度的定义）: $I=\dfrac{\Delta Q}{\Delta t}$

Resistance of a wire（导线中的电阻）: $R=\rho\dfrac{L}{A}$

Ohm's law（部分电路和全电路欧姆定律）: $I=\dfrac{V}{R}$, $I=\dfrac{V}{R+r}$

Power in a circuit（电功率）: $P=IV=I^2R$

1 Current 电流强度

1. Current（电流强度）:

In a circuit, the current is the amount of charge passing a given point per unit time. In more mathematical terms, current is defined as follows:

* 该知识点为 AP 物理 1 和 2 所共有。在 AP 物理 1 中，只涉及纯电阻电路，在 AP 物理 2 中，还涉及电容。因为这是一个知识体系，所以这里统一介绍。

$$I = \frac{\Delta Q}{\Delta t}$$

2. The unit:

The unit of current is Coulombs/second. 1 C/s =1 Ampere, abbreviated as 1 A.

2 Resistance and Ohm's law
电阻和欧姆定律

1. Resistance（电阻）:

A property of a circuit that resists the flow of current.

2. The unit:

The unit of resistance is ohms. 1 ohm is abbreviated as 1 Ω.

3. The calculation of resistance for a conduct wire（导线电阻的计算）:

If we have some length of wire, then the resistance of that wire can be calculated. Three physical properties of the wire affect its resistance:

- The material the wire is made out of the resistivity ρ which is an intrinsic property of that material. Good conducting materials, like silver, have low resistivities.
- The length of the wire, L.
- The cross-sectional area A of the wire.

We put all of these properties together in the equation for resistance of a wire:

$$R = \rho \frac{L}{A}$$

4. Resistor（电阻器）:

Resistor is that something you put it in a circuit to change the circuit's resistance.

5. Ohm's law（部分电路的欧姆定律）: $I = \frac{V}{R}$

V is the voltage across the part of the circuit you are looking at, I is the current flowing through that part of the circuit, and R is the resistance in that part of the circuit. Ohm's law is the most important equation.

6. Electric power（电功率）:

Electric power can be defined as the product of current and voltage:

$$P = VI$$

This equation says that the power, P dissipated in part of a circuit equals the current flowing through that part of the circuit multiplied by the voltage across that part of the circuit.

Using Ohm's law, it can be easily shown that $IV = I^2R = V^2/R$.

3 Resistors in series and in parallel
电阻的串联和并联

In a circuit, resistors can either be arranged in series with one another or parallel to one another. Before we take a look at each type of arrangement, though, we need first to familiarize ourselves with circuit symbols, as shown in figure 2—11—1:

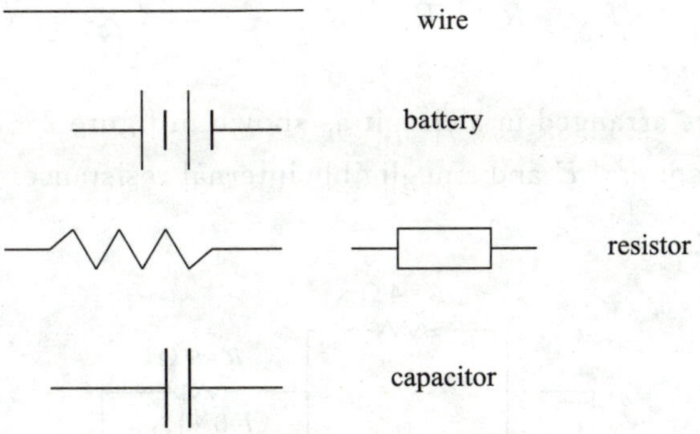

图 2—11—1 the symbols of electric circuits

注意：我国教科书上用的电阻符号为上图电阻符号右侧的小矩形，AP 物理中用左侧的折线。

1. The characteristics of series circuit（串联电路的特点）：

(1) When two resistors are connected in SERIES, the amount of current that flows through one resistor equals the amount of current that flows through the other resistor.

(2) The total voltage across all resistors is equal to the sum of voltage across individual resistor（图 2—11—2).

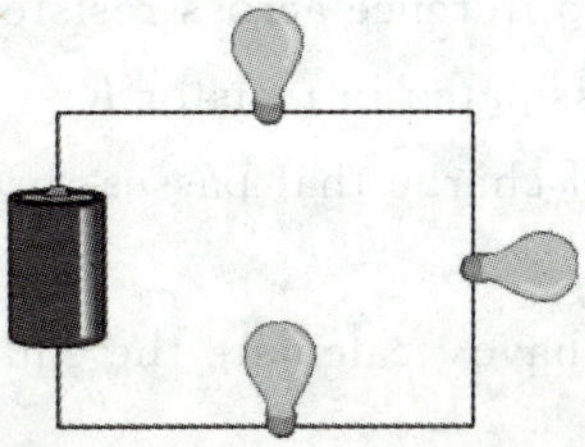

图 2—11—2 the resistors in series

To find the equivalent resistance of series resistors, we just add up all the individual resistors.

$$R_{eq}=R_1+R_2+R_3+\cdots=\sum_{i=1}^{n}R_i$$

2. The characteristics of parallel circuit（并联电路的特点）：

(1) When two resistors are connected in PARALLEL, the voltage across one resistor is the same as the voltage across the other resistor.

(2) The total current through all resistors is equal to the sum of current through individual resistor（图 2—11—3).

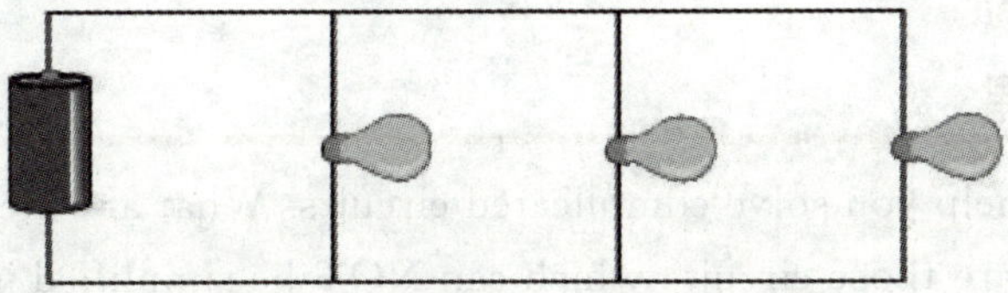

图 2—11—3 the resistors in parallel

The equivalent resistance of parallel resistors is found by this formula:

$$\frac{1}{R_{eq}}=\frac{1}{R_1}+\frac{1}{R_2}+\frac{1}{R_3}+\cdots=\sum_{i=1}^{n}\frac{1}{R_i}$$

Example:

Three resistors are arranged in a circuit as shown in figure 2—11—4. The battery has an unknown but constant emf E and a negligible internal resistance. The current I in resistor R_3 is 0.40 ampere.

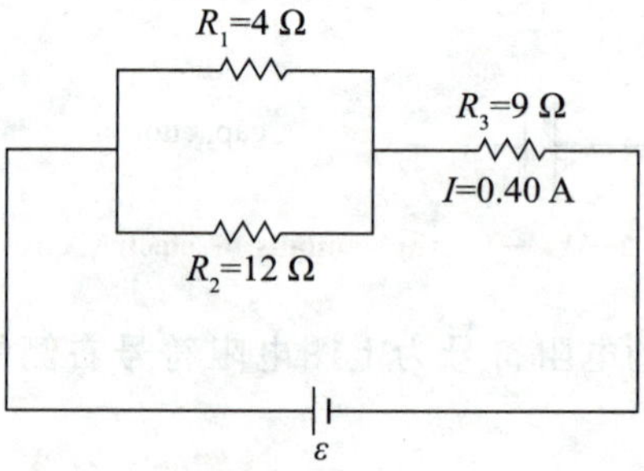

图 2—11—4

(a) Determine the equivalent resistance of the three resistors.

(b) Determine the emf (Voltage) of the battery.

(c) Determine the potential difference across resistor R_1

(d) Determine the power dissipated in resistor R_1

(e) Determine the amount of charge that passes through resistor R_3 in one minute.

Solution:

(a) Using the equations we have, calculate the equivalent resistance of the three resistors as follows:

(i) The equivalent resistance of R_1 and R_2: $\frac{1}{R_{para}}=\frac{1}{R_1}+\frac{1}{R_2}$, so, $R_{para}=3\ \Omega$.

(ii) The equivalent resistance of R_{para} and R_3: $R_{equ}=R_{para}+R_3=3+9=12\ (\Omega)$.

(b) Because the current through the circuit is 0.4 A, the emf of the battery is 4.8 V.

(c) The potential difference across resistor R_1 is $V_1=0.4\times3=1.2$ (V).

(d) The power dissipated in resistor R_1: $P=I^2R=\left(\frac{1.2}{4}\right)^2\times4=0.36$ (W).

(e) The amount of charge that passes through resistor R_3 in one minute$=0.4\times60=24$ (C).

4 Kirchoff's laws 基尔霍夫定律

Kirchoff's laws help you solve complicated circuits. What are the complicated circuits? Complicated circuits are those circuits which can NOT be simplified into a one loop circuit. There are two rules: one is about the junction, the other is about the loop.

1. Junction rule（节点定律）:

At any junction, the total current that enters a junction must equal the total current that leaves the junction.

2. Loop rule（回路定律）:

The sum of the potential differences (positive and negative) that traverse any closed loop in a circuit must be 0.

The junction rule is a restatement of the law of conservation of the charge.

Usually, in junction rule, the current out of the junction is considered as positive, the current into the junction is considered as negative.

The most important things to remember about loop rule are as follows:

- When going across a resistor in the same direction as the current, the potential drops by IR.
- When going across a resistor in the opposite direction from the current, the potential increases by IR.
- When going from the negative to the positive terminal of a source of emf, the potential increases by ε. (+)
- When going from the positive to the negative terminal of a source of emf, the potential decreases by ε. (−)

Example:

Use Kirchoff's rules to determine the current through R_2 in the circuit (图 2—11—5).

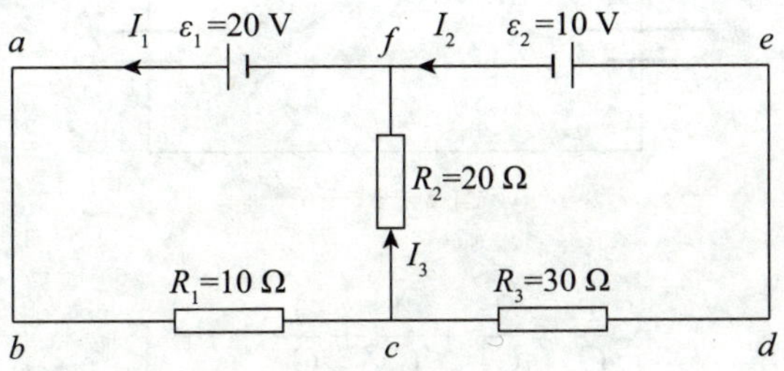

图 2—11—5

Solution:

The points a and f are junctions, but only one is independent. Suppose the currents I_1, I_2, and I_3 are flowing as shown in the diagram.

First use the first rule at junction f: $I_1 - I_2 - I_3 = 0$. Pay attention to the sign by the rule.

Then consider the two independent loops, $abcfa$ and $cdefc$:

The direction of $abcfa$ is counterclockwise: using the rules in loop, we have

$$-I_1R_1 - I_3R_2 + \varepsilon_1 = 0$$

The direction of $cdefc$ is also counterclockwise: using the rules in loop again, we have

$$-I_2R_3 + I_3R_2 - \varepsilon_2 = 0$$

Now that we have three equations and three unknown currents, we can get the current

I_3 through R_2.

$$I_3=\frac{7}{11}\text{A}=0.64\text{A}$$

注意：在利用基尔霍夫定律解决复杂电路问题时，可按下述步骤进行：

（1）首先要找到独立节点和独立回路。根据电路“假定”电流方向（这里强调“假定”，是指估计的方向，如果解出电流为正，说明真实电流方向与假定方向相同；为负，说明真实电流方向与假定方向相反）。

（2）按第一定律建立独立节点方程（如有 n 个节点，只有 $n-1$ 个独立节点方程）。

（3）确定绕行方向，即建立回路方程时考查电势变化的方向。

（4）再按照规定的关于在电阻和电源上电压降落的正、负号建立回路方程。

（5）未知数的个数和方程的个数相同，一般为多元一次方程，可用代入法或行列式法解。

5 Capacitors in series and in parallel 电容器的串并联

1. Capacitors in series（电容器串联）：

The charge stored on each capacitor in series（图 2—11—6）must be the same，and the total voltage is equal to the sum of individual voltage. So：

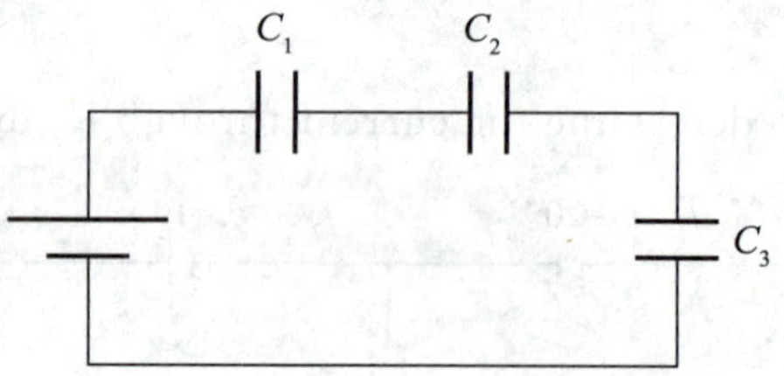

图 2—11—6

$$\frac{1}{C}=\frac{1}{C_1}+\frac{1}{C_2}+\frac{1}{C_3}+\cdots$$

2. Capacitors in parallel（电容器并联）：

The voltage across each capacitor in parallel（图 2—11—7）must be the same，and the total charge is equal to the sum of the charge on each capacitor. So：

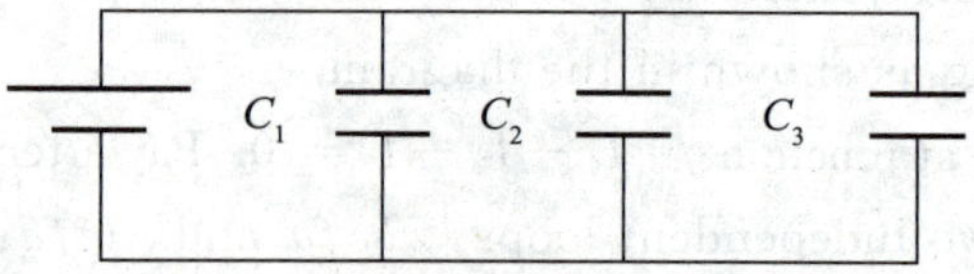

图 2—11—7

$$C=C_1+C_2+C_3+\cdots$$

3. The energy stored in a capacitor（电容器储存的能量）：

$$E=\frac{1}{2}CV^2$$

Once the circuit has been connected for a long time, capacitors stop current from flowing. To find the charge stored on or the voltage across a capacitor, just use the equation for capacitors, $Q = CV$.

6 Practice questions
练习题

Multiple-Choice（选择题）

Single-Select（单选题）

1. A set of 100 Ω, 120 Ω, and 150 Ω resistors are connected to a 9 V battery in the circuit as shown in figure 2—11—8. Which of the three resistors dissipates the most power?

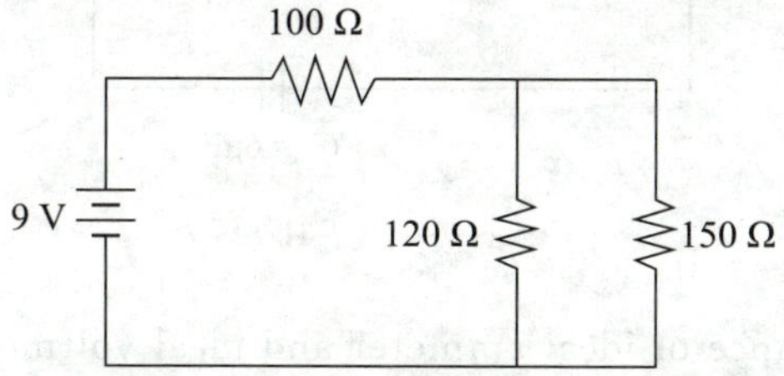

图 2—11—8

(A) the 100 Ω resistor
(B) the 120 Ω resistor
(C) the 150 Ω resistor
(D) both the 120 Ω and 150 Ω

2. A 1.0 F capacitor is connected to a 12 V power supply until it is fully charged. The capacitor is then disconnected from the power supply, and used to power a toy car. The average drag force on this car is 2 N. About how far will the car go?

(A) 36m (B) 72m
(C) 144m (D) 24m

3. Three resistors are connected to a 10 V battery as shown in the diagram (图 2—11—9). What is the current through the 2.0 Ω resistor?

(A) 0.25A (B) 0.50A
(C) 1.0A (D) 2.0A

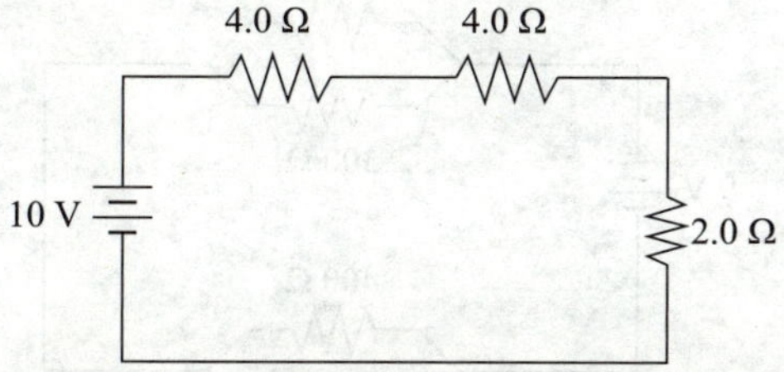

图 2—11—9

4. Three capacitors are connected as shown in the diagram（图 2—11—10）. $C_1 = 2\mu F$, $C_2 = 4\mu F$, $C_3 = 6\mu F$. If the battery provides a potential of 9V, how much charge is stored by this system of capacitor?

(A) 3.0μC　　(B) 30μC

(C) 2.7μC　　(D) 27μC

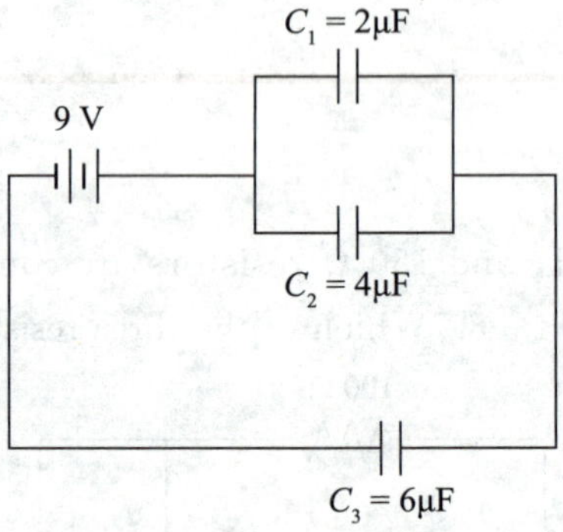

图 2—11—10

5. What is the resistance of ideal ammeter and ideal voltmeter?

	ideal ammeter	ideal voltmeter
(A)	zero	infinite
(B)	infinite	zero
(C)	zero	zero
(D)	infinite	infinite

Multi-Select（多选题）

6. If the temperature is kept as constant, the resistance of a metal wire will depend on

(A) The length of the wire only

(B) The cross section of the wire only

(C) Both the length and cross section of the wire

(D) The material of the wire

Free-Response（计算题）

7. (a) Simplify the circuit（图 2—11—11）so that it consists of one equivalent resistor and the battery.

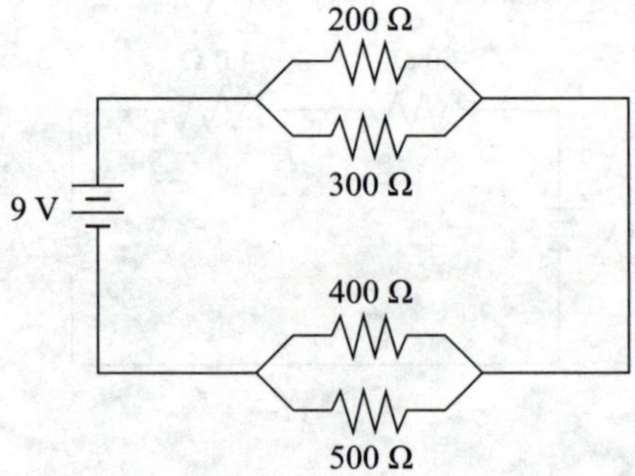

图 2—11—11

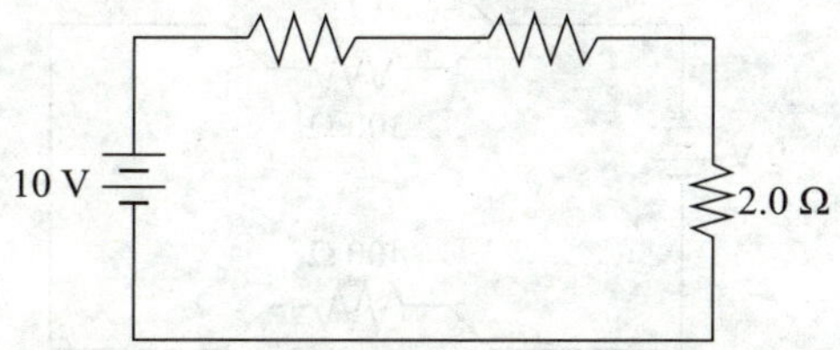

(b) What is the total current through this circuit?

(c) Find the voltage across each resistor. Record your answers in the spaces below.

voltage across the 200Ω resistor: ____________

voltage across the 300Ω resistor: ____________

voltage across the 400Ω resistor: ____________

voltage across the 500Ω resistor: ____________

(d) Find the current through each resistor. Record your answers in the spaces below:

current through the 200Ω resistor: ____________

current through the 300Ω resistor: ____________

current through the 400Ω resistor: ____________

current through the 500Ω resistor: ____________

(e) The 500Ω resistor is now removed from the circuit. State whether the current through the 200Ω resistor would increase, decrease, or remain the same. Justify your answer.

题解

单选题

1. 选（A）。先用欧姆定律计算出每个电阻上流过的电流，再用 I^2R 计算出消耗的功率。另外一种估算方法为：120Ω 和 150Ω 两电阻并联后的电阻值小于 100Ω，它们和 100Ω 的电阻串联，电流相同，所以 100Ω 电阻消耗的功率大。

2. 选（A）。电容器储藏能量为 $\frac{1}{2}CV^2$，把这个能量供给汽车，将电能转换成机械能。可以解出汽车位移为 36m。

3. 选（C）。应用欧姆定理，电路简化为 10V 的电池和 10Ω 等效电阻连接，电路电流为 1.0 A，三个电阻串联，所流过的电流相同。

4. 选（D）。先计算出等效电容：C_1，C_2 并联后的电容值为 6μF，然后和 C_3 串联，等效电容为 3μF，于是由 $Q=CV$，计算出储存电荷为 27 μC。

5. 选（A）。电流表和其他电路元件串联，为了不影响电路中的电流，其电阻越小越好。但伏特表和所测元件并联，为了不影响电路，其电阻越大越好。

多选题

6. 选（C）和（D）。根据导线电阻公式，电阻和导线长度及材料电阻率成正比，和导线截面积成反比。

计算题

7. (a) 两组并联电阻的等效电阻分别为 120Ω 和 222Ω，等效电路如图 2—11—12。

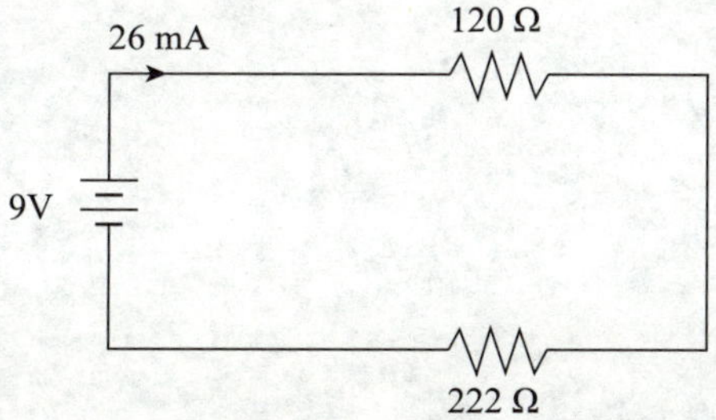

图 2—11—12

这两个等效电阻串联后的总电阻为 342Ω。

(b) 应用欧姆定律，总电压 9 V 除以总电阻 342Ω，得到线路电流为 0.026 A 或 26 mA。

(c) 用总电流乘以两个并联等效电阻可得它们的端电压分别为：3.1 V and 5.8 V。

即 3.1 V 在 200Ω 和 300Ω 并联电阻的两端。5.8V 在 400Ω 和 500Ω 并联电阻的两端。

(d) 应用欧姆定律，经过 200Ω 电阻的电流为 16 mA；经过 300Ω 电阻的电流为 10 mA。类似方法可求流经 400Ω 电阻的电流为 15 mA，流经 500Ω 电阻的电流为 12 mA。

(e) 将 500Ω 电阻移去后，电路总电阻将增加，总电流将变小，流经 200Ω 电阻的电流也将变小。

Chapter 12

Fluid mechanics

第十二章 流体力学

该知识点为参加 AP Physics 2 的学生准备，包括流体静力学和动力学两部分。学生要重点掌握帕斯卡定律和伯努利方程，会应用它们解题。

Main points（主要知识点）

- 流体静力学三原则

(a) The pressure in a static column depends on surface pressure and the column's height.

(b) The buoyant force on an object is equal to the weight of the displaced fluid.

(c) Pascal's principle states that pressure changes in an enclosed fluid are the same throughout, leading to the utility of hydraulic lifts.

- 流体动力学两原则

(a) The continuity equation is a statement of conservation of mass—what flows into a pipe must flow out if a fluid is incompressible.

(b) Bernoulli's equation is a statement of energy conservation.

Relevant equations（相关公式）

Pressure in a static column（静止液柱的压强）：$P = P_0 + \rho gh$

Buoyant force (Archimedes' law)（浮力）：$F_B = \rho_{fluid} V_{submerged} g$

Definition of pressure（压强的定义）：$P = \dfrac{F}{A}$

Fluid flow continuity（流体连续性方程）：$A_1 v_1 = A_2 v_2$

Bernoulli's equation（伯努利方程）：$P_1 + \rho g y_1 + \frac{1}{2}\rho v_1^2 = P_2 + \rho g y_2 + \frac{1}{2}\rho v_2^2$

1 Pressure and density 压强和密度

1. Pressure（压强）：

The pressure exerted by a fluid is equal to the amount of force with which it pushes on the walls of its container divided by the surface area of those walls. The units of pressure in SI system are Newtons per square meter，which are also called Pascals.

$$pressure = \frac{force}{area}$$

2. Density（密度）：

Density is defined as the mass per unit volume and it is abbreviated as the Greek letter ρ.

$$\rho = \frac{mass}{volume}$$

The units of density are kg/m^3.

2 Specific gravity 比重

It is useful to know the density of water，just because water is such a common fluid. What is the density of water? Water has a density of 1 000 kg/m^3.

Because water is so ubiquitous，the density of a substance is often compared to water. For instance，alcohol is eight tenths as dense as water；lead is eleven times as dense. We call this comparison the specific gravity of a substance—lead's specific gravity is 11 because it is 11 times more dense than water. So its density in standard units is 11 000 kg/m^3.

3 Density and mass 密度和质量的关系

When dealing with fluids，we do not generally talk about mass—we prefer to talk about density and volume. But，of course，mass is related to these two measurements：$mass = density \times volume$.

So the mass of any fluid equals ρV. And because weight equals mg，the weight of a fluid can be written as $\rho V g$.

4 Pressure in a static fluid column 静流体液柱的压强

Consider a column of water in a container，such as the one shown in figure 2—12—1：

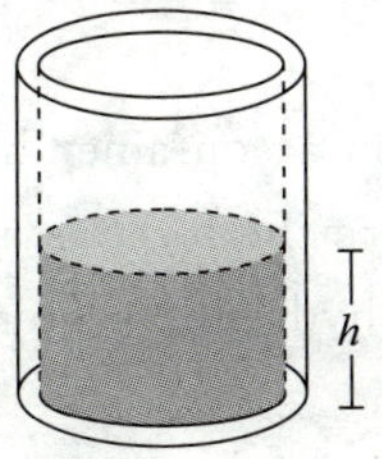

图 2—12—1 a column of water

The pressure due to the column of fluid of density ρ is given by this formula:

$$P = P_0 + \rho gh$$

In this formula, P_0 is the pressure at the top of the column, and h is the vertical height of the column. This formula actually makes intuitive sense—the pressure due to a column of fluid equals the pressure at the top of the column, P_0, plus the weight of the fluid divided by area ρgh.

Note that h is a vertical height, and it does not depend on the width of the container or even the shape of the container. P_0 represents the pressure at the top of the column, which is often atmospheric pressure. But not always! For example, if you put your finger on top of a straw and lift up some water, the pressure at the bottom of the straw is atmospheric (because that part of the water is open to the atmosphere). So, the pressure at the top of the column P_0 must be less than atmospheric pressure by $P = P_0 + \rho gh$.

The pressure $P = P_0 + \rho gh$ is called the absolute pressure of a fluid. Gauge pressure refers to just the ρgh part of the absolute pressure. These are terms to know because they might show up on the AP exam.

5 Buoyancy and Archimedes' principle 浮力和阿基米德定律

When an object is submerged wholly or partially in a fluid, the object experiences an upward force called a buoyant force. This leads us to Archimedes' principle, which says, in words:

" The buoyant force on an object is equal to the weight of the fluid displaced by that object."

In mathematics, Archimedes' principle can be expressed as:

$$F_B = (\rho_{fluid})\,(V_{submerged})\,g$$

Where ρ_{fluid} refers to the density of the fluid (not the submerged mass) and V represents only the submerged volume of an object.

6 Pascal's principle 帕斯卡定律

When fluid is in a closed container, it exerts a pressure outward on every surface of its container. But what happens when an external force is applied on the container? This is where Pascal's principle comes in.

If a force is applied somewhere on a container holding a fluid, the pressure increases everywhere in the fluid, not just where the force is applied.

This principle is most useful for hydraulic systems in machines.

7 Flowing fluids 流动的流体

So far we've only talked about stationary fluids. But what if a fluid is moving through a pipe? There are two questions that then arise: first, how do we find the velocity of the flow, and second, how do we find the fluid's pressure? To answer the first question, we will turn to the continuity principle, and to answer the second, we will use Bernoulli's equation.

1. Continuity principle (连续性原理)

Consider water flowing through a pipe. Imagine that the water always flows such that the entire pipe is full. Now imagine that you are standing in the pipe—how much water flows past you in a given time?

We call the volume of fluid that flows past a point every second the volume flow rate measured in units of m^3/s. This rate depends on two characteristics: the velocity of the fluid's flow, v, and the cross-sectional area of the pipe, A. Clearly, the faster the flow, and the wider the pipe, the more fluid flows past a point every second. To be mathematically precise:

$$\text{Volume flow rate} = A\,v$$

If the pipe is full, then any volume of fluid that enters the pipe must eject an equal volume of fluid from the other end. So, by definition, the volume flow rate is equal at all points within an isolated stream of fluid. This statement is known as the principle of continuity. For positions "1" and "2" in figure 2—12—2, we can write $A_1 v_1 = A_2 v_2$.

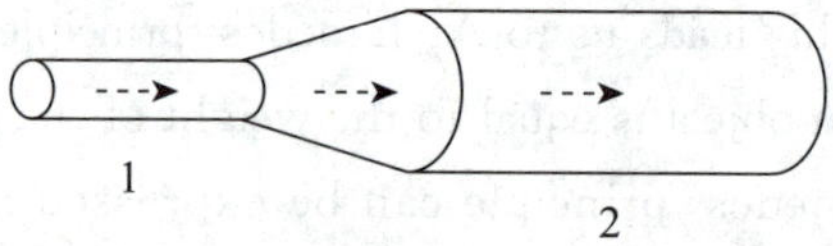

图 2—12—2 the principle of continuity

The most obvious physical consequence of continuity is that where a pipe is narrow, the flow is faster. So in the figure, the fluid must move faster in the narrower section labeled "1".

2. Bernoulli's equation (伯努利方程)

Bernoulli's equation is an application of conservation of energy in fluid mechanics. It can be written as follows:

$$P_1 + \rho g y_1 + \frac{1}{2}\rho v_1^2 = P_2 + \rho g y_2 + \frac{1}{2}\rho v_2^2$$

Bernoulli's equation is useful whenever you have a fluid flowing from one point, such as "1", to the other point, such as "2". The flow can be fast or slow; it can be vertical or horizontal; it can be open to the atmosphere or enclosed in a pipe. P is the pressure of the fluid at the specified point, y is the vertical height at the specified point, ρ is the fluid's density, and v is the speed of the flow at the specified point.

3. Principal consequence of Bernoulli's equation

The most important result that can be derived from Bernoulli's equation is that, where flow is faster, pressure is slower.

Take two pieces of paper and hold them parallel to each other, a couple of centimeters apart. Now get someone with strong lungs—a trombone player, perhaps—to blow a tast, focused stream of air between the sheets of paper. You will see that the pieces of paper actually move together. Why? Let us look at the terms in Bernoulli's equation. Take point "1" to be between the sheets of paper; take point "2" to be well away from the sheets of paper, out where the air is hardly moving, at the same level, so that y_1 and y_2 are the same (see figure 2—12—3). We can cancel the y terms and solve for the pressure between the sheets of paper:

$$P_1 = P_2 + \left(\frac{1}{2}\rho v_2^2 - \frac{1}{2}\rho v_1^2\right)$$

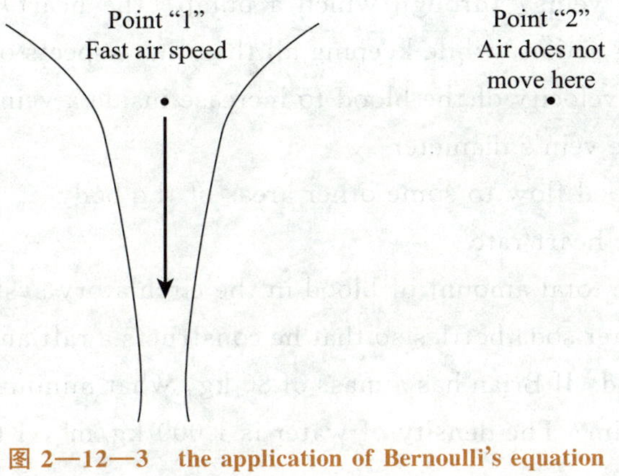

图 2—12—3 the application of Bernoulli's equation

Far away from the pieces of paper, the air is barely moving (so v_2 is just about zero), and the pressure P_2 must be the ambient atmospheric pressure. But between the sheets, the air is moving quickly. So, v_1 is substantially bigger than v_2. The pressure between the pieces of paper is less than atmospheric pressure, so the air pushes the pieces of paper together.

This effect can be used in the working principle in airplane. An airplane is pushed up when the air rushes fast over top of the wing than below the wing.

8 Practice questions
练习题

Multiple-Choice（选择题）

Single-Select（单选题）

1. A large tank is filled with water to a depth of 6 m. If point x is 1 m from the bottom and point y is 2 m from the bottom, how does P_x, the hydrostatic pressure due to the water at point x, compare to P_y, the hydrostatic pressure due to the water at point y?

(A) $P_x = 2P_y$ (B) $2P_x = P_y$

(C) $5P_x = 4P_y$ (D) $4P_x = 5P_y$

2. A person sips a drink through a straw. At which of the following three positions is the pressure lowest?

Ⅰ. Inside the person's mouth

Ⅱ. At the surface of the drink

Ⅲ. At the bottom of the drink

(A) only at position Ⅰ

(B) only at position Ⅱ

(C) only at position Ⅲ

(D) both at positions Ⅰ and Ⅲ

3. The circulatory system can be modeled as an interconnected network of flexible pipes (the arteries and veins) through which a pump (the heart) causes blood to flow. Which of the following action, while keeping all the other aspects of the system the same, would NOT cause the velocity of the blood to increase inside a vein?

(A) expending the vein's diameter

(B) cutting off blood flow to some other areas of the body

(C) increasing the heart rate

(D) increasing the total amount of blood in the circulatory system

4. Brian saves 2 liter soda bottles so that he constructs a raft and float out onto Haverford Collge's Duck Pond. If Brian has a mass of 80 kg, what minimum number of bottles is necessary to support him? The density of water is 1 000 kg/m^3, 1 000 L=1 m^3.

(A) 1 600 bottles

(B) 800 bottles

(C) 200 bottles

(D) 40 bottles

Multi-Select（多选题）

5. Water flows through the pipe shown in figure 2—12—4. At the larger end, the pipe has diameter D or radium R and the speed of the water is v_1. What is the speed of the water at the smaller end, where the pipe has diameter d or radium r?

(A) $\left(\frac{d}{D}\right)v_1$ (B) $\left(\frac{R^2}{r^2}\right)v_1$

(C) $\left(\frac{d^2}{D^2}\right)v_1$ (D) $\left(\frac{D^2}{d^2}\right)v_1$

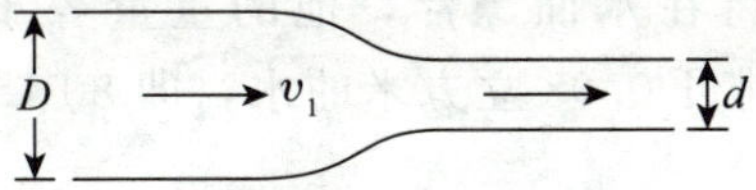

图 2—12—4

Free-Response（计算题）

6. The water tower in the drawing（图 2—12—5）is drained by a pipe that extends to the ground. The amount of water in the top spherical portion of the tank is significantly greater than the amount of water in the supporting column. Information you may need: Density of water=1 000 kg/m^3.

(a) What is the absolute pressure at the position of the valve if the valve is closed? Assuming that the top surface of the water at point P is at atmospheric pressure.

(b) Now the valve is opened; thus, the pressure at the valve is forced to be atmospheric pressure. What is the speed of the water past the valve ?

(c) Assuming that the radius of the circular valve opening is 10 cm, find the volume flow rate out of the valve.

(d) Considering that virtually all of the water is originally contained in the top spherical portion of the tank, estimate the initial volume of water contained by the water tower. Explain your reasoning thoroughly.

(e) Estimate how long it would take to drain the tank completely using this single valve.

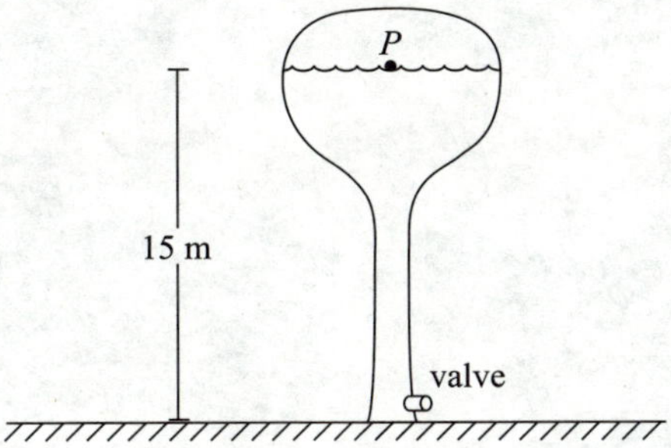

图 2—12—5

题解

单选题

1. 选（D）。由于点 x 在水面下 5 米，x 点的压强为 $5\rho g$，由于点 y 在水面下 4 米，y 点的压强为 $4\rho g$，因此，$\frac{P_x}{P_y}=\frac{5}{4}$，所以，$4P_x=5P_y$.

2. 选（A）。液体表面是大气压力，只有口腔的压力低于大气压力时，液体才会被大气压力通过吸管压入口腔。

3. 选（A）。流量（每秒流过血管截面的体积）是血管截面和流速的乘积。如果选 B、C、D，增加流量，就要增大血液的流速。选 A，扩大静脉的截面，就可以降低血液流速。

4. 选（D）。由于 Brian 将在水面漂浮，他的重量等于浮力。即 $\rho_{water} V_{submerged} g = 800N$，解出 $V_{submerge}$，得到要排开 0.08 立方米的水，即 80 L 的水，即需要 40 个瓶子。

多选题

5. 选（B）和（D）。根据流体连续性方程可判定。

计算题

6.（a）如果阀门关闭，由静止液柱的压强公式可得：$P = P_0 + \rho gh = 10^5 N/m^2 + (1\,000 kg/m^3)(10 N/kg)(15 m) = 250\,000 N/m^2$。

（b）应用 Bernoulli's 方程，选 P 点和阀门作为两个点。这两点均有大气压，因此，可以抵消。选阀门高度为零，液面速度近似为零，于是 Bernoulli's 方程的六项可以简化为：$\rho g y_{top} = \frac{1}{2}\rho v_{bottom}^2$，解出

$$v_{bottom} = 17 m/s$$

（c）因为流量定义为 Av，流管的截面为：$\pi(0.1 m)^2 = 0.031 m^2$，所以流量为：$0.53 m^3/s$。

（d）球形容器的体积为 $(4/3)\pi r^3$，其中罐的半径为 2～3 米，体积约为 100 m^3。罐仅仅充满 3/4 约为 70 m^3。

（e）由于流量为 $0.53 m^3/s$，共有 70 m^3 的液体，于是需要用 70/0.53 =140 秒的时间，大约是两分钟。

Chapter 13

Thermodynamics and ideal gas

第十三章 热学和理想气体

该知识点内容包括物体温度、热量、内能的概念和它们之间的关系，理想气体定律和理想气体状态方程，热力学第一定律。要了解 PV 图的意义和它描述的等温、等压、等容过程，并了解热机效率。

Main points（主要知识点）

- 内能是由分子振动决定的能量，宏观上内能只和温度有关
- 热量是热能的传输
- 物体温度升高，物体会膨胀
- 理想气体状态方程是描述理想气体的体积、压强和温度关系的方程
- 用气体分子运动论解释宏观量

热力学第一定律

（1）用理想气体 PV 图和状态方程计算内能

（2）理想气体做功可用 PV 图中热力学过程曲线下的面积决定

（3）系统吸收或放出的热量可用热力学第一定律求得

- 热机效率和卡诺循环

Relevant equations（相关公式）

Thermal length expansion（热膨胀）：$\Delta L = \alpha L_0 \Delta T$

Ideal gas law（理想气体方程）：$PV = nRT$

Internal energy of an ideal gas（理想气体内能）：$U = \frac{3}{2}nRT$

RMS speed of a gas molecule（分子的方均根速率）：$v_{rms} = \sqrt{\frac{3kT}{m}}$

First law of thermodynamics（热力学第一定律）：$\Delta U = Q + W$

Efficiency of any heat engine（热机效率）：$\eta = \frac{W}{Q_H}$

Efficiency of an ideal (Carnot) engine (卡诺循环效率): $\eta = \frac{T_H - T_C}{T_H}$

Rate of heat transfer (热传导率): $H = \frac{kA\Delta T}{L}$

1 Heat, temperature and power
热、温度和功率

Heat, represented by the variable Q, is a type of energy that can be transferred from one body to another. Heat is measured in joules, because it is a form of energy.

1. Internal Energy (内能): The sum of the energies of all of the molecules in a substance.

Internal energy, represented by the variable U is also measured in joules.

2. Temperature (温度): Related to the average kinetic energy per molecule of a substance.

Temperature is measured in Kelvins or degrees Celsius. To convert from Kelvins to degrees Celsius or vice versa, use the following formula:

$$T_{Celsius} = T_{Kelvin} - 273.$$

So, if room temperature is about 23°C, then room temperature is also about 296 K.

注意：在应用气体状态方程解热学问题时，一定要将温度用绝对温度表示。

3. Power (功率) is the work done per unit time.

Power is measured in joules/second, or watts:

$$1\text{W} = 1\ \text{J/s}.$$

2 Thermal expansion
热膨胀

The equation describing thermal expansion as follows:

$$\Delta L = \alpha L_0 \Delta T$$

This formula describes what is called linear expansion. It says that the change in an object's length, ΔL, is equal to some constant, α, multiplied by the object's original length, L_0, multiplied by the change on the object's temperature.

The constant α is determined experimentally: it is different for every material. Metals tend to have high values for α, meaning they expend a lot when heated. Such insulators as plastic are tend to have lower values for α.

3 Rate of heat transfer
热传导率

Heat will flow across a material when there is a temperature difference between two positions on that material. Heat will flow until the material has the same temperature everywhere, when the material is said to have reached thermal equilibrium. The bigger the temperature difference, the larger the heat transfer rate. Following parameters affect the rate of heat transfer:

- the nature of the material (metals carry heat better than non-metals)
- the cross-sectional area of the substance (more area makes it easier for heat to flow)
- the distance across the material between the two positions with a temperature difference (the longer this distance, the slower the heat transfer)

This can be summed up in the equation for the heat transfer rate H:

$$H = \frac{kA\Delta T}{L}$$

Where k is the thermal conductivity of the substance, A is the cross-sectional area, ΔT is the temperature difference, and L is the length of the material.

4 Ideal gas laws
理想气体定律

The state equation of idea gas (理想气体状态方程)

This is probably one of your favorite laws, because you used it so much in chemistry class.

It can be written two different ways; our preference is

$$PV = nRT$$

P is the pressure of the gas, V is the volume of the gas, n is the number of moles of gas you are dealing with, R is a constant ($R = 8.31$ J/mol·K), and T is the temperature of the gas in Kelvins. And in case you were wondering, the units of pressure are N/m^2, also known as Pascals, and the units of volume are m^3. (Yes, there are other units for pressure and volume, but don't use them unless you want to be really confused.)

The ideal gas law can also be written as

$$PV = Nk_BT$$

In this version, N is the number of molecules of gas, and k_B is called Boltzmann's constant ($k_B = 1.38\times10^{-23}$ J/K).

If you ever need to convert moles to molecules, remember that one mole contains 6.02×10^{23} molecules. This conversion is on the constant sheet.

Example:

Standard atmospheric pressure is 1.013×10^5 Pa, and at 0℃ = 273.15 K, one mole of any ideal gas occupies the same volume. What is this volume?

Solution:

Use the ideal gas equation: $PV = nRT$

$$V = \frac{nRT}{P} = \frac{1.00\ \text{mol} \times 8.31\text{J/mol} \cdot \text{K} \times 273.15\text{K}}{1.013 \times 10^5\ \text{Pa}} = 22.4 \times 10^{-3}\ \text{m}^3$$

The quantity 10^{-3} m^3 is the defination of a liter (L), so, the volume occupied by any gas at Standard Temperature and Pressure (0° C and 1 atm) (STP) is 22.4 L.

5 Kinetic theory of gases 气体的动力学理论

The following assumptions are used to make predictions about the macroscopic behavior of the gas:

- Molecules move in continuous, random motion.
- There are an exceedingly large number of molecules in any container of gas.
- The separation between individual molecules is large.
- Molecules do not act on one another at a distance; that is, they do not exert electrical or gravitational forces on other molecules.
- All collisions between molecules, or between a molecule and the walls of a container, are elastic (i. e. , no energy lost in these collisions).

There are two important equations as follows:

(1) Internal energy: The relationship between the internal energy of a gas and its temperature is

$$U = \frac{3}{2}nRT$$

This equation allows you to find the temperature of a gas given its internal energy, or vice versa.

(2) The root-mean-square (rms) speed of gas molecules is

$$v_{rms} = \sqrt{\frac{3kT}{m}}$$

Here, T is the temperature of the gas (always in Kelvins!) and m is the mass of one molecule of the gas (always in kilograms). Think of the rms speed as a sort of average speed of molecules in a gas. For most gases, this speed will be in the hundreds or thousands of meters per second.

6 The first law of thermodynamics 热力学第一定律

The first law of thermodynamics is simply a statement of conservation of energy.

$$\Delta U = Q + W$$

U is the internal energy of the gas in question. Q is the heat added to the gas, and W is the work done on the gas when the gas is compressed.

注意：应用热力学第一定律时的符号规定：

The signs of each term in the first law are extremely important.

(1) If heat is added to the Gas, Q is positive, and if heat is taken from the gas, Q is negative.

(2) If the gas's internal energy (and, thus, its temperature) goes up, then ΔU is positive; if the gas's temperature goes down, then ΔU is negative.

(3) The sign of W is more difficult to conceptualize. Positive W represents work done on a Gas. However, if a gas's volume expands, then work was not done on the gas; work was done by the gas. In the case of work done by a gas, W is given a negative sign.

Example:

Four samples of ideal gas are each initially at a pressure P_0 and volume V_0, and a temperature T_0 is shown in the figure (图 2—13—1). The samples are taken in separate experiments from this initial state to the final states Ⅰ, Ⅱ, Ⅲ, and Ⅳ along the processes shown in the figure

(a) One of the processes is isothermal. Identify this one and explain.

(b) One of the processes is adiabatic. Identify this one and explain.

(c) In which process (or processes) does the gas do work? Explain.

(d) In which process (or processes) is heat removed from the gas? Explain.

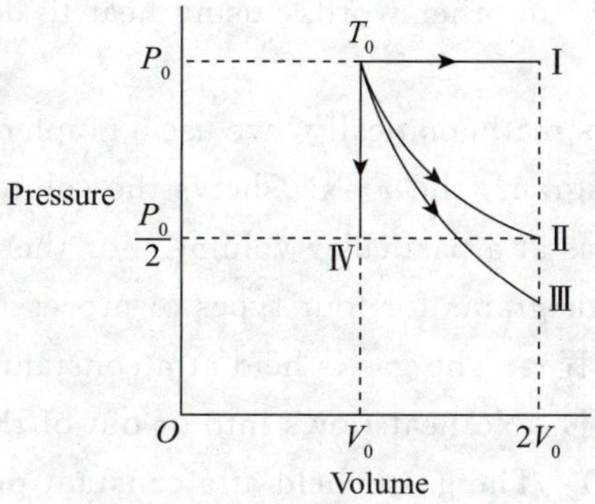

图 2—13—1

Solution:

(a) Process Ⅱ is isothermal. Because in isothermal process, the temperature is kept as a constant:

$$P_0V_0 = \left(\frac{P_0}{2}\right)(2V_0)$$

(b) Process Ⅲ is adiabatic. In adiabatic process, both P and V must change. So, processes Ⅰ and Ⅳ are eliminated, and Ⅱ is isothermal, and process Ⅲ is the only possible adiabatic process.

(c) Work is done by the gas in the processes Ⅰ, Ⅱ and Ⅲ because the volume of gas expends.

(d) Heat is given off by the gas in process Ⅳ. In process Ⅳ, the volume is constant, hence no work is done. The temperature is directly proportional to the pressure. Since the pressure decreases, the temperature decreases.

7 PV diagrams 压力—体积图

Imagine a gas in a cylinder with a moveable piston on top, as shown in figure 2—13—2:

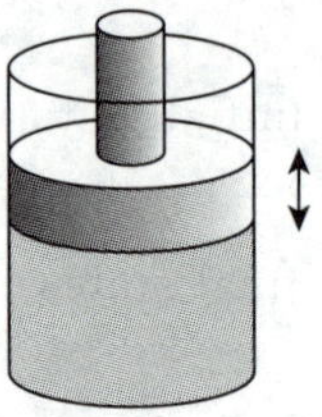

图 2—13—2 a cylinder with a moveable piston on top

When the gas expands, the piston moves up, and work is done by the gas. When the piston compresses, work is done on the gas. The point of a heat engine is to add heat to a gas, making the gas expand—in other words, using heat to do some work by moving the piston up.

To visualize this process mathematically, we use a graph of pressure vs volume, called a *PV* diagram. On a *PV* diagram, the x-axis shows the volume of the gas, and the y-axis shows the pressure of the gas at a particular volume. For the AP exam, you are expected to be familiar with the *PV* diagrams for four types of processes:

- Isothermal (等温过程): The gas is held at a constant temperature.
- Adiabatic (绝热过程): No heat flows into or out of the gas.
- Isobaric (等压过程): The gas is held at a constant pressure.
- Isochoric (等容过程): The gas is held at a constant volume.

We look at each of these processes individually.

1. Isothermal processes (等温过程)

Look at the ideal gas law: $PV = nk_BT$. In virtually all of the processes you will deal with, no gas will be allowed to escape. So n will be constant. Boltzmann's constant is, of course, always constant. And, in an isothermal process, T must be constant. So the entire right side of the equation remains constant during an isothermal process. This means that the product of P and V must also be constant. A *PV* diagram for an isothermal process is shown in figure 2—13—3:

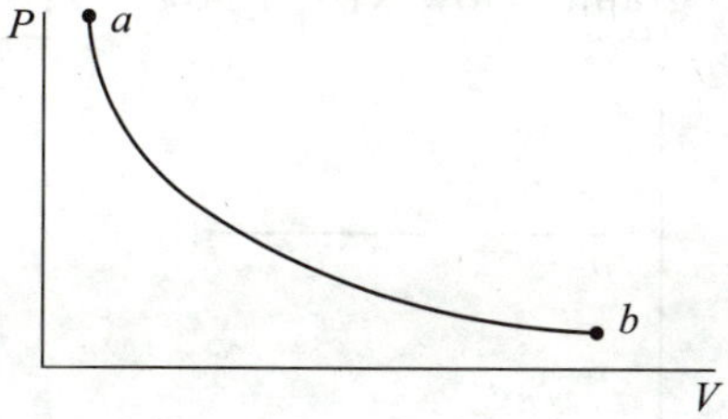

图 2—13—3 **isothermal process**

What this diagram shows is the set of all P—V values taken on by a gas as it goes from point "a" to point "b". At point "a" the gas is held at high pressure but low volume. We can imagine that there's a lot of gas squished into a tiny cylinder, and the gas is trying its hardest to push up on the piston—it is exerting a lot of pressure.

As the volume of the gas increases—that is, as the piston is pushed up—the pressure of the gas decreases rapidly at first, and then slower as the volume of the gas continues to increase. Eventually, after pushing up the piston enough, the gas reaches point "b" where it is contained in a large volume and at a low pressure.

Since the temperature of the gas does not change during this process, the internal energy of the gas remains the same, because $U=(3/2)\ nRT$. So, for an isothermal process, $\Delta U=0$.

2. Adiabatic processes (绝热过程)

In an adiabatic process, no heat is added to or removed from the system. In other words, $Q=0$. By applying the first law of thermodynamics, we conclude that $\Delta U=W$. This means that the change in internal energy of the gas equal the work done on the system. In an adiabatic process, you can increase the temperature of the gas by pushing done on the piston, or the temperature of the gas can decrease if the gas itself push up on the piston. The PV diagram for an adiabatic process looks similar, but not identical, to the diagram for an isothermal process. Please note that when a gas is expanded adiabatically, as from "a" to "b" in the following picture (图 2—13—4), the final state is on a lower position of isothermal process. (An isotherm is an curve on a PV diagram for which the temperature is constant.) In other words, in an adiabatic expansion, the temperature of the gas decreases.

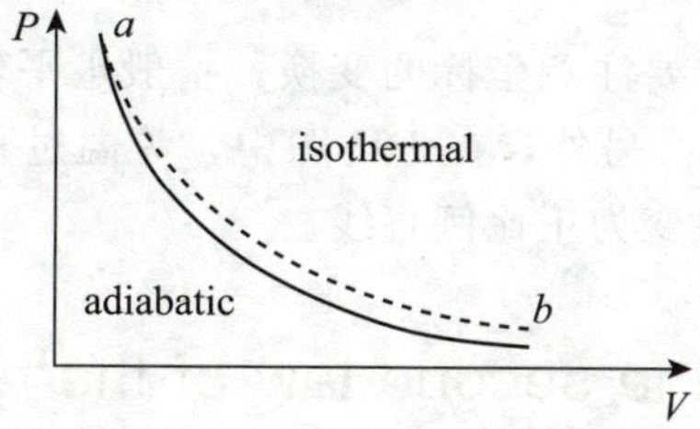

图 2—13—4 ***PV* diagram of an adiabatic process**

3. Isobaric processes (等压过程)

In an isobaric process, the temperature is kept as a constant. So, PV diagram for a isobaric process looks like the graph below（图 2—13—5）：

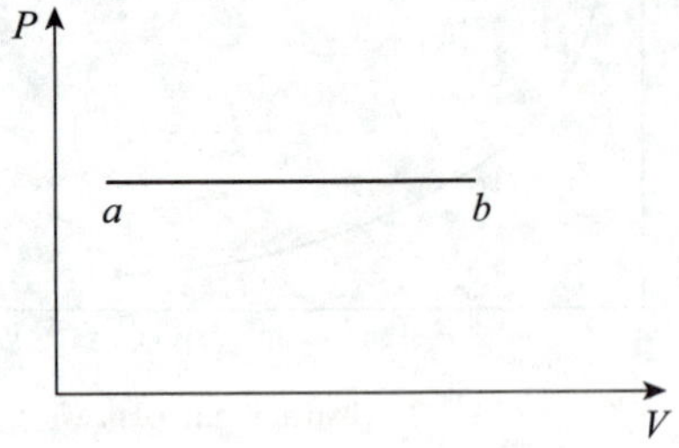

图 2—13—5　*PV* diagram of an isobaric process

In an isobaric process, the pressure of the gas is the same no matter what the volume of the gas is. Using $PV=Nk_BT$, we can see that, if we increase V but not P, then the left side of the equation increases. So the right side must also does, and the temperature must increase. Therefore, in an isobaric process, the temperature of the gas increases as the volume increases, but the pressure does not change.

4. Isochoric processes（等容过程）

In an isochoric process, the volume remains constant, as shown in figure 2—13—6:

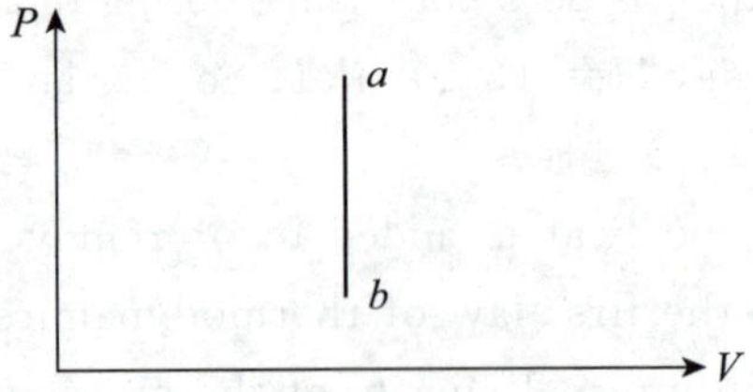

图 2—13—6　an isochoric process

As with an isobaric process, the temperature of a gas undergoing an isochoric process must change as the gas goes from point "*b*" to point "*a*". The temperature of the gas must increase as the pressure increases, and it must decrease as the pressure of the gas decreases.

用 *PV* 图解题有以下技巧：

（1）*PV* 图曲线下面的面积是气体做的功或外界对气体做的功。

（2）内能是温度的函数。

（3）利用 *PV* 图解题时，要注意坐标的变换，一般水平轴代表体积 V，纵轴代表压力 P。但有时，题目会调换位置。另外，在 *PV* 图中，等温过程是双曲线，但如果将体积坐标 V 转换为 $1/V$，等温过程将变为正比例直线。

8 Heat engines and the second law of thermodynamics 热机和热力学第二定律

1. The second law of thermodynamics（热力学第二定律）

The second law of thermodynamics is less mathematical than the first law, and it is also probably more powerful. One of many ways to state this law says:

Heat flows naturally from a hot object to a cold object.

Heat does NOT flow naturally from cold to hot.

Heat engines and their cousins, refrigerators, work by this principle.

In a heat engine, heat must flow from hot stuff to cold stuff... some of this flowing heat is used to do work. The rest of the input heat flows into the cold stuff as exhaust. A car engine is a heat engine: you start with hot gas in a cylinder, that hot gas expands to push a pistons (it does work), and you are left with colder exhaust gas.

2. The efficiency of a heat engine (热机的效率)

We define the efficiency of a heat engine by the ratio of the useful work done by the engine divided by the amount of heat that was put in (图 2—13—7). This is stated by the following equation:

$$\eta = \frac{W}{Q_H}$$

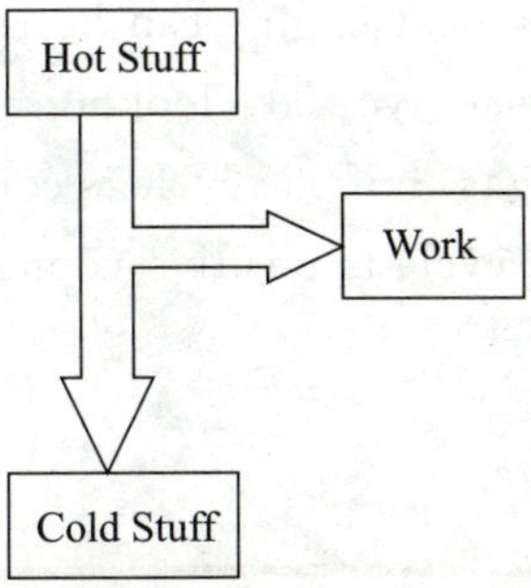

图 2—13—7 heat engine

The efficiency, η of an engine is always less than 1. Since all heat engines operate at less than 100% efficiency, it is useful to calculate what the maximum possible efficiency might be, assuming the engine is as well made as possible to eliminate energy losses to friction, insufficient insulation, and other engineering concerns. It is found that this ideal efficiency for a heat engine depends only on the temperatures of the hot and cold reservoirs:

$$\eta_{ideal} = \frac{T_H - T_C}{T_H}$$

T_H is the temperature of the hot stuff you start with, and T_C is the temperature of the colder exhaust. These temperatures need to be measured in Kelvins.

That a heat engine is "ideal" doesn't mean its efficiency is 1. The same argument we used above still applies: efficiency is always less than 1. And real, nonideal engines operate at only a fraction of their ideal efficiency.

A refrigerator like the backward cousin of a heat engine. Physicists draw a refrigerator as shown as follows (图 2—13—8):

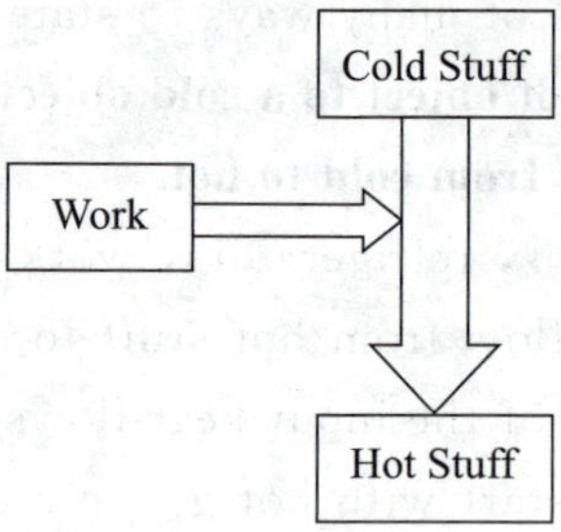

图 2—13—8　a refrigerator

In a refrigerator, work is done in order to allow heat to flow from cold stuff to hot stuff. This sound like a violation of the second law of thermodynamics. It isn't, though, because the work that is done to the system must have been the product of some sort of heat engine. Heat will not spontaneously flow from cold stuff to hot stuff; it is necessary to do net work on a gas to get heat to flow this way.

By the way, the behavior of a heat engine can be represented on a *PV* diagram by a closed cycle. If the cycle is clockwise, you are looking at a heat engine, and the network per cycle is negative (done by the gas); if the cycle is counterclockwise, you are looking at a refrigerator and the network per cycle is positive (done on the gas).

9 Practice questions
练习题

Multiple-Choice（选择题）

Single-Select（单选题）

1. A brass rod 5 m long and 0.01 m in diameter increases in length by 0.05 m, when its temperature is increased by 500℃. A similar brass rod of length 10 m has diameter of 0.02 m. By how much will this rod increase if the temperature increased by 1 000℃?

(A) 4×10^{-2} m　　(B) 4×10^{-3} m

(C) 4×10^{-4} m　　(D) 4×10^{-5} m

2. A gas is enclosed in a metal container with a moveable piston on top. Heat is added to the gas by placing a candle flame in contact with the container's bottom. Which of the following is true about the temperature of the gas?

(A) The temperature must go up if the piston remains stationary

(B) The temperature must go up if the piston is pulled out dramatically

(C) The temperature must go up no matter what happens to the piston

(D) The temperature must go down no matter what happens to the piston

3. A small heat engine operates using a pan of 100℃ boiling water as the high temperature reservoir and the atmosphere as a low temperature reservoir. Assuming ideal behavior, how much more efficient is the engine on a cold, 0℃ day than on a warm, 20℃ day:

(A) 1.2 times as efficient

(B) 2 times as efficient

(C) 20 times as efficient

(D) infinitely more efficient

4. A $1m^3$ container contains 10 moles of ideal gas at room temperature. At what fraction of atmospheric pressure is the gas inside the container ?

(A) 1/40 atm (B) 1/20 atm

(C) 1/10 atm (D) 1/4 atm

Multi-Select (多选题)

5. In time t, an amount of heat Q flows through the solid door of area A and thickness d (图 2—13—9). The temperatures on each side of the door are T_2 and T_1, respectively. Which of the following changes would be certain to decrease Q?

(A) Increasing A only

(B) Decreasing d only

(C) Decreasing $T_2 - T_1$

(D) Decreasing A

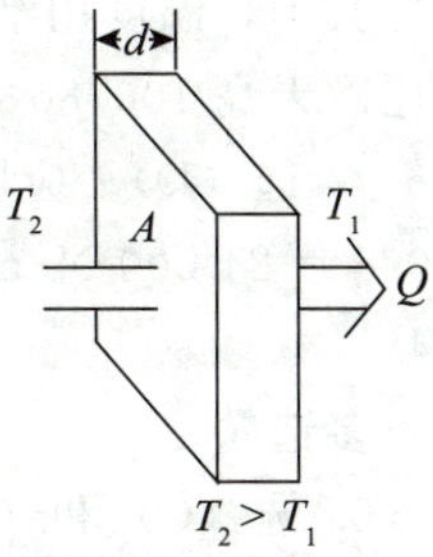

图 2—13—9

Free-Response (计算题)

6. A small container of gas undergoes a thermodynamic cycle. The gas begins at room temperature. First, the gas expands isobarically until its volume has doubled. Second, the gas expands adiabatically. Third, the gas is cooled isobarically. Finally, the gas is compressed adiabatically until it returns to its original state.

(a) The initial state of the gas is indicated on the PV diagram (图 2—13—10). Sketch this process on the graph.

(b) Is the temperature of the gas greater right before or right after the adiabatic expansion? Justify your answer.

(c) Is heat added to or removed from the gas in one cycle ?

(d) Does this gas act as a refrigerator or a heat engine?

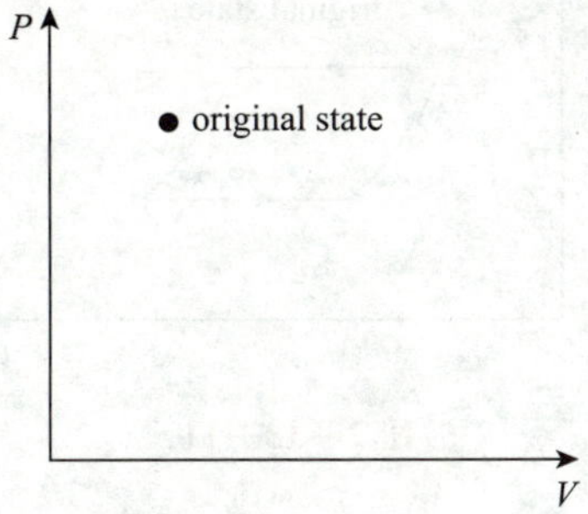

图 2—13—10

题解

单选题

1. 选 (C)。利用热膨胀方程:

$$\Delta T = \frac{\Delta L}{\alpha L}$$

先解出 $\alpha = 2\times10^{-5}℃^{-1}$，再用此公式解出 ΔL 为：4×10^{-4}m。

2. 选（A）。应用热力学第一定律，$\Delta U = Q + W$。蜡烛加热空气，因此 Q 为正。A 选项中，活塞静止不动，所以 $W=0$。因为内能和温度有关，因此如果 ΔU 是正值，则温度升高。

3. 选（A）。根据热机效率公式：

$$\eta = \frac{T_H - T_C}{T_H}$$

其中，温度用绝对温标。因为分母不变，热天为 100 K，冷天为 80 K，所以，热天效率比冷天高 100/80≈1.2 倍。

4. 选（D）。应用理想气体状态方程估算，$P = nRT/V$，代入数值：10 moles×8×300/1= 24 000 N/m^2。由于大气压强为 100 000 N/m^2，所以，容器压强约为大气压强的 1/4。

多选题

5. 选（C）和（D）。热量传输和温度差、传输截面成正比。

计算题

6. (a) 见图 2—13—11。

(b) 根据绝热过程定义，系统和外界没有热交换。因为是绝热膨胀，系统对外做功，W 为负值，于是，根据 $\Delta U = Q + W$，因为 $Q = 0$，W 为负，所以 ΔU 也是负值。因为内能降低，温度也降低。

(c) 由于整个过程中，气体起始和终了状态相同，内能不变化。考虑气体对外和外界对气体做功，在第一、二过程中，气体膨胀，曲线下面面积大于第二、三过程曲线下面的面积，所以气体膨胀做的功比气体被压缩做功多。所以，总功为负，系统吸热。

(d) 这是一个热机。热机工作时，气体吸热，然后对外做功。（对制冷机来说，外界对工作气体做功，热量由工作气体排出。）

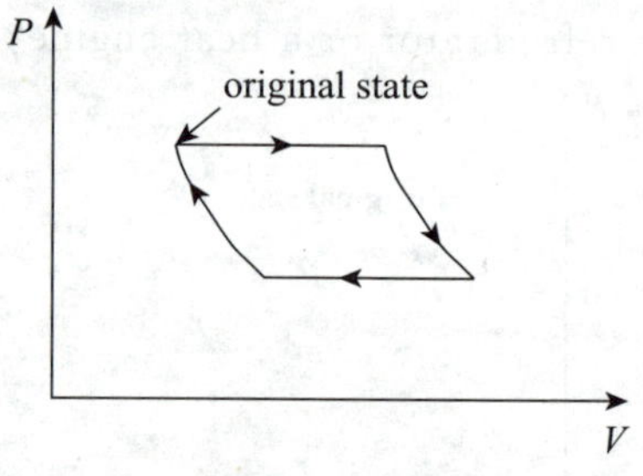

图 2—13—11

Chapter 14

Current and magnetic field

第十四章 电流和磁场

该部分讨论电流产生的磁场、载流直导线以及运动电荷在磁场中所受的力。在此基础上，定义了磁感应强度。

Main points（主要知识点）

- 电流周围激发磁场
- 在磁场中，载流导线受磁场力的作用
- 在磁场中，运动电荷受磁场力的作用。作用在一个以速率 v 垂直射入磁场 B 的带电粒子 q 的磁场力为 $F_m = qvB$，其方向由右手定则确定（一般情况下可用矢量叉乘确定）

Relevant equations（相关公式）

Force on a current-carrying wire（作用在载流直导线上的磁场）：$F = ILB$

Force on a charged particle in a magnetic field（磁场中运动电荷受力作用）：$F = qvB$

Magnetic field due to a long, straight current-carrying wire（距离载流强度为 I 的长直导线 r 处的磁感应强度）：$B = \dfrac{\mu_0 I}{2\pi r}$

Magnetic flux（匀强磁场中通过面积为 A 的闭合线圈的磁通）：$\varphi_B = BA$

1 Magnetic field 磁场

1. The definition of magnetic field strength（磁感应强度的定义）

All magnets are dipoles, which means that they have two "poles," or ends. One is called the north pole, and the other is the south pole. Opposite poles attract, and like poles repel. You can never create a magnet with just a north pole or just a south pole.

The magnetic field strength: The value of a magnetic field is a vector quantity, and it is abbreviated with the letter B. The value of a magnetic field strength can be defined as:

$$B=\frac{F_m}{dl \cdot I}$$

Where the $dl \cdot I$ is the elementary segment of a current, the F_m is the maximum magnetic force exerted on the current. The unit of a magnetic field is measured in teslas (T).

2. The magnetic field lines (磁感应线)

A magnet creates a magnetic field (see figure 2—14—1). Unlike electric field lines, which either go from a positive charge to a negative charge or extend infinitely into space, magnetic field lines form loops.

图 2—14—1　a magnet creates a magnetic field

These loops point away from the north end of a magnet, and toward the south end. Near the magnet the lines point nearly straight into or out of the pole.

When a magnetic field line is directed out of the page, it is drawn as shown in figure 2—14—2 (a). When a magnetic field line is directed into the page, it is drawn as shown in figure (b).

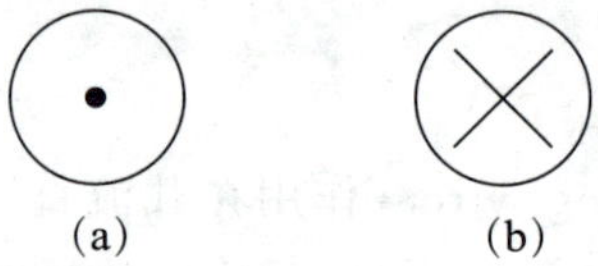

图 2—14—2

2 Long straight current-carrying wires 长直载流导线

The magnetic field created by a long, straight, current-carrying wire loops around the wire in concentric circles. The direction in which the magnetic field lines loop is determined by a right-hand rule.

Right-hand rule (右手定则):

Pretend you are holding the wire with your right hand. Point your thumb in the direction of the current. Your fingers wrap around your thumb in the same way as that the magnetic field wraps around the wire (图 2—14—3).

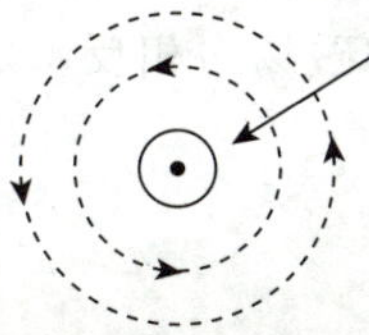

图 2—14—3 the magnetic field created by a long straight current-carrying wire

The formula that describes the magnitude of the magnetic field created by a long straight current-carrying wire is the following:

$$B=\frac{\mu_0 I}{2\pi r}$$

In this formula, B is the magnitude of the magnetic field, μ_0 is a constant called the "permeability of free space"（真空中的磁导率） ($\mu_0 = 4\pi\times10^{-7}$ T·m/A), I is the current flowing in the wire, and r is the distance from the wire. It can be proved in AP Physics C.

3 The moving charged particles in magnetic field 在磁场中运动的带电粒子

1. Lorentz force（运动带电粒子在磁场中受的力）

A magnetic field exerts a force on a charged particle if that particle is moving perpendicular to the magnetic field. $F=qvB$

The direction of the force can be determined by right-hand rule.

2. Right-hand rule（判断洛仑兹力的右手定则）

To find the force on a charged particle.

Point your right hand, with fingers extended, in the direction that the charged particle is traveling. Then, bend your fingers so that they point in the direction of the magnetic field.

- If the particle has a POSITIVE charge, your thumb points in the direction of the force exerted on it.
- If the particle has a NEGATIVE charge, your thumb points opposite the direction of the force exerted on it.

3. Lorentz force in vector form（矢量叉乘形式）

In general, the force exerted on a moving particle can be expressed by the cross product of the velocity v and the magnetic field B:

$$\vec{F}=q\vec{v}\times\vec{B}$$

This equation is sometimes written as $F = qvB\sin\theta$. The θ refers to the angle formed between the velocity vector of your particle and the direction of the magnetic field. So, if a particle moves in the same direction as the magnetic field lines $\theta = 0°$, $\sin\theta = 0$, and that particle experiences no magnetic force.

要特别注意：1. 粒子带正电与带负电所受的磁场力相反。2. 粒子沿磁感应线入射时，不受力。

4 Magnetic force on a wire 载流导线在磁场中受力

The formula for the force on a long straight current-carrying wire in the presence of a magnetic field is: $F = IBL$

When there is an angle θ between the current and magnetic field, the formula can be expressed as:

$$F = ILB \sin\theta$$

The direction of the force on a current-carrying wire is given by the same right-hand rule as for the force on a charged particle because current is simply the flow of positive charge.

In AP Physics C, the force exerted on an elementary current $I dl$ can be expressed as the cross product: $d\vec{F} = I d\vec{l} \times \vec{B}$

The ant current can be divided into many elementary segments, and the whole force exerted on a current is the integral of dF.

Example:

Two wires are placed parallel to each other. The direction of current in each wire is indicated in figure 2—14—4. How will these wires interact?

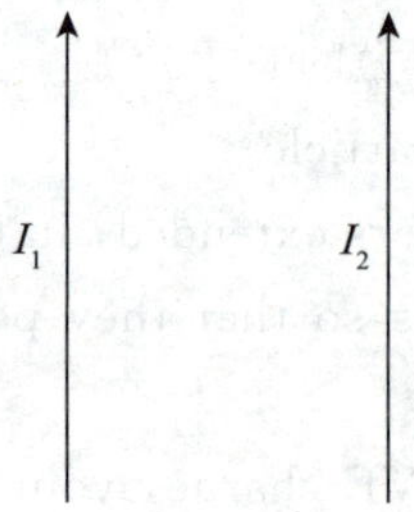

图 2—14—4 two parallel straight lines with current I_1 and I_2

(A) They will attract each other

(B) They will repel each other

(C) They will not affect each other

(D) This question cannot be answered without knowing the length of each wire

Solution:

Let's follow our advice and draw the magnetic field created by the left-hand wire (图 2—14—5).

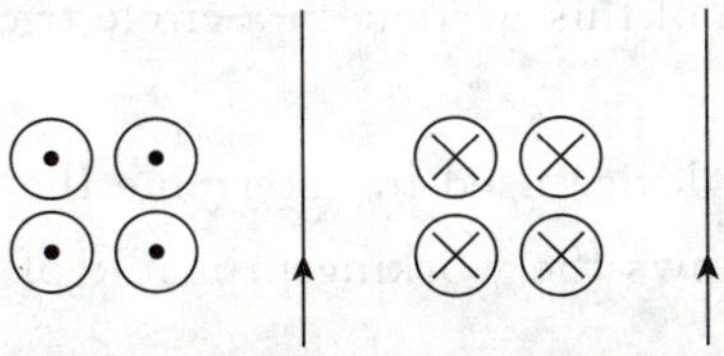

图 2—14—5 the magnetic field created by two parallel straight currents

Now, a wire's field cannot produce a force on itself. The field that we drew is caused by the left wire, but produces a force on the right-hand wire, which direction is that force? Use the right-hand rule for the force on a charged particle. The charges are moving up, in the direction of the current. So point up the page, and curl your fingers toward the magnetic field, into the page. The right wire is forced to the LEFT. Newton's third law says that the force on the left wire by the right wire will be equal and opposite. So, the wires attract, answer A.

5 Mass spectrometry
质谱仪

A magnetic field can make a charged particle travel in a circle.

Let's say you have a proton traveling through a uniform magnetic field coming out of the page, and the proton is moving to the right. The magnetic field exerts a downward force on the particle (use the right-hand rule). So the path of the particle begins to bend downward, as shown in figure 2—14—6.

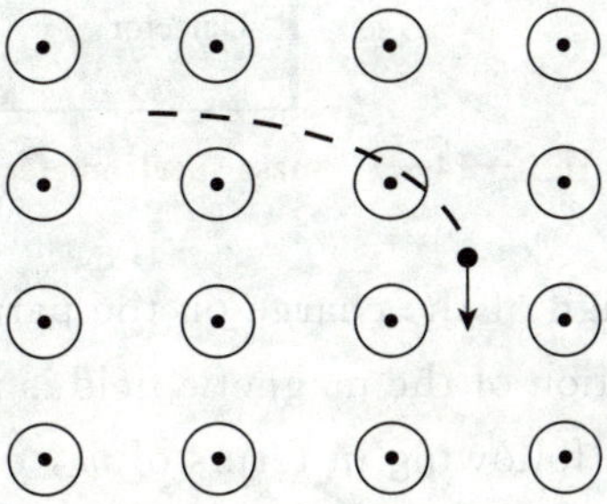

图 2—14—6 a charged particle travel in a circle in a magnetic field

We can figure out the radius of this path with some basic math. The force of the magnetic field is causing the particle to go in a circle, so this force must cause centripetal acceleration. That is,

$$qvB = \frac{mv^2}{r}$$

We can now solve for the radius of the particle's path:

$$r = \frac{mv}{qB}$$

The real-world application of this particle-in-a-circle trick is called a mass spectrometer (质谱仪).

A mass spectrometer is a device used to determine the mass of a particle.

The following example shows the working principle of mass spectrometer:

Example:

A particle of mass m and charge q is accelerated from rest in the plane of the page through a potential difference V between two parallel plates as shown in figure 2—14—7. The particle is injected through a hole in the right hand plate into a region of space containing a uniform magnetic field of magnitude B oriented perpendicular to the plane of the page. The particle curves in a semicircular path and strikes a detector. Neglect relativistic effects throughout this problem.

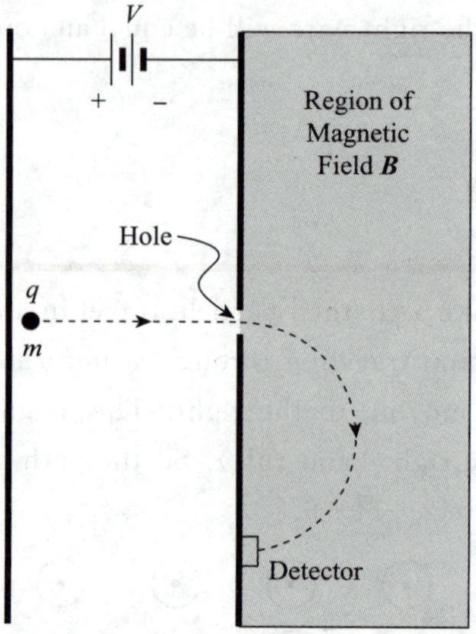

图 2—14—7 mass spectrometer

(a) i. State whether the sign of the charge on the particle is positive or negative.

ii. State whether the direction of the magnetic field is into the page or out of the page.

(b) Determine each of the following in terms of m, q, V, and B.

i. The speed of the charged particle as it enters the region of the magnetic field B.

ii. The force exerted on the charged particle by the magnetic field B.

iii. The distance from the point of injection to the detector.

Solution:

(a) i. Since the particle is accelerate toward negatively charged plate, it must be positively charged.

ii. According to the right hand rule, the magnetic field must point out of the page.

(b) i. The work done by the electric field in accelerating the particle equals the kinetic energy of the particle when it reaches the hole: $qV=\frac{1}{2}mv^2$, so the velocity of the particle should be:

$$v = \sqrt{\frac{2qV}{m}}$$

ii. The force exerted on the charged particle by the magnetic field B is:

$$F = qvB = qB\sqrt{\frac{2qV}{m}}$$

iii. The distance d from the point of injection to the detector is equal to two times of the radium r of the reflection of the particle:

$$d = 2r = \frac{2mv}{qB} = \frac{2m}{qB}\sqrt{\frac{2qV}{m}} = \frac{2}{B}\sqrt{\frac{2Vm}{q}}$$

Based on the above example, if the potential difference V and magnetic field B are kept as a constant, the distance from the point of injection to the detector d is directly proportional to the square root of the ratio between the mass of the particle and the charge:

$$d = \frac{2\sqrt{2V}}{B}\sqrt{\frac{m}{q}}$$

So, if the ratio changes, the distance changes. This is the working principle of the mass spectrometer.

6 Several important results of magnetic fields
关于磁场的几个重要结论

There are several important results of magnetic fields as follows:

(1) The magnetic field produced by a very long straight current (长直导线周围的磁场):

$$B = \frac{\mu_0 I}{2\pi r}.$$

(2) A current-carrying solenoid produces a magnetic field (直螺线管内部的磁场):

$$B_{solenoid} = \mu_0 nI\ .$$

Here I is the current in the solenoid, and n is the number of coils per meter in the solenoid.

7 Practice questions
练习题

Multiple-Choice (选择题)

Single-Select (单选题)

1. Two long, straight, parallel wires in the plane of the page carry equal currents I in the same direction, as shown in figure 2—14—8. Which of the following correctly describes the forces acting on the wires and the resultant magnetic field at points along the dotted line midway between the wires?

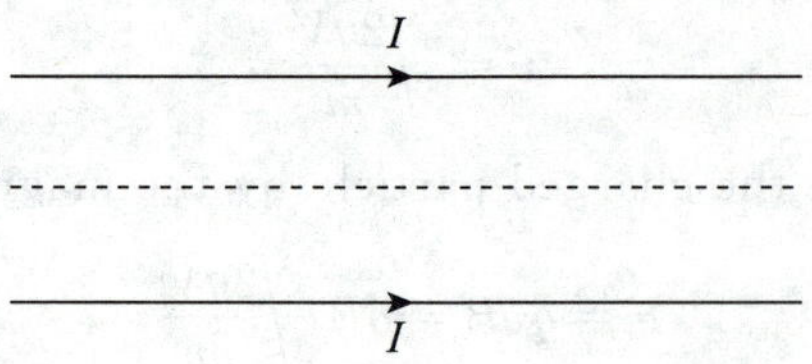

图 2—14—8

forces	field
(A) attractive	not zero
(B) attractive	zero
(C) zero	zero
(D) repulsive	not zero

2. One section of wire in the circuit (图 2—14—9) lies in a region containing a magnetic field B directed into the page. The magnetic force on the section of wire is directed

(A) into the page

(B) toward the right

(C) toward the left

(D) toward the bottom of the page

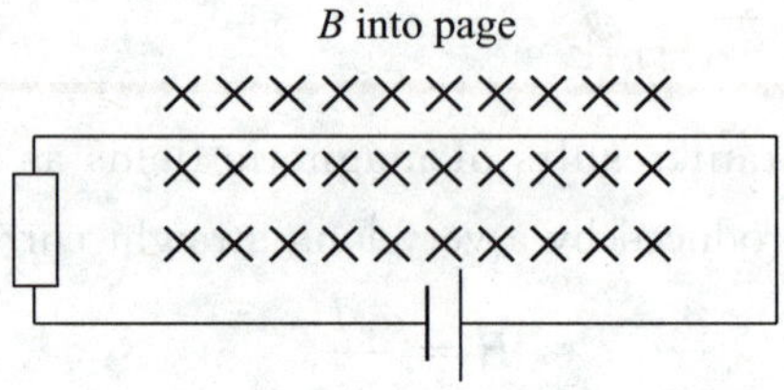

图 2—14—9

3. An electron moves to the right in a uniform magnetic field that points into the page (图 2—14—10). What is the direction of the electric field that could be used to cause the electron to travel in a straight line?

(A) down toward the bottom of the page

(B) up toward the top of the page

(C) into the page

(D) out of the page

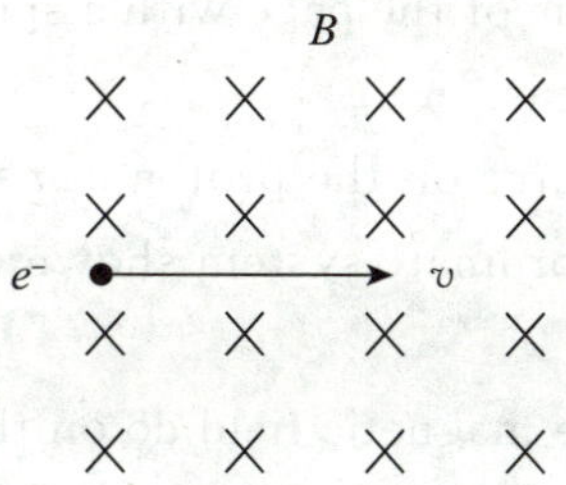

图 2—14—10

4. There is a counterclockwise current I in a circular loop of wire situated in an external magnetic field directed out of the page as shown in figure 2—14—11. The effect of the forces that act on this current is to make the loop

(A) expand in size

(B) contract in size

(C) rotate about an axis perpendicular to the page

(D) rotate about an axis in the plane of the page

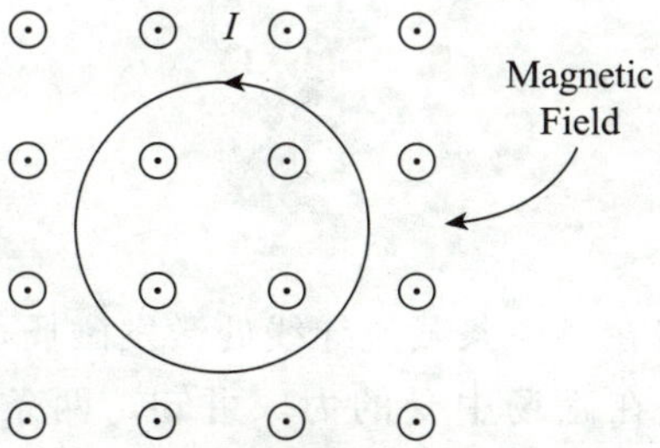

图 2—14—11

Multi-Select (多选题)

5. A beam of electrons with the velocity v is flying to the right and into a uniform electric field shown in figure 2—14—12. Which two of following ways have to be satisfied so that the electron beam can keep a straight line moving to the right?

(A) Put a magnetic field B perpendicular into the page

(B) Put a magnetic field B perpendicular out of the page

(C) The magnitude of magnetic field B should be equal to E/v

(D) The magnitude of magnetic field B should be equal to v/E

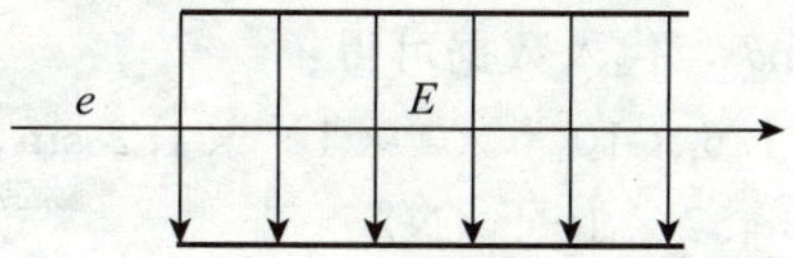

图 2—14—12

Free-Response (计算题)

6. A uniform magnetic field of magnitude $B = 1.2$ Teslas is directed toward the bottom of the page in the y direction, as shown in figure 2—14—13. At time $t = 0$, a proton

p in the field is moving in the plane of the page with a speed $v_0 = 4\times10^7$ meters per second in a direction 30° above the $+x$ axis.

(a) Calculate the magnetic force on the proton at $t=0$.

(b) With reference to the coordinate system shown on the right, state the direction of the force on the proton at $t=0$.

(c) How much work will the magnetic field do on the proton during the interval from $t=0$ to $t=0.5$ second?

(d) Describe (but do not calculate) the path of the proton in the field.

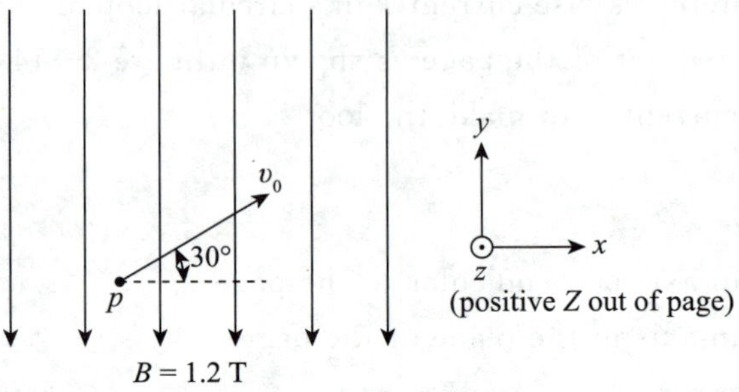

图 2—14—13

题解

单选题

1. 选 (B)。一条载流导线在另一条载流导线处激发磁场，用右手定则可以判断磁场的方向，再用左手定则判断电流在磁场中受的力，可知，两条导线互相吸引。在中间虚线处，两条导线产生的磁场互相抵消，因此磁场强度为零。

2. 选 (D)。电路中的电流方向是逆时针的，因此该段电流受力向下。

3. 选 (A)。带电粒子受的磁场力垂直向下。由于电场力必须抵消磁场力才能使粒子沿直线运动，所以电场力必须垂直向上。由于粒子带负电，电场应垂直向下。

4. 选 (A)。因为磁场力处处垂直于电流向外，因此使电流线圈扩张。

多选题

5. 选 (A) 和 (C)。该束电子水平射入电场方向是自上而下的电场中，受垂直向上的电场力，要保持电子束不向上偏转，就要加入垂直纸面向里的磁场，使电子束受向下的磁场力，磁场强度的大小应为 E/v。

计算题

6. (a) 应用 $F_m = qvB\sin\theta$，代入数据可得：

$$F_m = 1.6\times10^{-19}\times4\times10^7\times1.2\sin120^\circ$$
$$\approx 4.7\times10^{-12}\ (\text{N})$$

(b) 磁场力沿 z 轴负方向。

(c) 由于磁力永远和运动方向垂直，因此，不做功。

(d) 质子的运动是围绕 y 轴，向 y 轴正向平移的螺旋线。

Chapter 15 Electromagnetic induction

第十五章 电磁感应

该知识点讨论了变化的磁通量产生感生电动势（法拉第定律）和导体切割磁感应线产生动生电动势的物理规律。要求学生能用这些规律解决实际问题。

Main points（主要知识点）

- 当通过闭合导线的磁通量发生变化时，闭合导线中产生感生电动势（法拉第定律）
- 感生电动势方向的判定（楞次定律）
- 当导线切割磁感应线时，导线两端产生动生电动势
- 当电感线圈中电流发生变化时，产生自感电动势，当电流稳定时，电感线圈相当于短路

Relevant equations（相关公式）

Magnetic flux：

匀强磁场中通过面积为 A 的平面的磁通：$\varphi_B = BA$

非匀强磁场中通过面积为 A 的曲面的磁通：$\varphi_B = \iint_A \vec{B} \cdot \mathrm{d}\vec{A}$

Induced EMF（感生电动势）：$\varepsilon = -N\dfrac{\Delta\varphi}{\Delta t}$

Motional EMF（导体切割磁感应线产生的动生电动势）：

$\varepsilon = BLv$ for AP Physics 2 students

$\varepsilon = \int (\vec{v} \times \vec{B}) \cdot \mathrm{d}\vec{l}$ for AP Physics C students

1 Induced EMF 感生电动势—法拉第定律

1. Induced EMF（感生电动势）：The potential difference created by a changing mag-

netic field that causes a current to flow in a wire. EMF stands for Electro-Motive Force.

怎样表述穿过线圈的磁场变化？下面我们引进磁通量的概念：

2. Magnetic flux（磁通量）：The number of magnetic field lines that pass through an area. The units of flux are called webers；1 Weber = 1 T·m^2. The equation for magnetic flux is：

$$\varphi_B = BA$$

Where，φ_B is the magnetic flux，B is the magnitude of the magnetic field，A is the area of the region that is penetrated by the magnetic field.

In figure 2—15—1，a magnetic field points to the right，and a circular loop of wire lay down on the page. No field lines go through the loop. So we know that our flux should equal zero.

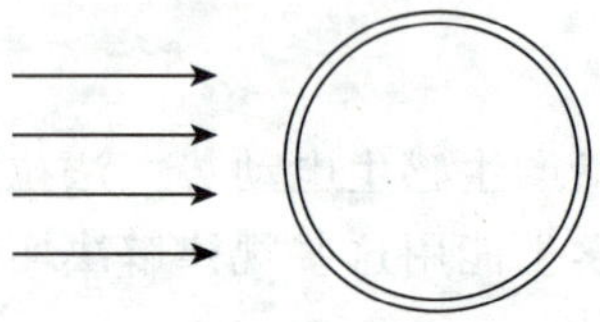

图 2—15—1 the surface of a circular loop is parallel with the magnetic field

Now，we will orient the field lines so that they pass through the middle of the loop. We will also specify the loop's radius = 0.2 m，and that the magnetic field is that of the Earth，$B = 5\times10^{-5}$ T. This situation is shown in figure 2—15—2：

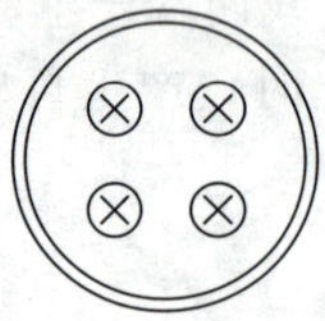

图 2—15—2 the surface of a circular loop is perpendicular to the magnetic field

The flux here is：$\varphi_B = 5\times10^{-5}\times\pi\times0.2^2 = 6.2\times10^{-6}$ (T·m^2)

Generally，when there is an angle between the normal line of the area and the magnetic field，the magnetic flux should be $BA\cos\theta$.

The magnetic flux can be expressed by the dot product of B and A：

$$\varphi_B = \vec{B}\cdot\vec{A}$$

Generally，if the surface is not a plane，we can divide the surface into many elementary areas as dA. Then we can integrate the elementary magnetic flux to obtain the whole magnetic flux as follows：

$$\varphi_B = \iint_A \vec{B}\cdot \mathrm{d}\vec{A}$$

3. Faraday's law（法拉第定律）：The induced EMF in a circuit is directly proportional to the rate of the change of magnetic flux through the circuit：

$$\varepsilon = \frac{\Delta \psi_B}{\Delta t} = N\frac{\Delta \varphi_B}{\Delta t}$$

Where ε is the induced EMF. $\Delta\varphi_B$ is the magnetic flux, N is the number of the loop.

4. Lenz's law (楞次定律): The direction of the induced current opposes the increase in flux, that is, the current that is induced will flow in such a way that the magnetic field it creates points opposite to the direction in which the already existing magnetic flux is changing.

As shown in figure 2—15—3: Initially, the magnetic flux through the loop is zero.

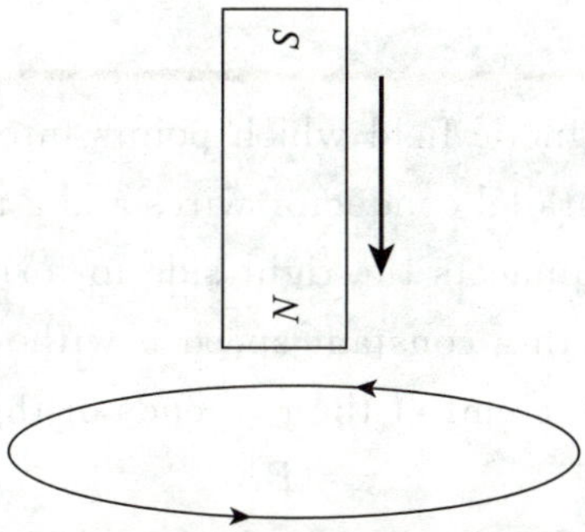

图 2—15—3 Lenz's law

Now, we will move the magnet into the loop. So, the magnetic flux will increase. Lenz's law says that the induced current will create a magnetic field that opposes this increase in flux. So the induced current will create a magnetic field up, against the magnet. By the right-hand rule, the current will flow counterclockwise (look down). This situation is shown in the figure.

After a while, the magnet will be entirely into the loop. Once it enters this loop, there will no longer be a changing flux, So, there is no induced EMF and current. Next, we can pull the magnet out of the loop, and the magnetic flux change again. The induced current in the loop just in the opposite direction is shown in figure 2—15—4.

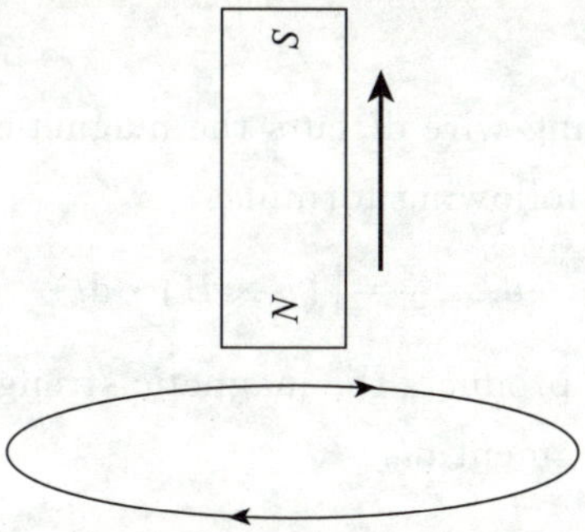

图 2—15—4 Lenz's law

To solve a problem that involves Lenz's law, you can follow the following steps:

- Point your right thumb in the initial direction of the magnetic field.

- Ask yourself, "Is the flux increasing or decreasing?"
- If the flux is decreasing, then just curl your fingers (with your thumb still pointed in the direction of the magnetic field). Your fingers show the direction of the induced current.
- If the flux is increasing in the direction you are pointing, then flux is decreasing in the other direction. So, point your thumb in the opposite direction of the magnetic field, and curl your fingers. Your fingers show the direction of the induced current.

2 Motional EMF 动生电动势

Consider in a uniform magnetic field which points into the paper, a rectangular wire frame which consists of two parallel conductor wires and a resistor R connects to the left end, a moving conductor bar connects the right side to form a rectangular loop. The conductor bar can slide to right with a constant speed v without friction, as shown in figure 2—15—5. It can be seen that the emf at the two ends of the conductor bar is

$$\varepsilon = Blv$$

Where, l is the length of the conductor bar.

v is the constant speed of the moving conductor bar.

B is the density of magnetic flux which is perpendicular to the circuit plane.

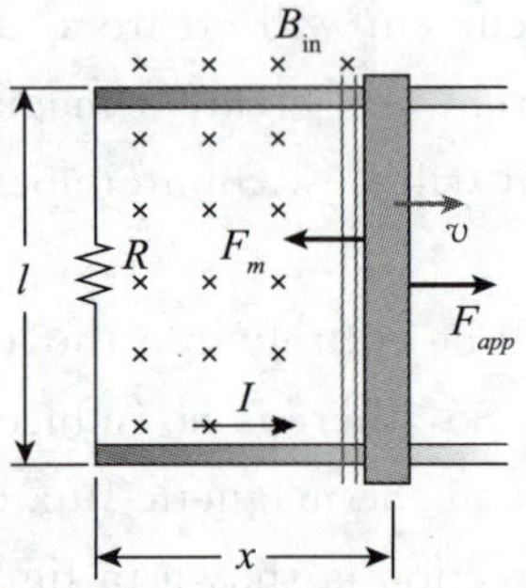

图 2—15—5 motional EMF

If any elementary conducting wire dl cuts the magnetic lines B in velocity v, the motional emf could be written as following formula:

$$\varepsilon_{motional} = \int (\vec{v} \times \vec{B}) \cdot d\vec{l}$$

First, the velocity v cross products the magnetic strength B, then the resultant vector dot products the elementary segment dl.

3 Inductors in circuits 电感电路

An inductor makes use of induced EMF to resist changes in current in a circuit. If part of a circuit is coiled, then the magnetic field produced by the coils induces a "back EMF" (反电动势) in the rest of the circuit. The EMF depends on how fast the current is changing, by Faraday's law.

$$V = L\frac{dI}{dt}$$

Where L is called the inductance of the inductor. Inductance is measured in units of henrys (H). Because H is too big, usually use $mH = 10^{-3}H$, or $\mu H = 10^{-6}H$.

Once the current goes into the coil and reaches stable value, the energy stored in the inductor is:

$$W_m = \frac{1}{2}IL^2$$

4 Practice questions 练习题

Multiple-Choice (选择题)

Single-Select (单选题)

1. Figure 2—15—6 shows a rectangular loop of wire of width l and resistance R. One end of the loop is in a uniform magnetic field of strength B at right angles to the plane of the loop. The loop is pulled to the right at a constant speed v. What are the magnitude and direction of the induced current in the loop?

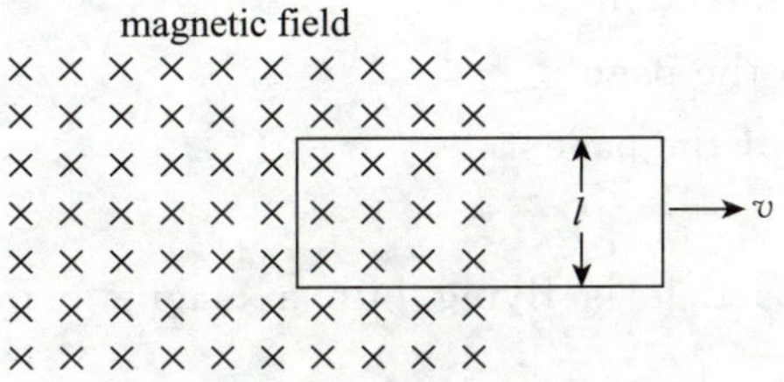

图 2—15—6

magnitude	direction
(A) $BlvR$	clockwise
(B) $BlvR$	counterclockwise
(C) Blv/R	clockwise
(D) Blv/R	counterclockwise

2. Two circular coils are situated perpendicularly to the z axis as shown in figure 2—15—7. There is a current in the primary coil. All of the following procedures will induce a current in the secondary coil EXCEPT

(A) rotating the secondary coil about the z axis

(B) rotating the secondary coil about a diameter

(C) moving the secondary coil closer to the primary coil

(D) varying the current in the primary coil

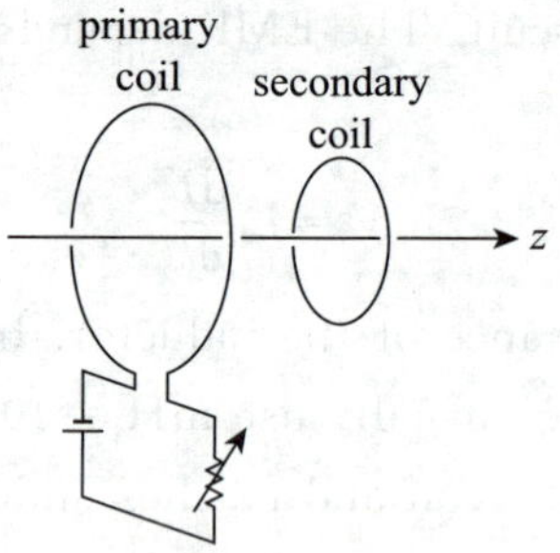

图 2—15—7

3. A magnetic field B that is decreasing with time is directed out of the page and passes through a loop of wire in the plane of the page, as shown in figure 2—15—8. Which of the following is true of the induced current in the wire loop?

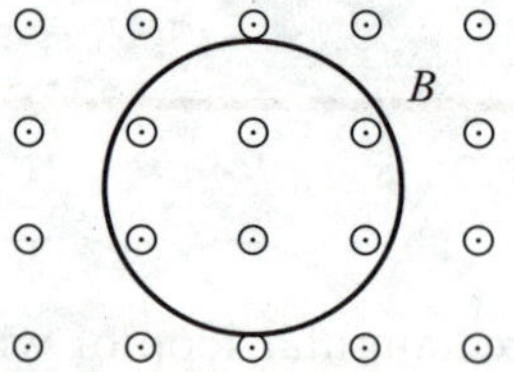

图 2—15—8

(A) It is counterclockwise in direction

(B) It is clockwise in direction

(C) It is directed into the page

(D) It is directed out of the page

Multi-Select (多选题)

4. When a charged particle is flying into a magnetic field B, the magnetic force exerted on the particle is

(A) Perpendicular to its velocity only

(B) Perpendicular to magnetic field

(C) Perpendicular both v and B

(D) Parallel to B

Free-Response (计算题)

5. A circular loop of wire of negligible resistanc and radius $R = 20$ cm is attached to the circuit shown in figure 2—15—9. Each resistor has resistance 10Ω. The magnetic field of the Earth points up along the plane of the page in the direction shown, and has magnitude $B = 5.0 \times 10^{-5}$ T.

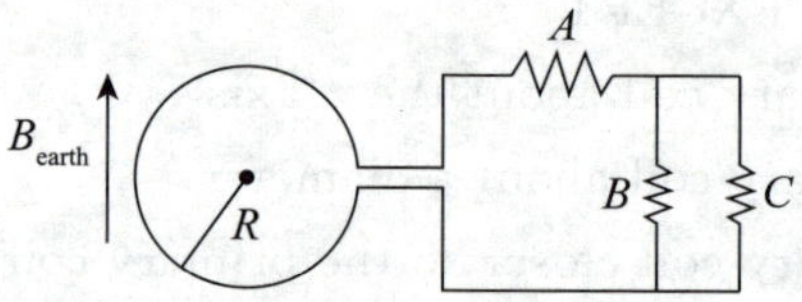

图 2—15—9

The wire loop rotates about a horizontal diameter such that after a quarter rotation the loop is no longer in the page, but perpendicular to it. The loop makes 500 revolutions per second, and remains connected to the circuit the entire time.

(a) Determine the magnetic flux through the loop when the loop is in the orientation shown.

(b) Determine the maximum magnetic flux through the loop.

(c) Estimate the average value of the induced EMF in the loop.

(d) Estimate the average current through resistor C.

题解

单选题

1. 选（C)。导线向右运动产生动生电动势，大小为 Blv， 电流大小为 Blv/R ，只可以选（C）或（D)。由楞次定律可以判定，感生电流的方向为顺时针。

2. 选（A)。因为第二个线圈绕 z 轴旋转不会改变通过第二个线圈的磁通量，因此不能产生感生电动势。

3. 选（A)。因为原磁场在减弱，根据楞次定律，感生电流激发的磁场应该补偿原磁场的减弱，应该是逆时针方向。

多选题

4. 选（B）和（C)。磁场力垂直于粒子速度方向和磁感应强度方向所构成的平面。

计算题

5. (a) 磁通为零，因为线圈的法线和磁场垂直。

(b) 当线圈旋转到垂直于纸面位置时，磁通最大，$BA = 5.0\times10^{-5}\ \mathrm{T}\times\pi\ (0.20\ \mathrm{m})^2 = 6.3\times10^{-6}\ \mathrm{Tm^2}$。

(c) 由于线圈每秒转 500 转，所以每一转用时为$\frac{1}{500}$秒。而每转 1/4 周时，线圈的磁通变化由 0 到最大值，即所用时间为$\frac{1}{2\,000}$秒。根据法拉第定律，感生电动势应为 $6.3\times10^{-6}\ \mathrm{Tm^2}\div 0.000\,5\mathrm{s} = 0.013\mathrm{V}$，即 13mV。

(d) 电路的电动势为 13 mV。并联电阻 B 和 C 阻值是 5Ω。电路总电阻是 15Ω。因此，电路总电流是 $0.013\mathrm{V}/15\ \Omega = 8.4\times10^{-4}\mathrm{A}$，或 $840\mu\mathrm{A}$。因为电阻 B 和 C 的阻值相等，因此每个支路电流为 $420\mu\mathrm{A}$。

Chapter 16 Waves and the wave property of light

第十六章　波和光的波动性

本知识点包括 AP Physics 1 关于波的基本概念，如波长、波速、频率及其相关公式，以及波的干涉、衍射及驻波。对于 AP Physics 2 的考生，本章主要介绍了电磁波和电磁波谱。光从本质上说是电磁波，因此存在干涉、衍射等性质。该知识点综合了 AP 物理 1 和 2 的内容。

Main points（主要知识点）

- 波是振动在介质中的传播。在一定介质中，波速为常数。当波从一种介质传到另一种介质时，波长和波速都发生变化，但频率保持不变
- 两个波相遇，在满足一定条件下，会产生干涉。学生要掌握声波形成驻波的条件和公式
- 当波源和观察者有相对运动时，接收到的波的频率发生变化，称为多普勒效应
- 杨氏双缝干涉实验可以观察到干涉条纹，这是光的波动性的有力佐证
- 波从薄膜反射时，有时会产生相位差，同样可以观察到干涉条纹

Relative equations（相关公式）

波速、波长和频率的关系：$v=\lambda f$

两端固定的弦振动的驻波公式：$\lambda_n=\frac{2L}{n}$，其中 L 是两个固定点之间的距离，n 是驻波的阶数，$n=1$ 时，为 1 阶驻波，波长等于 $2L$；$n = 2$ 时，为 2 阶驻波，波长等于 L

杨氏双缝实验公式：$x=\frac{m\lambda L}{d}$，其中 x 是干涉条纹间距，L 是双缝到屏的距离，d 是双缝之间的距离

单缝衍射的公式：$d\sin\theta=m\lambda$，其中 d 是单缝的宽度，m 是衍射条纹级数

介质的折射率：$n=\frac{c}{v}$

介质中的波长：$\lambda_n = \frac{\lambda}{n}$

薄膜干涉加强公式：$2t = m\lambda_n$，其中，t 为薄膜厚度，m 为干涉级数

1 Transverse and longitudinal waves
横波和纵波

1. Wave（波）

A rhythmic up-and-down or side-to-side motion. Energy is transferred from one particle to another in waves.

2. Transverse wave（横波）

The particles in the wave move perpendicularly to the direction of the wave's motion. When you jiggle a string up and down, you create a transverse wave. A transverse wave is shown in the figure（图 2—16—1）：

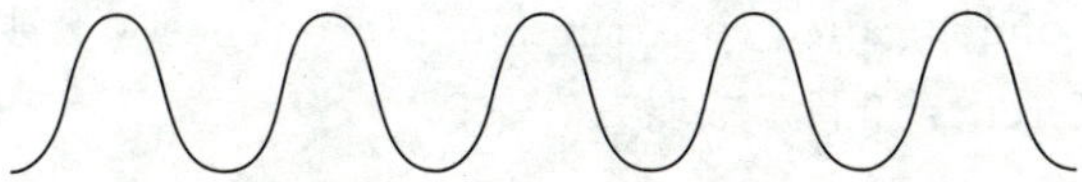

图 2—16—1 a transverse wave

3. Longitudinal wave（纵波）

When particles move parallel to the direction of the wave's motion, this wave is called a longitudinal wave. Sound waves are examples of longitudinal waves. The figure（图 2—16—2）shows a longitudinal wave.

图 2—16—2 a longitudinal wave

A crest (or a peak) is a high point on a wave, and a trough is a low point on a wave. The distance from peak-to-peak or from trough-to-trough is called the wavelength（波长）. This distance is abbreviated with the Greek letter λ. The distance that a peak or a trough is from the horizontal axis is called the amplitude（振幅）, abbreviated with the letter A, as shown in the figure（图 2—16—3）.

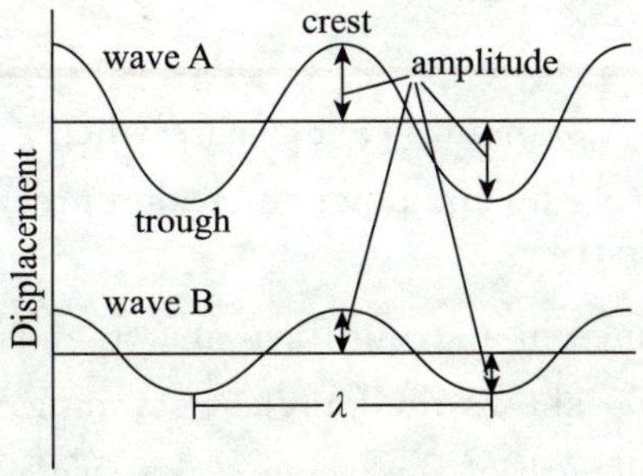

图 2—16—3 trough and crest in a wave

The time necessary for one complete wavelength to pass a given point is the period (周期), abbreviated as T. The number of wavelengths that passes a given point in one second is the frequency (频率), abbreviated as f. The period and frequency of a wave are related with a simple equation, which we first saw when we discussed simple harmonic motion:

$$T=\frac{1}{f}$$

The relationship among velocity, wavelength and frequency is:

$$v=\lambda f$$

For sound waves moving through air at room temperature, $v = 343$ m/s; for light waves in a vacuum, $v = 3.0\times10^{8}$ m/s, which is the maximum velocity in universe.

2 Interference 干涉

When two waves cross each other's path, they interact with each other. There are two ways they can do this—one is called constructive interference (干涉加强), and the other is called destructive interference (干涉减弱).

Constructive interference happens when the peaks of one wave align with the peaks of the other wave. The figure (图 2—16—4) (a)、(b)、(c) shows this process.

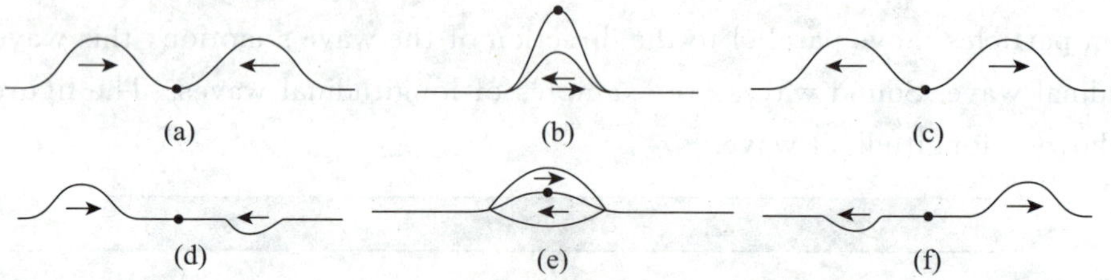

图 2—16—4

However, in (d)、(e)、(f), if the peaks of one wave align with the troughs of the other wave, we get destructive interference.

3 Standing waves 驻波

Standing wave (驻波): A standing wave occurs when a string is held in place at both ends. When you move the string up and down at certain precise frequencies, you produce a standing wave.

(1) Standing waves formed in a string fixed at two ends.

The figure (图 2—16—5) shows the simplest standing wave, called the fundamental frequency (基频). The wavelength of this wave λ_1 is twice as long as the distance between the two walls, L.

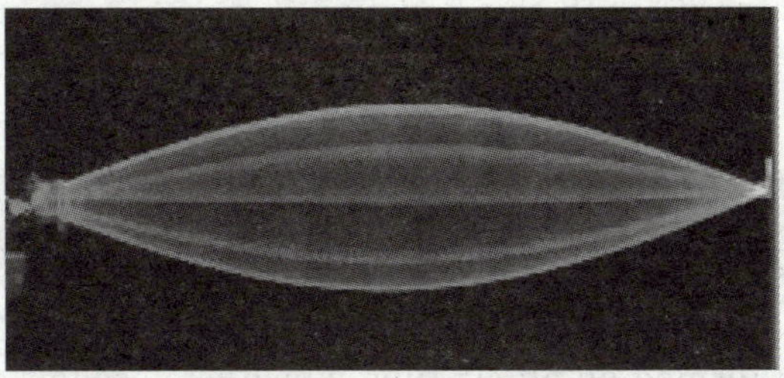

图 2—16—5 the simplest standing wave

Node（波节）: Point on a standing wave where the string (or whatever the wave is traveling in) does not move at all.

Antinode（波腹）: Point on a standing wave where the wave oscillates with maximum amplitude.

The figure（图 2—16—6）shows the next simplest standing wave:

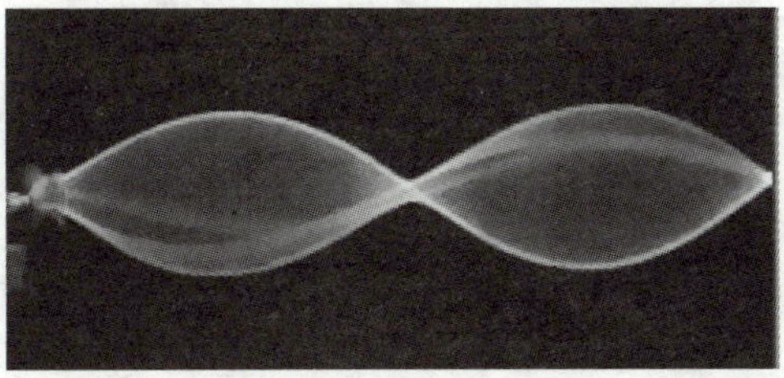

图 2—16—6 the second simplest standing wave

Here, the wavelength, λ_2, is exactly equal to the distance between the walls, L. The next standing wave in our progression is drawn in the figure（图 2—16—7）:

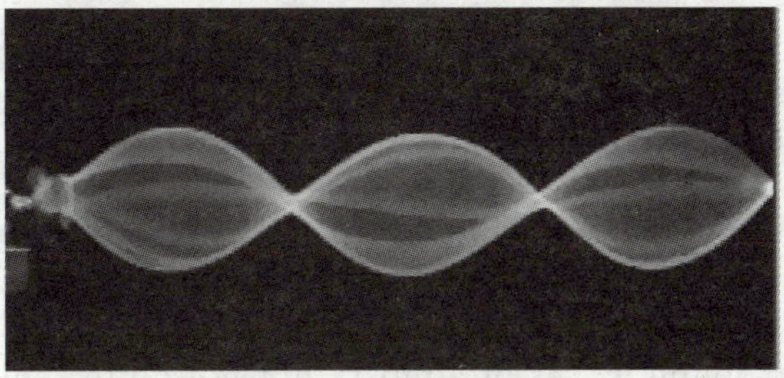

图 2—16—7 the third simplest standing wave

Notice that the relationship between L, the distance between the walls, and the wavelength of the standing wave is

$$L = \frac{n}{2}\lambda$$

We can rewrite it as:

$$\lambda_n = \frac{2L}{n}$$

And we know that the velocity of a wave is $v = \lambda f$. So we can do some more manipulation of variables and come up with this equation:

$$f_n = \frac{nv}{2L}$$

This equation says that the frequency of the 1st, 2nd, 3rd, or nth standing wave equals n multiplied by the velocity of the wave v and then divided by two times the distance between the walls, L.

(2) Standing waves formed on an air tube which is closed at one end.

If we have sound waves in a pipe that's closed at one end, the situation looks slightly different. The fundamental frequency is shown in figure 2—16—8.

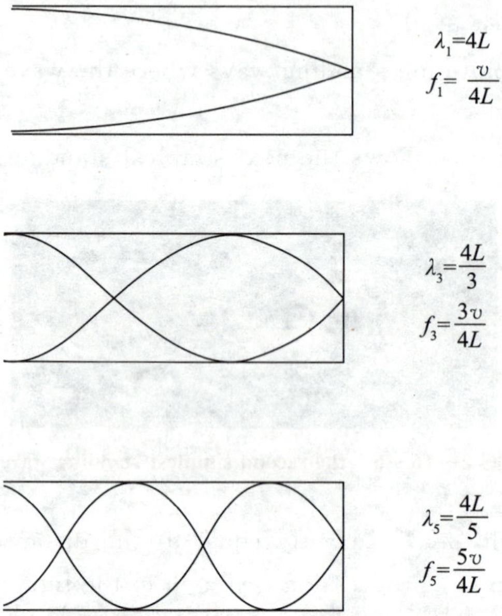

图 2—16—8 standing waves formed on an air tube which is closed at one end

The frequency of the nth standing wave in a closed pipe is:

$$f_n = \frac{nv}{4L},$$

where n must be odd

When you have two sound waves of almost, but not quite, equal frequency, you may hear beats.

Beats (拍): Rhythmic interference that occurs when two notes of unequal but close frequencies are played.

If you have a couple of tuning forks of similar—but not identical—frequency to play with you might enjoy generating some beats of your own. They make a wonderful "wa—wa" sound.

4 Doppler effect 多普勒效应

Doppler Effect（多普勒效应）：The apparent change in a wave's frequency that you observe whenever the source of the wave is moving toward or away from you.

To understand the Doppler effect, let's look at what happens as a fire truck travels past you.

As the fire truck moves toward you, its sirens are emitting sound waves. Let's say that the sirens emit one wave pulse when the truck is 50 meters away from you. It then emits another pulse when the truck is 49.99 meters away from you. And so on. Because the truck keeps moving toward you as it emits sound waves, it appears to you that these waves are getting scrunched together as shown in the figure（图 2—16—9）.

图 2—16—9 a fire truck moves toward a person

Then, once the fire truck passes you and begins to move away from you, it appears as if the waves are stretched out, you hear a lower frequency. As shown in the figure（图 16—10）.

图 2—16—10 a fire truck passes away from a person

Doppler effect can be calculated by the following equation:

$$f_D = \frac{v \pm v_D}{v \pm v_S} f_S$$

Where f_S is the frequency of waves that source emits, and f_D is the frequency that the detector hears. v is the velocity of the sound; v_D and v_S stand for the velocity of detector and source respectively.

For example, If the the source of sound is kept rest, and detector moves toward the source, that means the relative velocity between detector and source increases; choose positive sign for v_D. So, the formula is simplified as:

$$f_D = \frac{v + v_D}{v} f_S$$

So, $f_D > f_S$, the detector will hear higher frequency signal.

5 Electromagnetic waves 电磁波

When radio waves are beamed through space, or when X-rays are used to look at your bones, or when visible light travels from a light bulb to your eyes, electromagnetic waves are at work.

All these types of radiation fall in the electromagnetic spectrum, as shown in the figure (图 2—16—11):

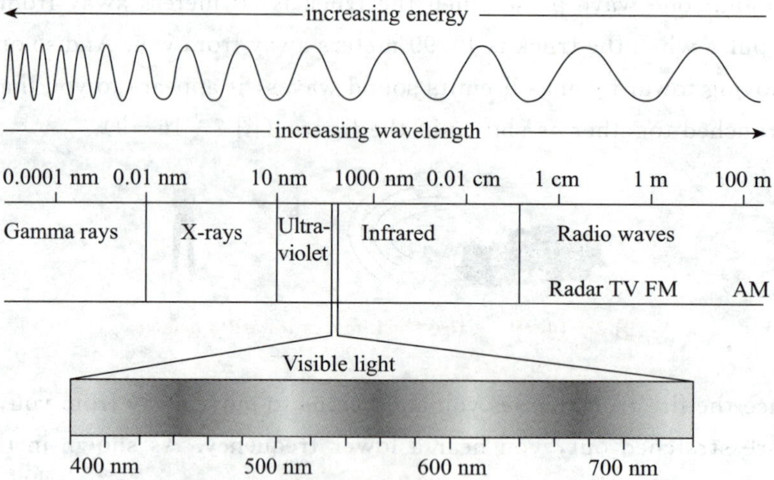

图 2—16—11 the electromagnetic spectrum

The unique characteristic about electromagnetic waves is that all of them travel at exactly the same speed through a vacuum—3×10^8 m/s. The more famous name for "3×10^8 m/s" is "the speed of light," or "c".

What makes one form of electromagnetic radiation different from another form is simply the frequency of the wave. AM radio waves have a very low frequency and a very long wavelength; whereas, gamma rays have an extremely high frequency and an exceptionally short wavelength. But they are all just varying forms of light waves.

6 Young's double-slits experiment 杨氏双缝干涉实验

The way that physicists showed that light behaves like a wave was through the famous Young's double-slits experiment. Consider light shining through two very small slits, located very close together—slits separated by tenths or hundredths of millimeters. The light shone through each slit and then hit a screen. The only way to explain this experimental phenomenon was to conclude that light behaves like a wave, as shown in the figure (图 2—16—12):

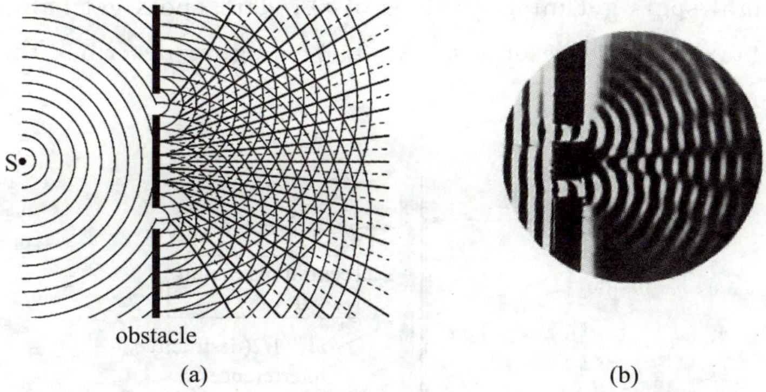

图 2—16—12 Young's double-slits experiment

So when the light waves hit the screen, at some places they constructively interfered with one another and in other places they destructively interfered with one another. In order to obtain the equation of interference fringes, the following figure（图 2—16—13) shows a good approximation.

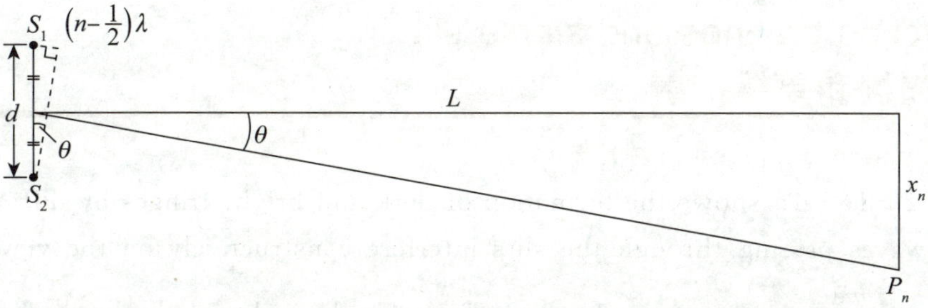

图 2—16—13 the geometric relation in Young's double-slits experiment

The location of bright spots is as follows:

$$d\sin\theta = m\lambda$$

Where, d is the distance between slits, λ is the wavelength of the light, and m is the "order" of the bright spot; we discuss m below. θ is the angle at which an observer has to look to see the bright spot. Usually, the bright spots are pretty close together, and almost directly across from the slits. In this case, the angle θ is small, so we can use the following equation instead. It describes where on the screen you would find patches of constructive or destructive interference.

$$x = \frac{m\lambda L}{d}$$

In this equation, x is the distance from the center of the screen where you would find the region you are seeking (either a bright region or a dark region), λ is the wavelength of the light shining through the slits, L is the distance between the slits and the screen, and d is the distance between the slits. The variable m represents the "order" of the bright or dark spot, measured from the central maximum as shown in the figure（图

2—16—14). Bright spots get integer values of m; dark spots get half-integer values of m. The central maximum represents $m = 0$. The figure describes the relationship of these quantities.

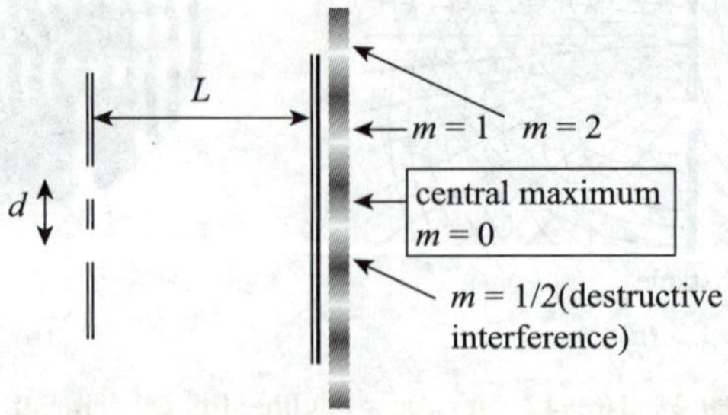

图 2—16—14 the interference patterns in Young's double-slits experiment

Usually, people are more interested in the two bright fringes or two dark fringes, the formula becomes: $\Delta x = \frac{\lambda L}{d}$

7 Diffraction gratings and single slit 衍射光栅和单缝衍射

1. Diffraction gratings (衍射光栅)

Figure 2—16—15 shows the formation of dark and bright fringes by diffraction gratings. The waves passing through the slits interfere constructively on the viewing screen when $\sin\theta_m = \frac{m\lambda}{d}$, where $m = 1, 2, 3, \cdots$, is the order of the bright line or maximum.

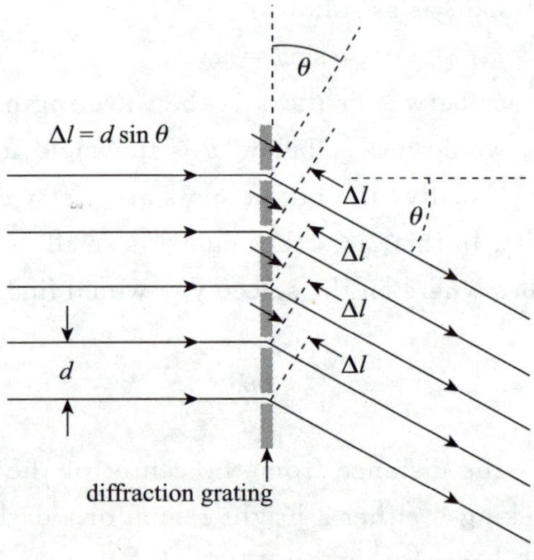

图 2—16—15 diffraction gratings

A diffraction grating consists of a large number of slits, not just two slits. The locations of bright and dark spots on the screen are the same as for a double slit, but the bright spots produced by a diffraction grating are very sharp dots.

2. Single slit（单缝衍射）

Figure 2—16—16 shows the mechanism of single slit interference. For a single slit, the central maximum is bright and very wide; the other bright spots are regularly spaced, but dim relative to the central maximum. According to the figure, the following results can be drawn:

For interference minima or dark fringes $\sin\theta_n = \frac{n\lambda}{W}$ (minima or dark, $n=1, 2, 3, \cdots$)

For interference maxima or bright fringes $\sin\theta_m = \frac{\left(m+\frac{1}{2}\right)\lambda}{W}$ (maxima or bright, $m=1, 2, 3, \cdots$)

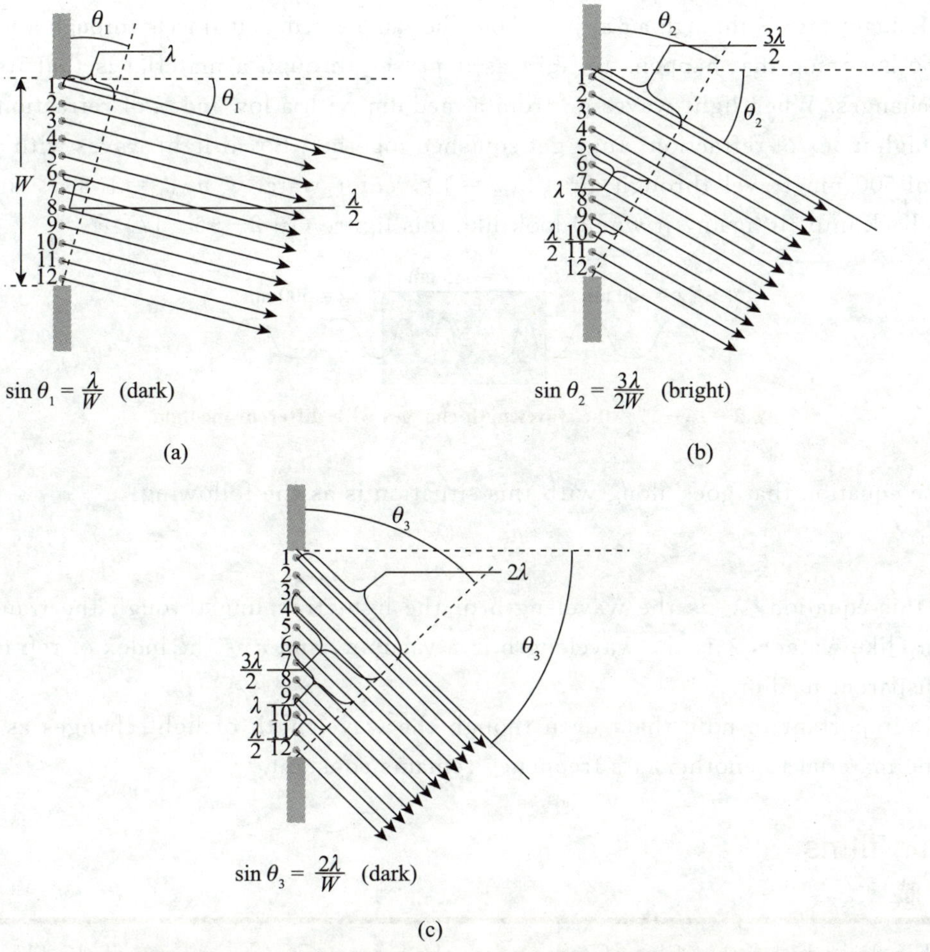

图 2—16—16 the dark and bright fringes formation in single slit

8 Index of refraction 折射率

Now let us examine how light behaves when it passes through different materials.

Light—or any electromagnetic wave—travels at the speed c, or 3×10^8 m/s. But it only travels at this speed through a vacuum, when there aren't any pesky molecules to get in the way. When it travels through anything other than a vacuum, light slows down. The amount by which light slows down in a material called the material's index of refraction.

Index of refraction (折射率): A number that describes by how much light slows down when it passes through a certain material, abbreviated as n.

The index of refraction can be calculated using this equation:

$$n = \frac{c}{v}$$

For example, the index of refraction of glass is about 1.5. This means that light travels 1.5 times faster through a vacuum than it does through glass. The index of refraction of air is approximately 1. Light travels through air at just about the same speed as it travels through a vacuum.

Another thing that happens to light as it passes through a material is that its wavelength changes. When light waves go from a medium with a low index of refraction to one with a high index of refraction, they get squished together. So, if light waves with a wavelength of 500 nm travel through air ($n_{air} = 1$), enter water ($n_{water} = 1.33$), and then emerge back into air again, it would look like this figure (图 2—16—17):

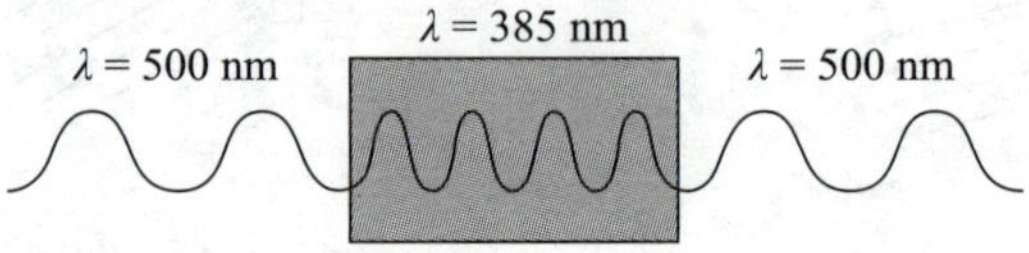

图 2—16—17 **the wavelength changes with different medium**

The equation that goes along with this situation is as the following:

$$\lambda_n = \frac{\lambda}{n}$$

In this equation, λ_n is the wavelength of the light traveling through the transparent medium (like water), λ is the wavelength in a vacuum, and n is the index of refraction of the transparent medium.

It is important to note that, even though the wavelength of light changes as it goes from one material to another, its frequency remains constant.

9 Thin films 薄膜

When light hits a thin film of some sort, the interference properties of the light waves are readily apparent.

Consider a situation where monochromatic light (单色光) (meaning "light that is all of the same wavelength") hits a thin film, as shown in figure 2—16—18. At the top surface, some light will be reflected, and some will penetrate the film. The same thing will happen at the bottom surface: some light will be reflected back up through the film, and some will keep on traveling out through the bottom surface. Notice that the two reflected light waves overlap; the wave that reflected off the top surface and the wave that traveled through the film and reflected off the bottom surface will interfere.

Actually, as long as you know the following working principles, you will be able to solve the problems involved in the thin film interference:

(1) Only the reflect light waves are inverted (180° phase change), when the light waves pass from fast medium (smaller n) to slower medium (bigger n). Transmitted waves are never inverted.

(2) The path difference between two light rays also causes phase change. One light ray has no phase change.

For thin films, according the two principles you can analyse the interference in different cases. The more levels thin films, the more complex. You must be very careful.

1. Interference from reflected light(反射光干涉)(图 2—16—18)

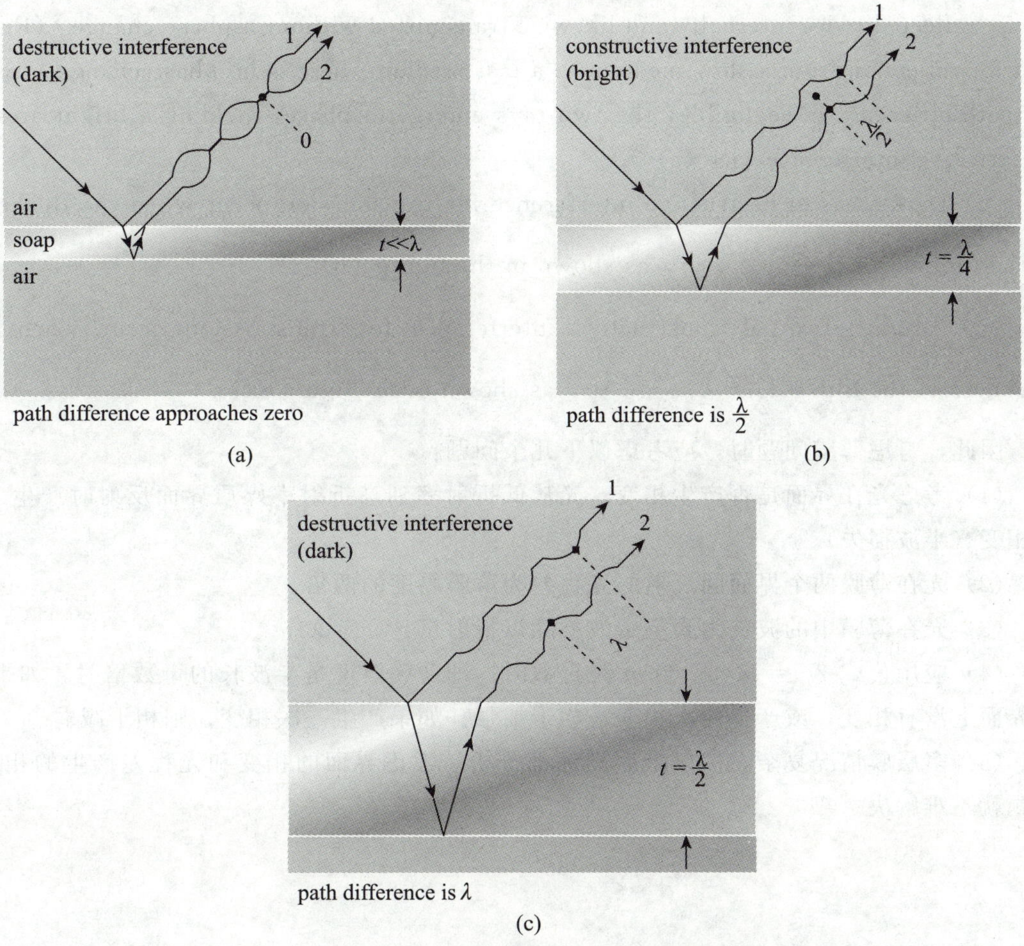

图 2—16—18 thin films interference for reflected light

(a) Since both rays originate from the same source, they are initially in phase. Ray 1 will be inverted when it is reflected, whereas ray 2 will not. Because the film is very thin ($t \ll \lambda$), the extra distance travelled by ray 2 is negligible, and the two rays, being 180° out of phase, interfere destructively (图 (a)).

(b) If the soap film is a little thicker, ray 2 will have an appreciable path difference in comparison with ray 1. If the thickness t of the film is $\frac{1}{4}\lambda$, the path difference is $\frac{1}{2}\lambda$, for nearly normal paths, yielding a 180° phase delay. The two rays, initially out of phase because of reflection, are now back in phase again. Constructive interference occurs, and a bright area is observed (图 (b)).

(c) When the thickness of the film is $\frac{1}{2}\lambda$ and the path difference is λ (图 (c)), the two reflected rays are again out of phase and there is destructive interference.

2. Interference from transmitted light (透射光干涉) (图 2—16—19)

(a) When the thickness t of the film is essentially zero ($t \ll \lambda$), the transmitted light can be considered to consist of two rays: ray 1, transmitted without any phase change, and ray 2, reflected twice internally and likewise transmitted without a phase change. (Recall that for reflection from a slow medium to a fast medium there is no phase change.) Since the path difference is negligible, the two rays emerge in phase, yielding a bright area of constructive interference (图 (a)).

(b) Dark areas of destructive interference for transmission occur when the thickness of the film is $\frac{1}{4}\lambda$, $\frac{3\lambda}{4}$, $\frac{5\lambda}{4}$, …, as shown in the figure (b).

(c) Bright areas of constructive interference for transmission occur when the thickness of the film is $\frac{\lambda}{2}$, λ, $\frac{3\lambda}{2}$, …, as shown in the figure (c).

因此，考虑薄膜问题时，应考虑以下几个问题：

(1) 考虑光在界面是否产生相变，光从低折射率到高折射率媒质界面反射时产生 180 度相变（半波损失）。

(2) 光在薄膜两个界面间反射的光程差为薄膜厚度的两倍。

(3) 光在薄膜中的波长为真空中波长除以折射率：$\lambda_n = \lambda / n$ 。

(4) 应用公式 $2t = m\lambda_n$。当 m 是奇数时，即薄膜厚度是半波长的奇数倍时，如果光在界面上没有相变，或产生两次相变，相干加强。如果产生一次相变，则相干减弱。

(5) 多层膜情况复杂，按以上原则仔细分析，考虑界面的相变和光程差产生的相变，问题就不难解决。

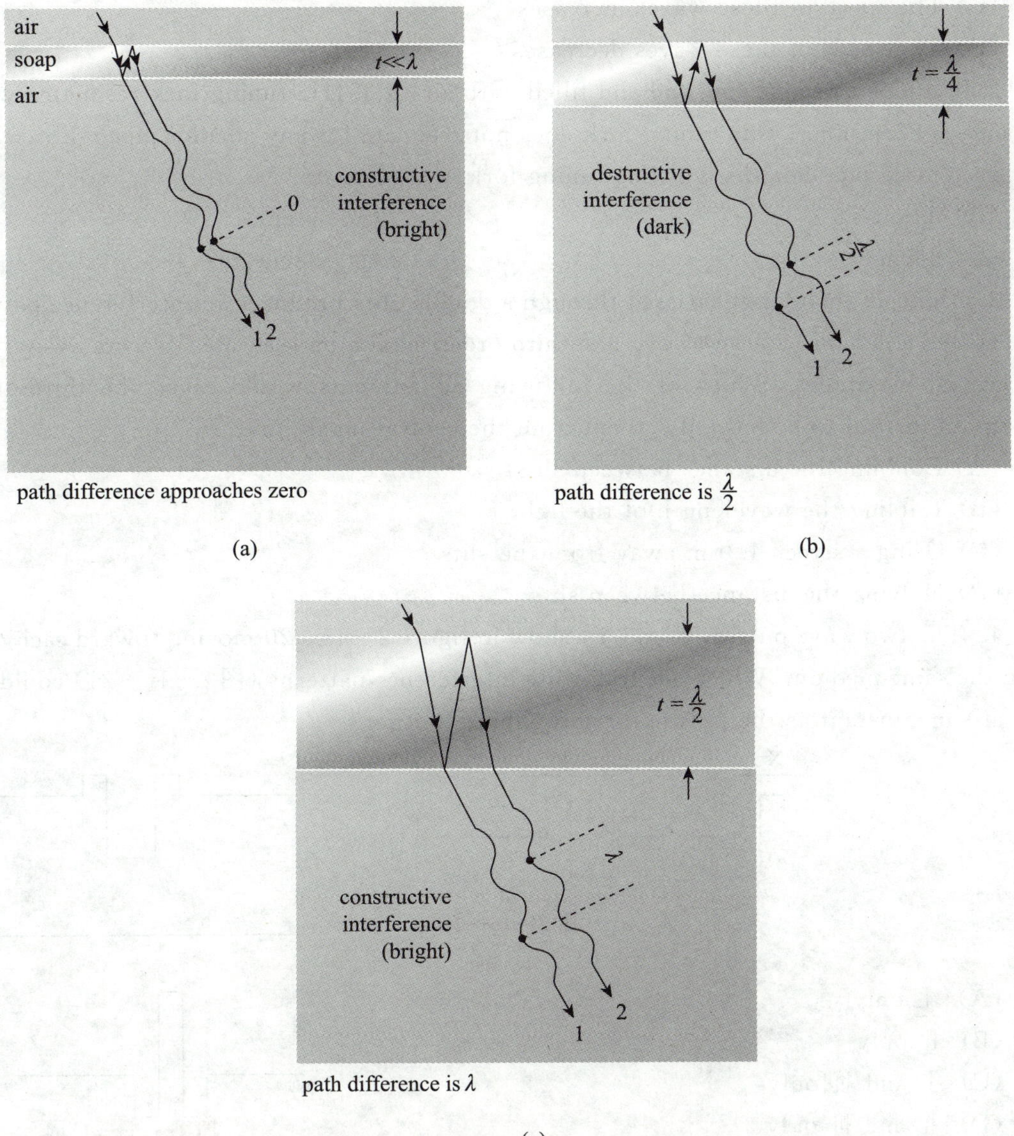

图 2—16—19　thin films interference for transmitted light

10 Practice questions 练习题

Multiple-Choice（选择题）

Single-Select（单选题）

1. A talk show host inhales helium; as a result, the pitch of his voice rises. What happens to the standing waves in his vocal cords to cause this effect?

(A) The wavelength of these waves increases

(B) The wavelength of these waves decreases

(C) The speed of these waves increases

(D) The speed of these waves decreases

2. In a pipe closed at one end and filled with air, a 384Hz tuning fork resonates when the pipe is 22cm long; this tuning fork does not resonate for any smaller pipes. For which of these closed pipe lengths will this tuning fork also resonate?

(A) 11cm (B) 44cm

(C) 66cm (D) 88cm

3. Monochromatic light passed through a double-slits produces an interference pattern on a screen a distance 2. 0 m away. The third order-maximum is located 1. 5 cm away from the central maximum. Which of the following adjustments would cause the third-order maximum instead to be located 3. 0 cm from the central maximum?

(A) Doubling the distance between slits

(B) Tripling the wavelength of the light

(C) Using a screen 1. 0 m away from the slits

(D) Halving the distance between slits

4. The two wave pulses, X and Y, show in figure 2—16—20 moving toward each other in the same medium. Which the following interference patterns (图 2—16—21) could result at some instant as the pulse pass through each other?

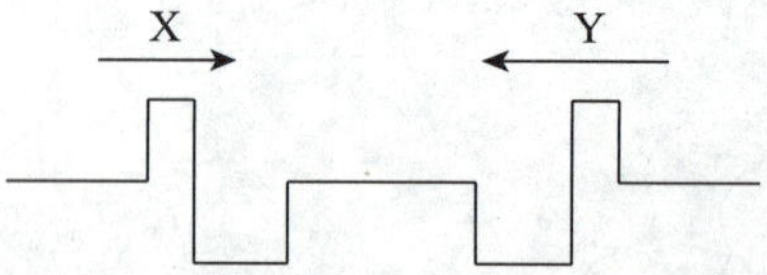

图 2—16—20

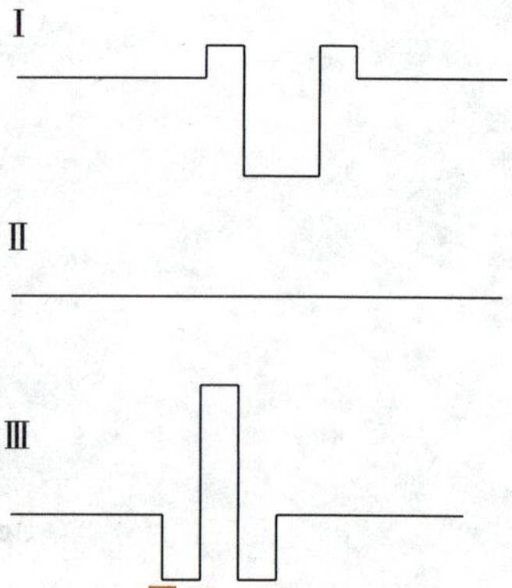

图 2—16—21

(A) Ⅰ only

(B) Ⅱ only

(C) Ⅰ and Ⅲ only

(D) Ⅱ and Ⅲ only

Multi-Select (多选题)

5. Which of the following experiments reviews the wave property of the light? Choose two of them.

(A) The diffraction experiment of the light

(B) The reflection law of the light

(C) Yong's double-slits interference experiment

(D) Photoelectric effect of the light

Free-Response (计算题)

6. Laser light passes through a diffraction grating with 7 000 lines per centimeter. Light is projected onto a screen far away. An observer by the diffraction grating observes the first order maximum 25° away from the central maximum.

(a) What is the wavelength of the laser?

(b) If the first order maximum is 40 cm away from the central maximum on the screen, how far away is the screen from the diffraction grating?

(c) How far, measured along the screen, from the central maximum will the second-order maximum be?

题解

单选题

1. 选（C）。由于声音的音调由频率决定，音调变高，则波的频率增加。由于主发音体发出的声波的波长只和其物理结构有关，所以主发音体发出的声波的波长不变，根据 $v=\lambda f$ ，可知该声波的波速增加。所以声音在氦气中的传播速度比在空气中快。

2. 选（C）。当波在一段封闭的气柱中传播时，形成驻波。封闭端为波节，敞开端为波腹。22cm 为管长，其振动如图 2—16—22 所示：

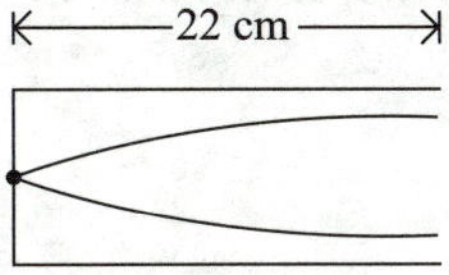

图 2—16—22

22cm 的管长对应 1/4 波长，所以波长为 88cm。下一个驻波产生时，管子一端仍为波节，另一端为波腹，如图 2—16—23 所示：

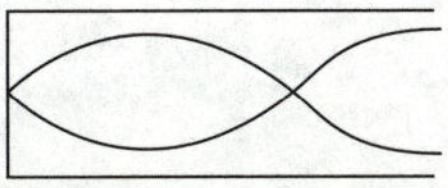

图 2—16—23

管子包括 3 个 1/4 波长，因此管长应为 66cm。

3. 选（D）。应用双缝干涉公式：

$$x=\frac{m\lambda L}{d}$$

要想加倍三级亮纹间距离，即使方程中 x 值加倍，就要减半双缝的距离 d。

4. 选（C）。当两个波的脉冲叠加时，其振幅相加，两个脉冲相遇不可能抵消，因此，Ⅱ不可能，只有Ⅰ和 Ⅲ 是可能的结果。

多选题

5. 选（A）和（C）。光的衍射和双缝干涉实验是光的波动性的最好证明。

计算题

6. (a) 应用光栅衍射公式：$d\sin\theta=m\lambda$，其中 d 代表光栅间距离。由于已知光栅每厘米 7 000 线，所以光栅间距为 $\frac{1}{7\,000}$ 厘米，即 1.4×10^{-4} cm，或 1.4×10^{-6} m。对于第一极大值，$m=1$，$\theta=25°$，代入公式，得到波长约等于 6×10^{-7} m，即 600 nm。

(b) 这是一个几何问题，如图 2—16—24。

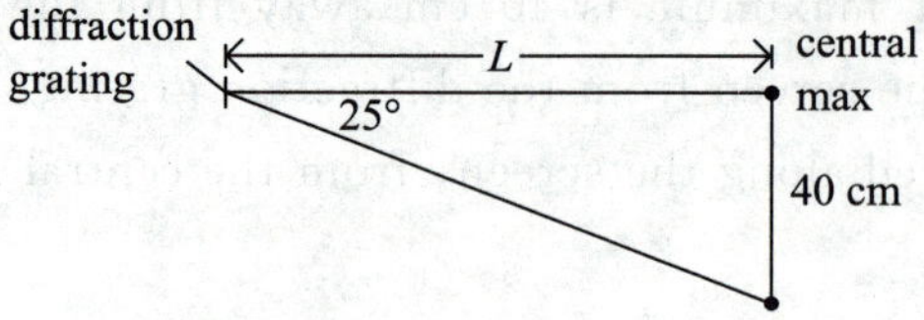

图 2—16—24

$\tan 25^\circ = \frac{40}{L}$，解出 L 等于 86cm。

(c) 应用 $d\sin\theta = m\lambda$；当 $m=2$ 时，解出 θ。即 $\sin\theta = \frac{2\times6\times10^{-7}}{1.4\times10^{-6}}$，得 $\theta = 59^\circ$。现在，应用 (b) 中同样的几何规律，可以解出所要求的距离，即 $\tan 59^\circ = \frac{x}{86}$，因此，$x = 143\text{cm}$。

Chapter 17

Geometric optics

第十七章 几何光学

这部分知识包括几何光学的主要内容，即：反射、折射定律，面镜和透镜的成像规律及其公式。

Main points（主要知识点）

- 当光线照射到两种媒质界面时，部分光反射，部分光折射。折射光遵从斯奈尔定律（Snell's law）
- 光线从高折射率媒质射到高折射率与低折射率媒质界面时，当入射角大于一定值时，会发生全反射
- 凸透镜和凹透镜是光学器件，它们能将光线汇聚到焦点，或从焦点发散
- 通过透镜成像公式，可以判定在什么条件下，透镜可成实像，在什么条件下，透镜可成虚像，何时成正立的像，何时成倒立的像

Relative equations（相关公式）

Snell's law（for refraction)(斯奈尔定律)：$n_1\sin\theta_1 = n_2\sin\theta_2$

Critical angle for total internal reflection（全反射的临界角）：

$$\sin\theta_c = \frac{n_2}{n_1}$$

透镜（面镜）成像公式：

$$\frac{1}{f} = \frac{1}{d_i} + \frac{1}{d_o}$$

Magnification equation（放大率公式）：

$$m = -\frac{d_i}{d_o} = \frac{h_i}{h_o}$$

1 Snell's law 斯奈尔定律

Light can change its direction when it travels from one medium to another. The way in which light changes its direction is described by Snell's law:

$$n_1 \sin\theta_1 = n_2 \sin\theta_2$$

Where θ_1 is the incident angle, θ_2 is the refract angle. Figure 2—17—1 describe the situation: a ray of light is going from air into water. The dotted line perpendicular to the surface is called the "normal". This line is not real; rather it is a reference, a line for use in Snell's law.

As the light ray enters the water, it is bent toward the normal. The angles θ_1 and θ_2 are marked on the figure, and the index of refraction of each material, n_1 and n_2 is also noted. If we knew that θ_1 equals 55°, for example, we could solve for θ_2 using Snell's law.

$$(1.00)(\sin 55°) = (1.33)(\sin\theta_2)$$

$$\theta_2 = 38°$$

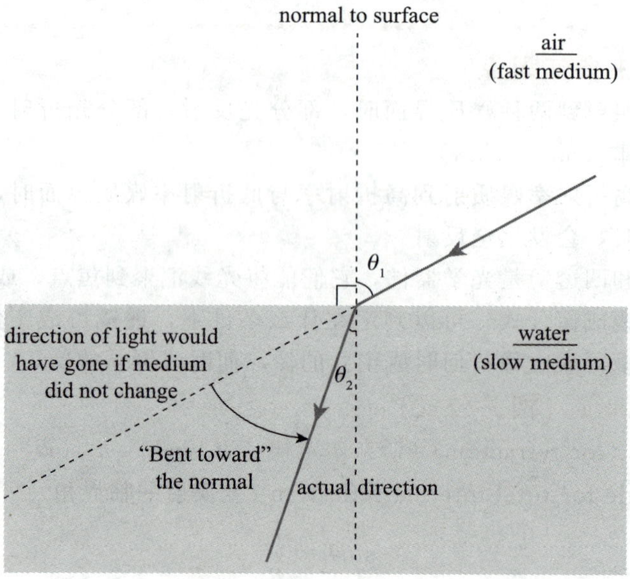

图 2—17—1 Snell's law

Whenever light goes from a medium with a low index of refraction to one with a high index of refraction—as in our drawing—the ray is bent toward the normal. Whenever light goes in the opposite direction—say, from water into air, the ray is bent away from the normal.

2 Total reflection 全反射定律

When light goes from a medium with a high index of refraction to one with a low index of refraction, we could get total internal reflection. For total internal reflection to occur, the light ray must be directed at or beyond the critical angle.

Critical angle (临界角): The angle past which rays cannot be transmitted from one material to another, abbreviated as θ_c.

Let's take a look at figure 2—17—2: a ray of light shines up through water. The critical angle for light going from water to air is 40°. However, the angle of the incident ray is greater than the critical angle. Therefore, the light cannot be transmitted into the air. Instead, all of it reflects inside water.

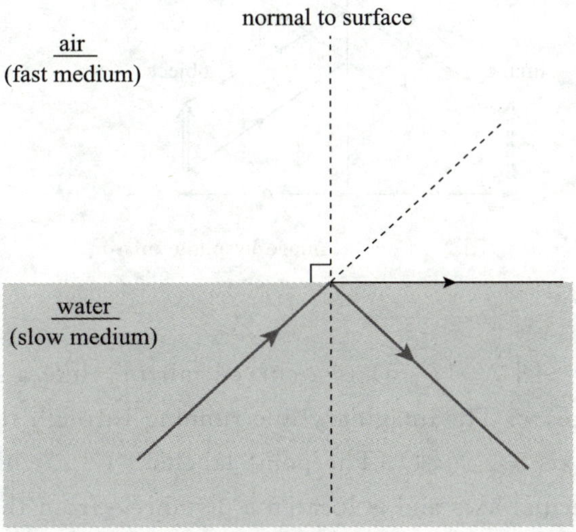

图 2—17—2 total reflection

The law of total reflection has many applications. The most important application is used in optical fiber communication.

3 Mirrors 面镜

1. Plane mirror (平面镜): The angle at which a ray of light hits the mirror equals the angle at which it bounces off, as shown in figure 2—17—3. In other words, or more accurately, $\theta_i = \theta_r$, where θ_i is the incident angle, and θ_r is the reflected angle.

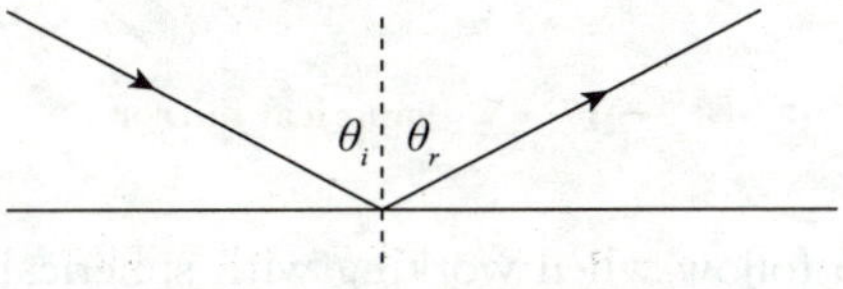

图 2—17—3 plane mirror

Whenever you are working with a plane mirror, follow these rules:

- The image is upright. Another term for an upright image is a virtual image.

- The image is the same size as the original object. That is, the magnification, m, is equal to 1.
- The image distance, d_i, equals the object distance, d_o. Figure 2—17—4 shows this result.

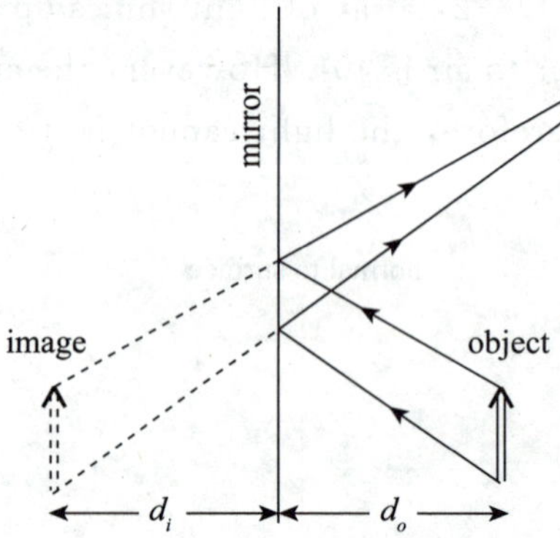

图 2—17—4 image by plane mirror

2. Spherical mirror（球面镜）:

A spherical mirror（图 2—17—5）is a curved mirror—like a spoon—that has a constant radius of curvature, r. The imaginary line running through the middle of the mirror is called the principal axis（主光轴）. The point labeled "C", which is the center of the sphere, lies on the principal axis and is located a distance, from the middle of the mirror. The point labeled "F" is the focal point（焦点）, and it is located a distance f where $f=\frac{r}{2}$（焦距）, from the middle of the mirror. The focal point is also on the principal axis.

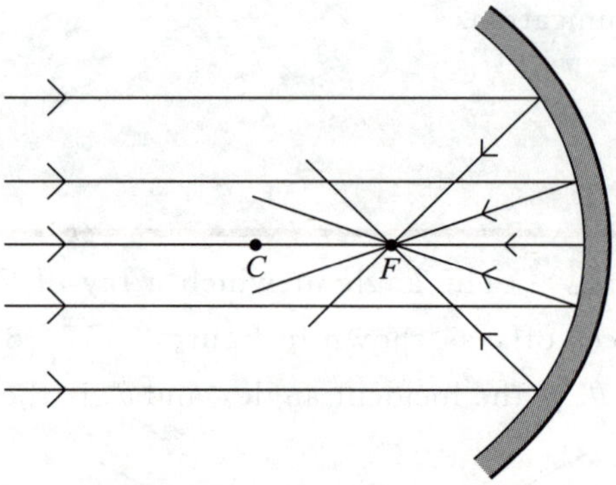

图 2—17—5 spherical mirror

There are several rules to follow when working with spherical mirrors（球面镜成像规律）:

- Incident rays that are parallel to the principal axis reflect through the focal point.
- Incident rays that go through the focal point reflect parallel to the principal axis.
- Any points that lie on the same side of the mirror as the object are a positive distance from the mirror. Any points that lie on the other side of the mirror are a negative dis-

tance from the mirror.

- $\frac{1}{f}=\frac{1}{d_i}+\frac{1}{d_o}$

The last rule is called the "mirror equation"(面镜方程).

To demonstrate these rules, we will draw three different ways to position our arrow with respect to the mirror. In the first scenario, we will place our arrow on the principal axis beyond point "C" as shown in figure 2—17—6.

(1) upside down image(凹面镜成倒立实像)

Notice that the image here is upside down. Whenever an image is upside down, it is called a real image (实像). A real image can be projected onto a screen, whereas a virtual image cannot.

The magnification, m, is found by:

$$m=-\frac{d_i}{d_o}$$

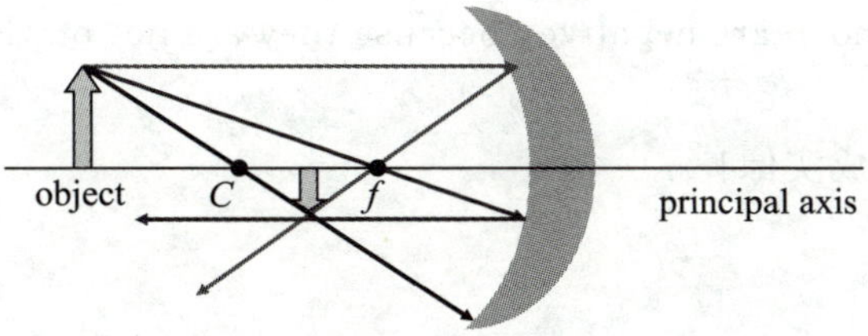

图 2—17—6

这里，放大率为负值，表示像是倒立的。

(2) upright virtual image (正立虚像)

When an object is placed between "F" and the mirror (图 2—17—7), the image created is a virtual image (虚像). It is upright.

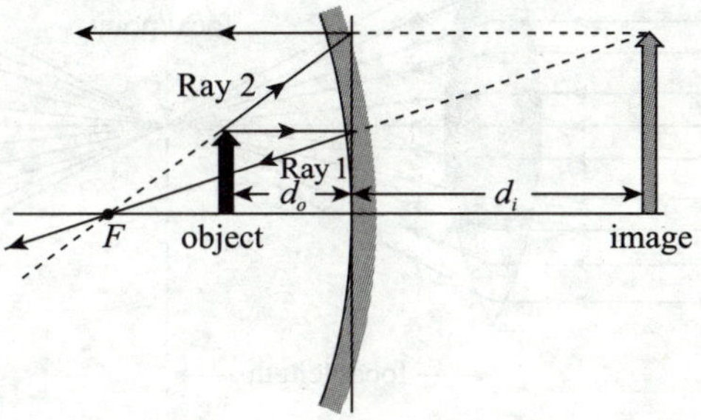

图 2—17—7 an upright virtual image formed by a concave mirror

(3) Convex mirror (凸面镜)

Finally we will place our object on the other side of the mirror, as shown in figure 2—17—8.

Now, we have a convex mirror, which is a diverging mirror—parallel rays tend to spread away from the focal point. In this situation, the image is again virtual.

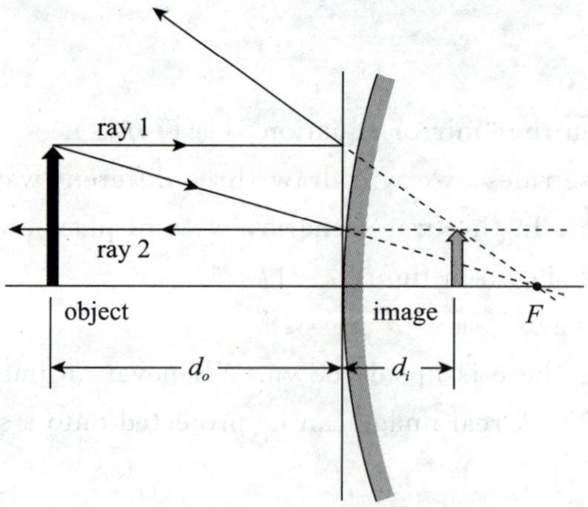

图 2—17—8 an upright virtual image formed by a convex mirror

When we use the mirror equation here, we have to be especially careful: d_o is positive, but both d_i and f are negative, because they are not on the same side of the mirror as the object.

注意：凸面镜不能成实像！

4 Lenses 透镜

1. Convex lens（凸透镜）

A convex lens, also known as a converging lens, is shown in figure 2—17—9:

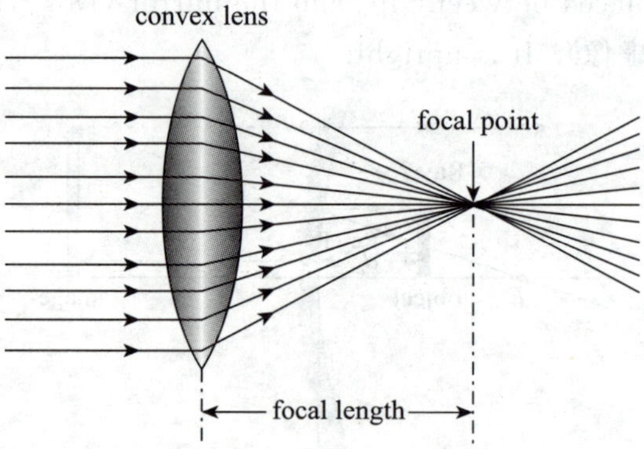

图 2—17—9 a convex lens

2. Concave lens（凹透镜）

A concave lens, or "diverging lens", is shown in figure 2—17—10:

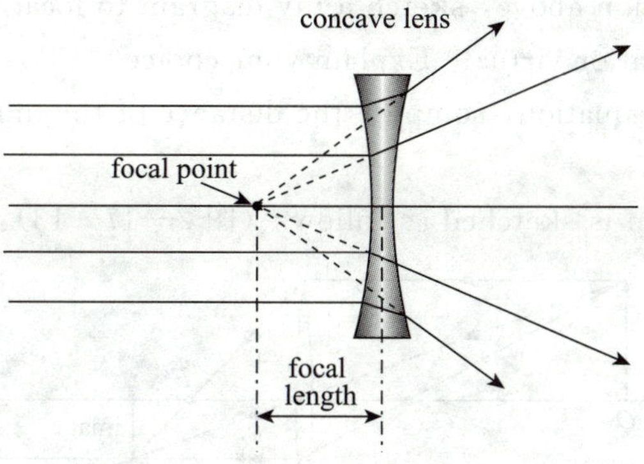

图 2—17—10 a concave lens

3. The rules to follow with convex lens（凸透镜成像规律）

- An incident ray that is parallel to the principal axis refracts through the far focal point.
- An incident ray that goes through near focal point refracts parallel to the principal axis.
- The lensmaker's equation and the equation to find magnification are the same as for mirrors. In the lensmaker's equation，f is positive.

We will start，in figure 2—17—11 by placing our arrow farther from the lens than the focal point.

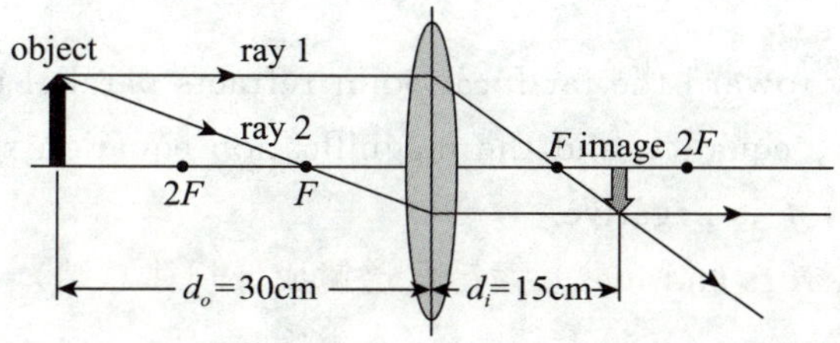

图 2—17—11 ray diagram formation by a convex lens

请考虑：如果将物体置于焦点和光心之间，应该怎样画光路图？请同学自己总结。

Example：

An object O is placed 18 centimeters from the center of a converging lens of focal length 6 centimeters as illustrated in figure 2—17—12：

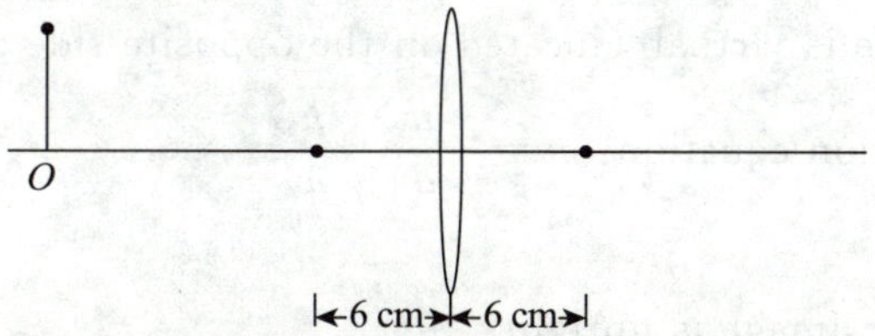

图 2—17—12

(a) On the illustration above, sketch a ray diagram to locate the image.

(b) Is the image real or virtual? Explain your choice.

(c) Using the lens equation, compute the distance of the image from the lens.

Solution:

(a) The ray diagram is sketched as follows (图 2—17—13):

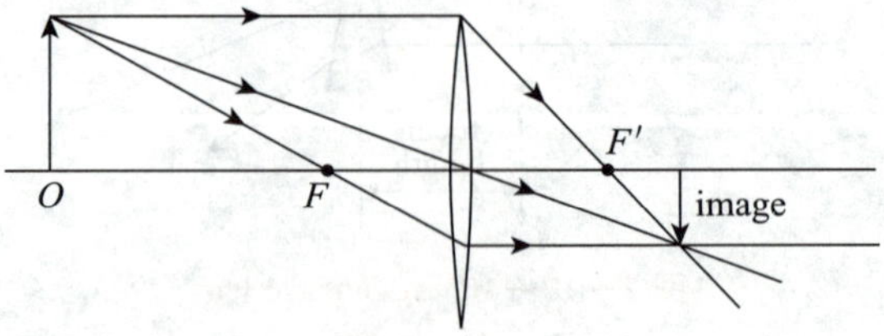

图 2—17—13

(b) The image is real, because the light rays actually meet at the image.

(c) The distance of the image from the lens can be calculated as:

$$\frac{1}{f}=\frac{1}{d_i}+\frac{1}{d_o}$$

Where $f=6$, $d_o=18$, so, $\frac{1}{d_i}=\frac{1}{f}-\frac{1}{d_o}=\frac{1}{6}-\frac{1}{18}=\frac{1}{9}$, Therefore, $d_i=9\text{cm}$

4. When working with diverging lenses, follow these rules (凹透镜成像规律)

- An incident ray parallel to the principal axis will refract as if it came from the near focal point.
- An incident ray toward the far focal point refracts parallel to the principal axis.
- The lensmaker's equation and the magnification equation still hold true. With diverging lenses, though, f is negative.

5. Summary of mirrors and lens (面镜和透镜成像规律总结)

(1) Mirrors (面镜)

Converging—concave → f is positive

Diverging—convex → f is negative

According to the equation: $\frac{1}{f}=\frac{1}{d_i}+\frac{1}{d_o}$

d_o always positive (real object)

d_i positive → image is real (located on the same side of mirror as object)

d_i negative → image is virtual (located on the opposite side of mirror from object)

Based on magnification equation: $m=-\frac{d_i}{d_o}=\frac{h_i}{h_o}$

h_o is positive

h_i and m positive → image is upright

h_i and m negative → image is inverted

(2) Lens（透镜）

Convex lens—converging → f is positive

Concave lens—diverging → f is negative

According to the equation: $\frac{1}{f}=\frac{1}{d_i}+\frac{1}{d_o}$

d_o always positive (real object)

d_i positive → image is real (located on the opposite side of lens from object)

d_i negative → image is virtual (located on the same side of lens as object)

Based on magnification equation: $m=-\frac{d_i}{d_o}=\frac{h_i}{h_o}$

h_o is positive

h_i and m positive → image is upright

h_i and m negative → image is inverted

5 Practice questions 练习题

Multiple-Choice（选择题）

Single-elect（单选题）

1. A ray of light in glass that is incident on an interface with ice, as shown in figure 2—17—14, is partially reflected and partially refracted. The index of refraction n for each of the two media is given in the figure. How do the angle of reflection and the angle of refraction compare with the angle of incidence θ ?

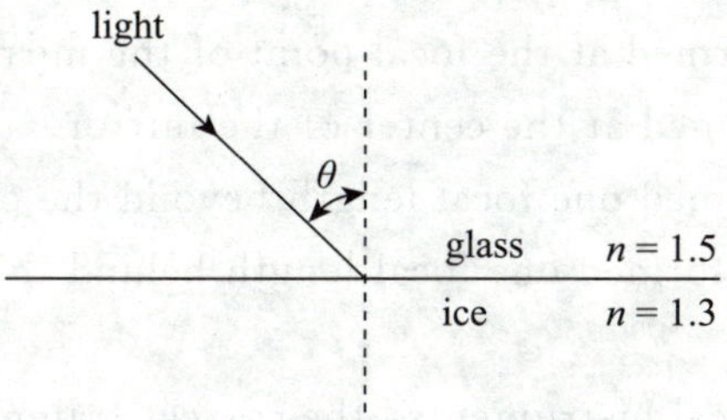

图 2—17—14

	angle of reflection	angle of refraction
(A)	same	larger
(B)	same	smaller
(C)	smaller	same
(D)	smaller	smaller

2. In an aquarium, light traveling through water (n=1. 3) is incident upon the glass container (n=1. 5) at an angle of 36° from the normal. What is the angle of transmission in

the glass?

(A) The light will not enter the glass because of total internal reflection

(B) 31°

(C) 36°

(D) 41°

3. Which of the following optical instruments can produce a virtual image with magnification 0.57?

Ⅰ. convex mirror

Ⅱ. concave mirror

Ⅲ. convex lens

Ⅳ. concave lens

(A) Ⅰ and Ⅳ only

(B) Ⅱ and Ⅳ only

(C) Ⅰ and Ⅱ only

(D) Ⅲ and Ⅳ only

4. Light waves traveling through air strike the surface of water at an angle. Which of the following statements about the light's wave properties is correct?

(A) The light's speed, frequency and wave length all stay the same

(B) The light's speed, frequency and wave length all change

(C) The light's speed and frequency change, but the wave length stays the same

(D) The light's wave length and speed change, but the frequency stays the same

5. An object is placed at the center of a concave spherical mirror. What kind of image is formed, and where is that image located?

(A) A real image is formed at the focal point of the mirror

(B) A real image is formed at the center of the mirror

(C) A real image is formed one focal length beyond the center of the mirror

(D) A virtual image is formed one focal length behind the mirror

Multi-Select (多选题)

6. In the following optical instruments, the correct statements about the function are:

(A) The vertical image can be formed by plane mirror

(B) The convex lens can only form an enlarged image

(C) The near-sighted glasses are concave lenses. The incident light from the concave lens is diverged when it passes through the lens

(D) In a camera, an enlarged real image can be formed on the film

Free-Response (计算题)

7. Light traveling through air encounters a glass aquarium filled with water (图 2—17—15). The light is incident on the glass from the front at an angle of 35°.

(a) At what angle does the light enter the glass?

(b) At what angle does the light enter the water?

(c) On the diagram, sketch the path of the light as it travels from air to water. Include all reflected and refracted rays; label all angles of reflection and refraction.

After entering the water, the light encounters the side of the aquarium, hence traveling back from water to glass. The side of the tank is perpendicular to the front.

(d) At what angle does light enter the glass on the side of the aquarium?

(e) Does the light travel out of the glass and into the air, or does total internal reflection occur? Justify your answer.

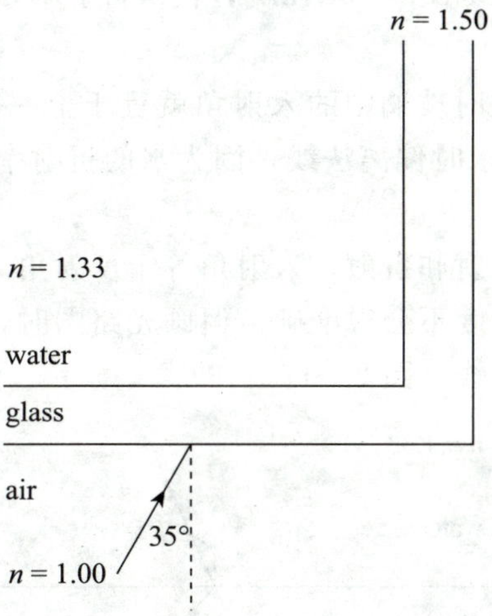

图 2—17—15

题解

单选题

1. 选（A）。光线从光密媒质射入光疏媒质时，反射角等于入射角，折射角大于入射角。

2. 选（B）。将水作为介质“1”，玻璃作为介质“2”，应用斯奈尔定律：$1.3 \cdot \sin 36° = 1.5 \cdot \sin\theta_2$，可得折射角为 31°。如果没有计算器，我们知道当光从光疏媒质射入光密媒质时，将向法线偏转。所以，选（B）。

3. 选（A）。从面镜成像理论可知，当物处于焦点之内时，只成虚像，而且像大于物。就像你在一个汤勺中看自己一样。但凸面镜和凹透镜总是成一个缩小正立像。

4. 选（D）。光速和传播的媒质有关，当光线进入水中时，速度变小，但频率不变，这就是为什么绿色的树叶在水中还是绿色的原因。根据公式 $v = \lambda f$，波长在不同媒质中不同。

5. 选（B）。可以画光路图估计，也可以用透镜成像公式计算：

$$\frac{1}{f} = \frac{1}{d_o} + \frac{1}{d_i}$$

因为球面镜的半径是焦距的两倍，将物置于光心，所以物距为 $2f$。

解出像距：

$$d_i = \frac{f d_o}{d_o - f}$$

代入物距 $2f$ 于上式右端得：$d_i=\frac{f\times 2f}{2f-f}$，恰为 $2f$，像距也是焦距的 2 倍。由于像距 d_i 是正值，所以是实像。

多选题

6. 选（A）和（C）。平面镜只能成等大小的虚像。近视镜是凹透镜，对入射光起发散作用。

计算题

7. (a) 用斯奈尔定律：$n_1 \sin\theta_1 = n_2 \sin\theta_2$，代入折射率 n_1，n_2 和 θ_1，得 $1.0\times\sin 35°=1.5\times\sin\theta_2$，解出 θ_2 为 22°。

(b) 再用斯奈尔定律，这时玻璃中的入射角就等于上一问解出的折射角 22°，得折射角为 25°。当光线从玻璃进入水时偏离法线，因为水的折射率比玻璃小。

(c) 画光路图的要点：

光在两个界面都会产生反射和折射，入射角等于反射角。

你没有量角器，所以，角度不会很准确。但画光路图时，当光线进入玻璃时，一定要偏向法线，进入水时要偏离法线（图 2—17—16）。

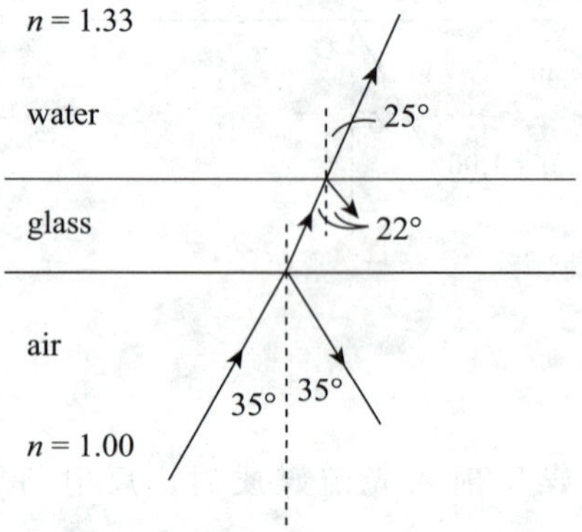

图 2—17—16

(d) 光线和界面的角度不是 25°，而是 90°−25°＝65°。应用斯奈尔定律：$1.33\sin 65°=1.50\sin\theta_2$。折射角是 53°。

(e) 根据公式 $\sin\theta_c=1.0/1.5$，光从玻璃到空气的临界角为 $\theta_c=42°$。由（d）的计算，光从玻璃到空气的入射角为 53°，大于全反射角 42°，所以发生全反射。

Chapter 18

Quantum physics, atomic and nuclear physics

第十八章 量子物理、原子物理和核物理

Main points (主要知识点)

- 原子中电子占有一定的能级，电子吸收或放出能量，实现不同能级的跃迁
- 光子的能量为 $E=\frac{hc}{\lambda}$，其中 $hc=1\ 240$ eV nm
- 光电效应及爱因斯坦解释
- 原子能级图
- 三种核衰变：Alpha，Beta，and Gamma
- 在核衰变过程中，电荷、质量数守恒
- 在核衰变过程中，质量—能量守恒。质能转换公式

Relevant equations (相关公式)

Energy of a photon (光子的能量)：$E=hf=\frac{hc}{\lambda}$

Photo electric effects (光电效应公式)：$E_k=hf-W$

Momentum of a photon (光子的动量)：$p=\frac{h}{\lambda}=\frac{E}{c}$

de Broglie wavelength (德布洛意波)：$\lambda=\frac{h}{p}$

Energy-mass conversion (质能转换公式)：$E=mc^2$

1 Subatomic particles 亚原子粒子

The nucleus of an atom is made up of two types of subatomic particles：protons and neutrons. Protons are positively charged particles, neutrons carry no charge, and electrons are negatively charged. For an atom to be electrically neutral, it must contain an equal number of protons and electrons. When an atom is not electrically neutral, it is called an ion; a positive ion has more protons than electrons, and a negative ion has more electrons than protons.

One particle we didn't introduce is the photon. A photon is a particle of light. Photons have no mass, but they can transfer energy from or to electrons.

Note that the mass of a proton (and, similarly, the mass of a neutron) equals one atomic mass unit, abbreviated as "1 amu." (表 2—18—1) This is a convenient unit of mass to use when discussing atomic physics.

表 2—18—1 **properties of the subatomic particles**

name	mass	charge
proton	1.67×10^{-27} kg = 1amu	positive
neutron	1.67×10^{-27} kg = 1amu	zero
electron	9.11×10^{-31} kg	negative
photon	0	zero

2 The electron-volt 电子伏特

The electron-volt, abbreviated as eV, is a unit of energy that is particularly useful for problems involving subatomic particles. One eV is equal to the amount of energy needed to change the potential of an electron by one volt.

$$1\text{eV}=1.6\times10^{-19}\text{J}$$

The conversion to joules shows that an eV is a very small unit of energy.

3 Energy levels in an atom 原子中的能级

According to quantum mechanics, the electrons in an atom can only have certain, specific amounts of energy.

In figure 2—18—1, we have drawn a few of the energy levels of a hypothetical atom. E_1 is the lowest energy level, and it is called the ground state energy of the atom (原子基态能级). When an electron is sitting as close to the nucleus as possible, and when it's completely unexcited, it will have the energy or E_1, −14 eV.

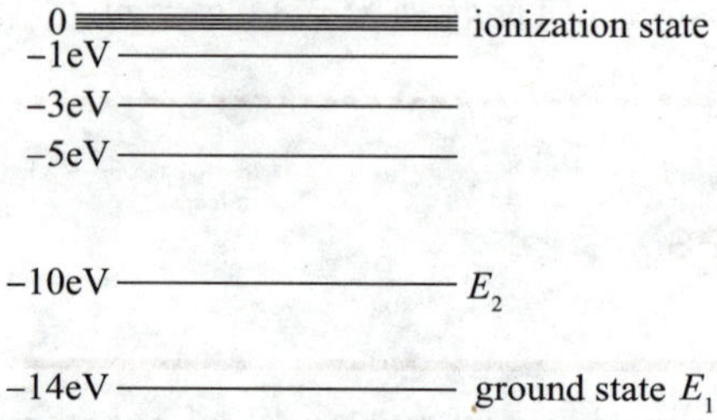

图 2—18—1 **energy level in an atom**

To increase the energy of the electron—to move the electron into a higher energy level—energy must somehow be transferred to the electron. This energy transfer is done by a photon. To jump up in energy, an electron absorbs a photon, and to drop down in energy, an electron emits a photon.

The energy diagram in the figure tells us that to get to E_2 from the ground state, an electron must absorb a photon carrying 4 eV—and only 4 eV! —of energy. If a photon with an energy of 3.9 eV comes along and knocks into the electron, nothing will happen.

Similarly, if a photon with an energy of 4.1 eV comes along, nothing will happen. It's all or nothing; either the electron gets just the right amount of energy to go from one energy level to the next, or it doesn't.

The energy of a photon can be calculated by the formula:

$$E = hf$$

Where h is Planck's constant (6.63×10^{-34} Js)

The frequency is related to the wavelength by the formula:

$$v = \lambda f$$

For light, the velocity is c, or 3×10^{8} m/s, so we can write:

$$c = \lambda f$$

So, the energy of a photon:

$$E = \frac{hc}{\lambda}$$

These formulas tell us that a photon with a high frequency, and therefore with a small wavelength, is higher in energy than a photon with a low frequency and long wavelength. So gamma rays, for example, are a lot higher energy than radio waves because gamma rays have a higher frequency.

Example:

An energy level diagram for a hypothetical atom is shown in figure 2—18—2.

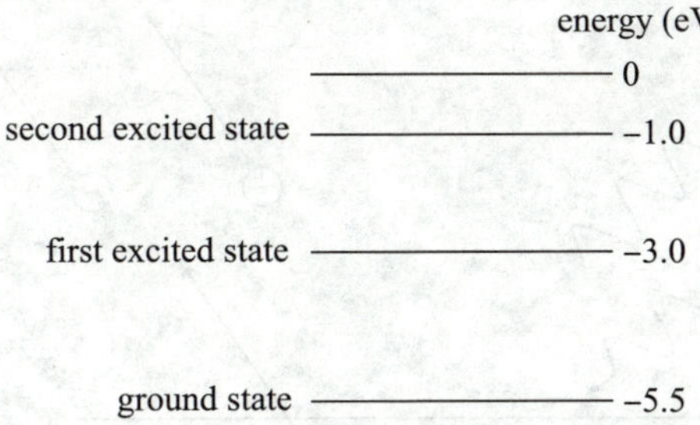

图 2—18—2

(a) Determine the frequency of the lowest energy photon that could ionize the atom, initially in its ground state.

(b) Assume the atom has been excited to the state at 1.0 electron volt.

i. Determine the wavelength of the photon for each possible spontaneous transition.

ii. Which, if any, of these wavelengths are in the visible range?

Solution:

(a) The lowest energy photon that could ionize the atom is 5.5 eV if the electron initially in ground state. So, we have $\Delta E = hf = 5.5\text{eV}$.

The frequency $f=\dfrac{\Delta E}{h}=\dfrac{5.5}{4.14\times 10^{-15}}=1.33\times 10^{15}$ (Hz)

(b) i.

The possible spontaneous transitions are 2.0eV, 2.5eV, 4.5 eV.

So the wavelengths of the photon for each possible spontaneous transition are:

$$\lambda_1=\frac{hc}{\Delta E}=\frac{(4.14\times 10^{-15})(3\times 10^{8})}{2.0}=\frac{12.4\times 10^{-7}}{2.0}=6.21\times 10^{-7}\ (\text{m})$$

$$\lambda_2=\frac{hc}{\Delta E}=\frac{(4.14\times 10^{-15})(3\times 10^{8})}{2.5}=\frac{12.4\times 10^{-7}}{2.5}=4.97\times 10^{-7}\ (\text{m})$$

$$\lambda_3=\frac{hc}{\Delta E}=\frac{(4.14\times 10^{-15})(3\times 10^{8})}{4.5}=\frac{12.4\times 10^{-7}}{4.5}=2.76\times 10^{-7}\ (\text{m})$$

ii. Because the range of wavelengths for visible lights is from 4×10^{-7} m to 7×10^{-7} m, the 2.0 and 2.5 transitions are within visible wavelengths.

注意：电子从高能级跃迁到低能级，放出光子；反之，电子从低能级跃迁到高能级，需要吸收光子。在考虑所有可能的跃迁时，特别注意不要丢掉任意一个可能的跃迁！

4 Photo electric effect 光电效应

Energy in the form of light can cause an atom to eject one of its electrons（图 2—18—3）, but only if the frequency of the light is above a certain value. This frequency became known, for obvious reasons, as the cutoff frequency（截止频率，或临界频率）. The cutoff frequency is different for every type of metal atom.

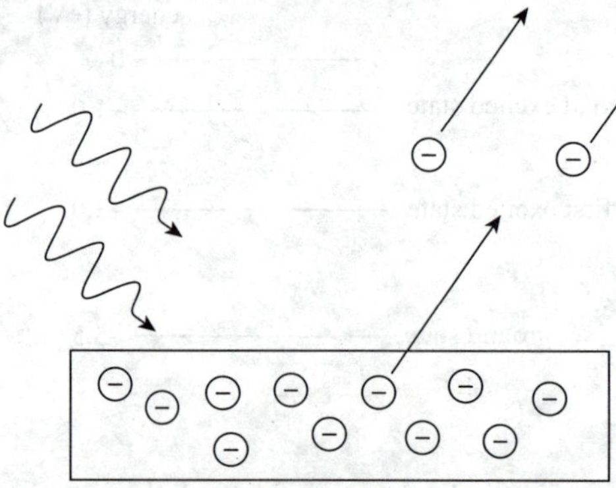

图 2—18—3 photo electric effect

Einstein equation（爱因斯坦公式）

The kinetic energy of a photoelectron is the energy of an incident photon substrates the energy of work function of the metal. The relation can be expressed as:

$$E_k = hf - W$$

It is called Einstein equation. So, if we choose the incident light with higher frequency and the metal with lower work function, a photoelectron with higher kinetic energy can be obtained.

When the E_k is equal to zero, the work function W is equal to hf.

For example:

A metal surface has a work function of 10 eV. What is the cutoff frequency for this metal?

Remembering that hc equals 1 240 eV · nm, we can easily find the wavelength of a photon with an energy of 10 eV:

$$\lambda = \frac{hc}{W} = \frac{1\,240\text{eV} \cdot \text{nm}}{10\text{eV}} = 124 \text{ nm}$$

Using the equation $c = f\lambda$, we find that $f = 2.42 \times 10^{15}$ Hz. So any photon with a frequency equal to or greater than 2.42×10^{15} Hz carries enough energy to eject a photon from the metal surface.

5 Momentum of a photon 光子的动量

The momentum of a photon can be found by this equation:

$$p = \frac{h}{\lambda}$$

The momentum of a photon equals Planck's constant divided by the photon's wavelength.

You can rearrange this equation a bit to show that the energy of a photon, which equals hc/λ, also equals the photon's momentum times c.

6 de Broglie wavelength 德布洛意波

According to de Broglie, a moving mass behaves like a wave. The wavelength of a moving massive particle, called its de Broglie wavelength, is found by this formula:

$$\lambda = \frac{h}{p}$$

So an object's de Broglie wavelength equals Planck's constant divided by the object's momentum. For a massive particle, momentum $p = mv$. Note that the formula above is valid even for massless photons—it is just a rearrangement of the definition of the photon's momentum.

Assume that your mass is approximately 60 kg and that your velocity when jogging is about 5 m/s. Plugging these values into de Broglie's equation, we find that your wavelength when jogging is about 10^{-36} m, which is too small to detect.

Only subatomic particles have de Broglie wavelengths large enough to actually be detected.

7 Three types of nuclear decay processes 三种核衰变

Three particles alpha, beta, and gamma are as follows:

- Alpha particle α: two protons and two neutrons stuck together.
- Beta particle β: an electron or a positron.
- Gamma particle γ: a photon.

A couple of observations: The alpha particle is by far the most massive; the gamma particle is both massless and chargeless.

1. Nuclear notation (原子核的表征)

The two properties of a nucleus that are most important are its atomic number and its mass number. The atomic number, Z, tells how many protons are in the nucleus: this number determines what element we are dealing with because each element has a unique atomic number. The mass number, A, tells the total number of nuclear particles a nucleus contains. It is equal to the atomic number plus the number of neutrons in the nucleus. Isotopes (同位素) of an element have the same atomic number, but different mass numbers.

A nucleus is usually represented using the following notation:

$$^{A}_{Z}\text{Symbol}$$

Where "symbol" is the symbol for the element we are dealing with. For example, $^{4}_{2}$He represents helium with two protons and two neutrons. A different isotope of helium might be $^{5}_{2}$He which contains three neutrons.

2. Alpha decay (α 衰变)

Alpha decay happens when a nucleus emits an alpha particle. Since an alpha particle has two neutrons and two protons, then the daughter nucleus (the nucleus left over after the decay) must have two fewer protons and two fewer neutrons than it had initially.

For example:

$^{238}_{92}$U undergoes alpha decay. Give the atomic number and mass number of the nucleus produced by the alpha decay.

The answer is found by simple arithmetic. The atomic number decreases by two, to 90. The mass number decreases by four, to 234. The element formed is thorium. This alpha decay can be represented by the equation below:

$$^{238}_{92}\text{U} \rightarrow {}^{234}_{90}\text{Th} + {}^{4}_{2}\alpha$$

3. Beta decay (β 衰变)

In beta decay, a nucleus emits either a positron (β^{+} decay) or an electron (β^{-} decay).

Because an electron is not a normal nuclear particle, the total mass number of the nucleus must stay the same after beta decay. But the total charge present must not change—charge is a conserved quantity. So, consider an example of neon (Ne) undergoing β^+ decay:

$$^{19}_{10}Ne \rightarrow ^{19}_{9}F + e^+$$

Here e^+ indicates the positron. The mass number stayed the same. But look at the total charge present. Before the decay the neon nucleus has a charge of +10. After the decay, the total charge must still be +10, and it is: +9 for the protons in the Fluorine (F), and +1 for the positron. Effectively, then, in β^+ decay, a proton turns into a neutron and emits a positron.

For β^- decay a neutron turns into a proton, as in the decay process important for carbon dating:

$$^{14}_{6}C \rightarrow ^{14}_{7}N + e^-$$

The mass number of the daughter nucleus did not change. But the total charge of the carbon nucleus was initially +6. Thus, a total charge of +6 has to exist after the decay as well. This is accounted for by the electron (charge −1) and the nitrogen (charge +7).

In beta decay, a neutrino or an antineutrino also must be emitted to carry off some extra energy. But that doesn't affect the products of the decay, just the kinetic energy of the products.

4. Gamma decay (γ 衰变)

A gamma particle is a photon, a particle of light. It has no mass and no charge. So a nucleus undergoing gamma decay does not change its outward appearance:

$$^{238}_{92}U \rightarrow ^{238}_{92}U + \gamma$$

However, the photon carries away some energy and momentum, so the nucleus recoils.

8 $E = \Delta mc^2$ 质能关系

We can imagine that the nuclei in the last few examples are at rest before emitting alpha, beta, or gamma particles. But when the nuclei decay, the alpha, beta, or gamma particles come out at an appreciable fraction of light speed. So, the daughter nucleus must recoil after decay in order to conserve momentum.

But now consider energy conservation. Before decay, no kinetic energy existed. After the decay both the daughter nucleus and the emitted particle have got kinetic energy. Where did this energy come from? It comes from mass!

The total mass present before the decay is very slightly greater than the total mass present in both the nucleus and the decay particle after the decay. That slight difference in mass is called the mass defect (质量亏损), often labeled Δm. This mass is destroyed and converted into kinetic energy.

Use Einstein's famous equation to find out how much energy can be converted. Multiply the mass defect by the speed of light squared:

$$E = (\Delta m)c^2$$

Although the mass defect is a small quantity, the light squared is huge, and the product is very large.

9 Practice questions 练习题

Multiple-Choice (选择题)

Single-Select (单选题)

1. Figure 2—18—4 shows a portion of the energy level diagram for hydrogen. Possible transitions among the energy levels are represented by arrows Ⅰ, Ⅱ and Ⅲ. Which of following lists the transitions in order of their associated wavelengths of light, from shortest wavelength to longest wavelength?

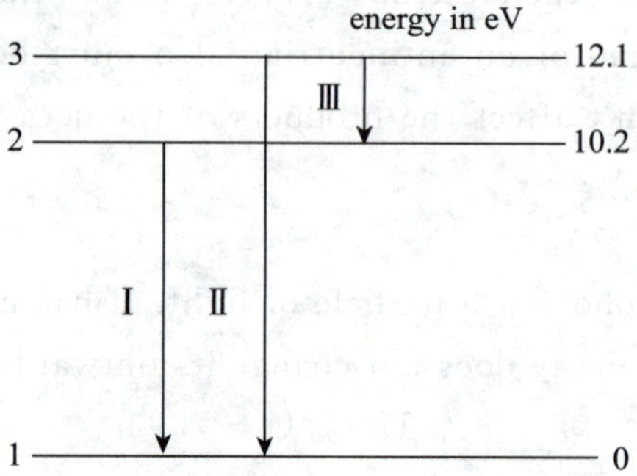

图 2—18—4

(A) Ⅰ, Ⅱ, Ⅲ　　(B) Ⅰ, Ⅲ, Ⅱ

(C) Ⅱ, Ⅰ, Ⅲ　　(D) Ⅱ, Ⅲ, Ⅰ

Questions 2 and 3:

In a nuclear reactor, uranium fissions into krypton and barium via the reaction:

$$n + {}^{235}_{92}\mathrm{U} \rightarrow {}^{141}_{56}\mathrm{Ba} + \mathrm{Kr} + 3n$$

2. What are the mass number A and atomic number Z of the resulting krypton nucleus?

	A	Z
(A)	92	36
(B)	90	36
(C)	94	36
(D)	92	33

3. How much mass is converted into the kinetic energy of the resulting nuclei?

(A) 1 ainu (B) 2 amu

(C) 3 amu (D) much less than 1 amu

4. ${}^{15}_{8}O$ decays via β^+ emission. Which of the following is the resulting nucleus?

(A) ${}^{15}_{9}F$ (B) ${}^{16}_{9}O$

(C) ${}^{15}_{7}N$ (D) ${}^{16}_{7}N$

Multi-Select（多选题）

5. In photo electric effect, in order to increase the kinetic energy of photoelectron, which of following quantities should be change?

(A) Choose the metal with higher work function

(B) Choose the metal with lower work function

(C) Increase the frequency of incident light

(D) Decrease the frequency of incident light

Free-Response（计算题）

6. A hypothetical atom has two energy levels, as shown in figure 2—18—5.

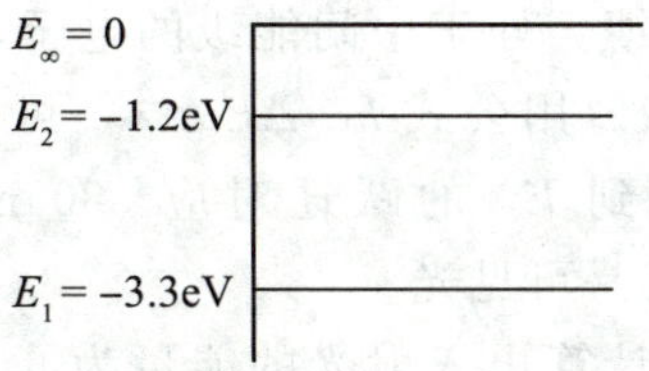

图 2—18—5

(a) What wavelengths of electromagnetic radiation can be absorbed by this atom? Indicate which of these wavelengths, if any, represents visible light.

(b) Now, monochromatic 180nm ultraviolet radiation is incident on the atom, ejecting an electron from the ground state. What will be

i. the ejected electron's kinetic energy?

ii. the ejected electron's speed?

iii. the incident photon's speed?

For parts (c) and (d), imagine that the 180nm radiation ejected an electron that, instead of being in the ground state, was initially in the −1.2 eV state.

(c) Would the speed of the ejected electron increase, decrease, or stay the same? Justify your answer briefly.

(d) Would the speed of the incident photon increase, decrease, or stay the same? Justify your answer briefly.

题解

单选题

1. 选（C）。由光子能量公式 $E = hf$ 可知，高频辐射（最短波长）对应高能量跃迁。所以，从短波长到长波长对应的跃迁顺序为Ⅱ，Ⅰ，Ⅲ。

2. 选（A）。根据质子和中子数守恒，聚变前质子和中子数总和为 235，另外有一个自由中子参加反应，总数为 236 amu。反应后得到质量数为 141 amu 的钡（barium）和 3 个自由中子，共 144amu，再加 92 amu 的氪。

根据电荷守恒，反应前铀元素的总电荷为+92，反应后，钯中有 56 个质子，氪必须有 36 个质子。

3. 选（D）。根据质能公式 $E = mc^2$，质量亏损转换为能量。由于反应后仍有 236 amu 质量，所以转换为能量的质量小于 1 amu.

4. 选（C）。在 β^+ 衰变中，原子核辐射出一个正电子（positron），这并不会改变质子和中子总数，所以质量数仍为 15。但根据电荷守恒，氧原子核带电为+8，放出一个正电子后应为+7。因此选 $^{15}_{7}N$。

多选题

5. 选（B）和（C）。根据光电效应的公式，光电子的动能等于入射光子能量和金属逸出功之差。所以，可以通过增加入射光频率和选择低逸出功金属提高光电子动能。

计算题：

6. (a) 由于 $\Delta E = hc/\lambda$，所以 $hc/\Delta E = \lambda$，$hc = 1\,240 \text{eV} \cdot \text{nm}$。$\Delta E$ 对应能级差。一个基态的电子可以产生两个跃迁：一是吸收 2.1eV 的光子跃迁到中间能级，另一个是吸收 3.3 eV 的光子逃逸出原子的束缚。处于中间能级的电子，吸收 1.2 eV 的光子也可以电离。这样就出现三个不同的能级。用公式 $hc/\Delta E = \lambda$，可以计算出对应的波长：从 E_1 到 E_2 的跃迁对应 590nm，从 E_1 到 E_∞ 的跃迁对应 380 nm，从 E_2 到 E_∞ 的跃迁对应 1 030nm。但只有 590nm 的波长是可见光。

(b) i. 用公式 $\Delta E = hc/\lambda$ 计算出入射光的能量为 6.9eV。这是电子吸收的总能量，但只要电子具有 3.3eV 能量就可以逸出原子，所以还剩 3.6eV 的动能。

ii. 由于动能等于 $\frac{1}{2}mv^2$，单位为焦耳，所以要把电子从原子中逸出后剩下的能量 3.6 eV 转换成焦耳（$1\text{eV} = 1.6 \times 10^{-19}$ J），为 5.8×10^{-19} J，再求出电子的速度为 1.1×10^6 m/s。

iii. 光子的速度为 3.0×10^8 m/s。

(c) 由于电子吸收同样的能量，但电离能仅为 1.2 eV，剩余 5.7 eV 转换为动能，因此射出的电子的动能比原来的大。

(d) 光子的速度仍为 3.0×10^8 m/s。

Chapter 19

Practice questions for AP Physics 1 & 2

AP 物理 1 和 2 练习题

改革以后的 AP 物理 1 和 2 的首次考试在 2015 年 5 月举行，在编撰本书时，还没有 AP 物理 1 和 2 考试原题公布。为了帮助参加考试的学生熟悉 AP 物理 1 和 2 的题型，我们编撰了以下三类练习题：

1. 选择题中的多选题

AP 物理 1 和 2 考题的一个重要改革，就是在选择题中增加了多选题（Multi-Select），即在列出的 4 个选项中选出两个正确答案，这样避免了在单选题中可获得概率分的可能。多选题需要对相关物理概念和定律有较深刻的理解，所以难度较单选题大。

2. 简答题（Short Answer Questions）

AP 物理 1 和 2 考题的另一个重要改革，就是在计算题（Free-Response）中增加了一项简答题（Short Answer Questions）。简答题主要是考查学生对物理概念和定律的理解。题目中，计算部分简单，但答题要求有一定的科技英语书写能力。因此，对中国学生是一个难点，需要平时加强对英语写作能力的训练。

3. 模拟考试题（Mock Examination）

我们编撰了两套模拟考题：

AP Physics 1 和 2 Practice Exam（1）是美国大学理事会 2014 年公布的标准试题练习，出题原则紧扣 6 个主题思想和 7 个实践要点。本书摘要编录该套练习的主要部分、试题和参考答案，供师生参考。由于篇幅有限，出题原则和答案解释请登录 AP 官网查询。

AP Physics 1 和 2 Practice Exam（2）是作者按照官网的出题原则，自行编撰的一套练习题，供参加 AP 物理 1 和 2 考试的学生和指导老师参考。

A Practice: Multi-Select Questions

Multi-Select Questions of AP Physics 1

Directions: Each of the questions or incomplete statements below is followed by four suggested answers or completions. Select two that are best in each case and place the letter of your choice in the corresponding box on the student answer sheet.

Note: To simplify calculations, you may use $g=10\ \text{m/s}^2$ in all problems.

1. A ball is thrown straight up by a student at rest on the surface of the Earth. A graph of the position y as a function of time t, in seconds, is shown in figure 1. Air resistance is negligible.

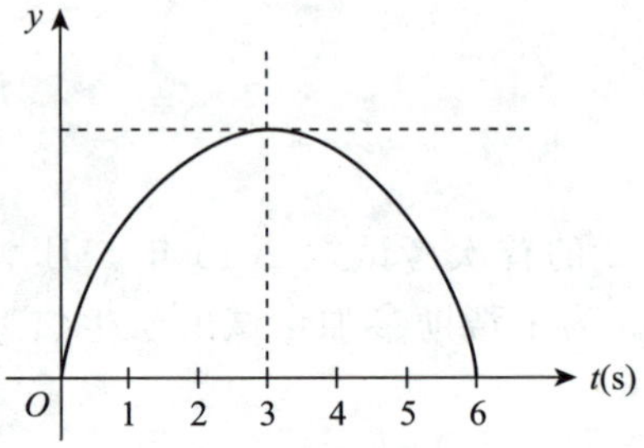

图 1

Two of the following statements describe the acceleration of the ball correctly are

(A) It is downward and constant from 0 to 6 s

(B) It is downward and increases in magnitude from 0 to 3 s, then decreases

(C) It is downward and decreases in magnitude from 0 to 3 s, then increases

(D) It is downward and constant from 0 to 3 s

2. An object is projected vertically upward from ground level. It rises to a maximum height H. If air resistance is negligible, which of the following must be true for the object when it is at a height $H/2$?

(A) Its speed is half of its initial speed

(B) Its kinetic energy is half of its initial kinetic energy

(C) Its potential energy is half of its maximum potential energy at H

(D) Its total mechanical energy is half of its initial value

3. An object of mass M travels along a horizontal air track at a constant speed v and collides elastically with an object of identical mass that is initially at rest on the track. Which two of the following statements are true for the two objects after the impact?

(A) The total momentum is Mv and the total kinetic energy is $0.5Mv^2$

(B) The total momentum is Mv and the total kinetic energy is less than $0.5Mv^2$

(C) The total momentum and kinetic energy are conserved

(D) The momentum of each object is $0.5Mv$

4. A 2 kg object initially moving with a constant velocity is subjected to a force of magnitude F in the direction of motion. A graph of F as a function of time t is shown in

figure 2. What does it increase, if any, during the time the force is applied?

(A) The momentum of the object increases 3.0 kg m/s

(B) The velocity of the object increases 2.0 m/s

(C) The velocity of the object increases 3.0 m/s

(D) The momentum of the object increases 6.0 kg m/s

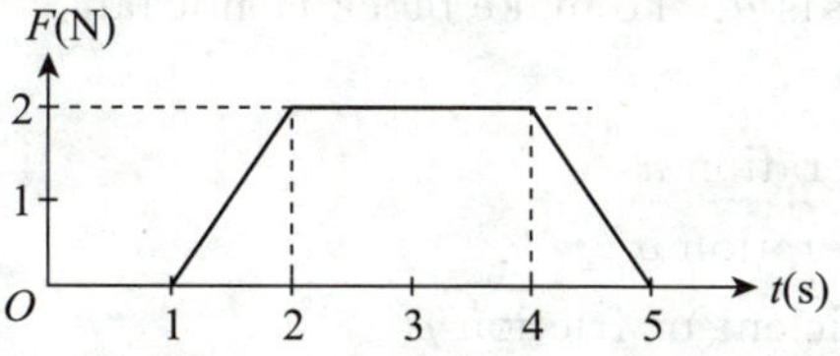

图 2

5. A spherical planet has mass greater than that of the Earth, but its density is unknown. The weight of an object on that planet compared with its weight on the Earth is which of the following?

(A) Larger

(B) It cannot be determined without information about the distance between the object and the centre of that planet

(C) Smaller

(D) It cannot be determined without information about the planet's size

6. Which two of the following statements are true for both sound waves and electromagnetic waves?

(A) They are both longitudinal waves

(B) They both can undergo diffraction

(C) They can produce a two-slit interference pattern

(D) They both are transverse waves

7. An object initially at rest is subjected to a constant net force. Measurements are taken of its velocity v at different distances d from the starting position. A graph of which of the following should exhibit a straight-line relationship?

(A) d^2 versus v^{-2} (B) $\sqrt{d}$ versus v (C) d versus v (D) d versus v^2

8. A meterstick of negligible mass is placed on a fulcrum at the 0.60 m mark, with a 2.0 kg mass hung at the 0 m mark and a 3.0 kg mass hung at the 1.0 m mark. The meterstick is released from rest in a horizontal position. If we replaced the fulcrum at 0.4 m mark , how could keep the balance again?

(A) Increase the mass at 0 m mark to 6 kg

(B) Increase the mass at 0 m mark to 4.5 kg

(C) Change the position of 2.0 kg and 3.0 kg each other

(D) Decrease the hung mass at 1.0 m mark to 1.0 kg

9. A standing wave pattern is created on a guitar string as a person tunes the guitar by changing the tension in the string. Which of the following properties of the waves on the string will change as a result of adjusting only the tension in the string?

(A) Speed of the traveling wave that creates the pattern

(B) Frequency of the standing wave

(C) Wavelength of the standing wave

(D) The amplitude of the wave

10. Figure 3 shows a cart of mass M accelerating to the right with a block of mass m held to the front surface only by friction. The coefficient of friction between the surfacesis μ. To make block m not fall, what are the possible ways?

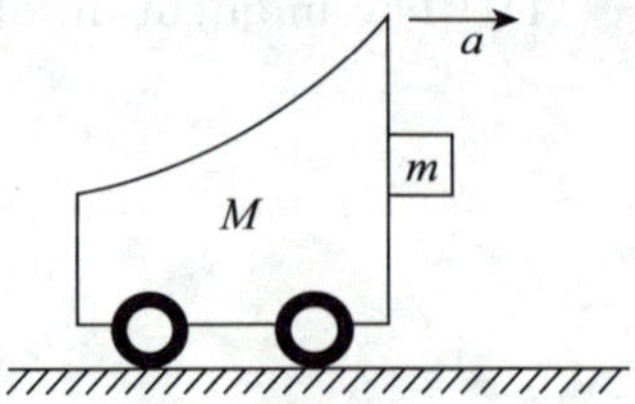

图 3

(A) Increase the acceleration a

(B) Decrease the acceleration a

(C) Decrease the coefficient of friction μ

(D) Increase the coefficient of friction μ

11. What quantities affect the acceleration of a satellite of mass m in a circular orbit of radius R around a planet of mass M ?

(A) The mass m of the satellite

(B) The mass M of the planet

(C) The distance between the satellite and the planet

(D) The mass m and the speed of the satellite

12. If three resistors with unequal resistance are connected in parallel in a DC circuit, which of the following are true for the total resistance?

(A) It is higher than the value of the highest resistance

(B) Its reciprocal is equal to the sum of the reciprocal of each resistance

(C) It is equal to the average of the three resistances

(D) It is lower than the value of the lowest resistance

13. In what motion the acceleration is always kept as a constant vector?

(A) Uniform circular motion

(B) Free fall motion

(C) Project motion

(D) Non-uniform circular motion

14. A sphere of mass m, which is suspended from a string of length l, is displaced to the right as shown in figure 4 and released from rest so that is swings as a simple pendulum with small amplitude. Which of the following statements are true ?

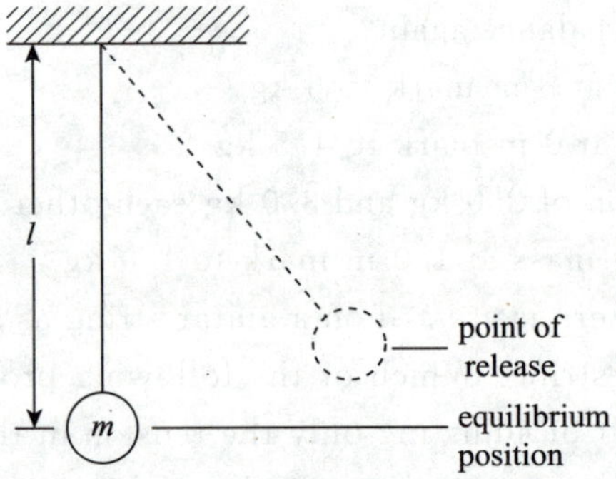

图 4

(A) The maximum kinetic energy is attained as the sphere passes through its equilibrium position

(B) The maximum kinetic energy is attained as the sphere reaches its point of release

(C) The minimum gravitational potential energy is attained as the sphere passes through its equilibrium position

(D) The maximum gravitational potential energy is attained as the sphere reaches its point of release

15. A block attached to the lower end of a vertical spring oscillates up and down. If the spring obeys Hooke's law, the period of oscillation depends on which of the following?

(A) Mass of the block

(B) Amplitude of the oscillation

(C) Restoring force of the spring

(D) Spring constant

16. A new planet is discovered; the mass is M, and the radium is R. The acceleration of an object with the mass of m on the surface of the planet depends on

(A) The mass of the object m

(B) The mass of the planet M

(C) The size of the object

(D) The radium R of the new planet

17. If the temperature is kept as constant, the resistance of a metal wire will depends on

(A) the length of the wire only

(B) the cross section of the wire only

(C) both the length and cross section of the wire

(D) the material of the wire

18. How does an air mattress protect a stunt person landing on the ground after a stunt?

(A) It reduces the average force acting on the stunt person

(B) It reduces the momentum change of the stunt person

(C) It increases the momentum change of the stunt person

(D) It lengthens the stopping time of the stunt person and reduces the force applied during the landing

19. A simple pendulum and a mass hanging on a spring both have a period of 1s when set into small oscillatory motion on the Earth. They are taken to Planet X, which has the same diameter as the Earth but twice the mass. Which of the following statements are true about the periods of the two objects on Planet X compared to their periods on the Earth?

(A) Both are shorter

(B) Both are longer

(C) The period of the pendulum is shorter

(D) The mass on the spring is the same

20. Two fire trucks have sirens that emit waves of the same frequency. As the fire trucks approach a person, the person hears a higher frequency from truck X than from truck Y. Which of the following statements about truck X and Y can be correctly inferred from this information?

(A) Truck X is traveling slower than truck Y

(B) Truck X is traveling faster than truck Y

(C) Truck X is speeding up, and truck Y is slowing down

(D) Truck Y is traveling slower than X

21. The two blocks of masses M and $2M$ shown in figure 5 initially travel at the same speed v but in opposite directions. They collide and stick together. Which of the following statements are true after the collision?

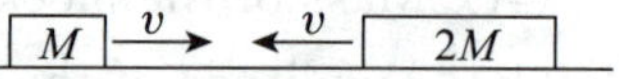

图 5

(A) They will move to the right together with speed of $\frac{1}{3}v$

(B) They will move to the left together with speed of $\frac{1}{3}v$

(C) The mechanical energy is conserved

(D) The mechanical energy is lost to another form

22. A satellite of mass m and speed v moves in a stable, circular orbit around a planet of mass M. The radius of the satellite's orbit

(A) is directly proportional to the mass of planet M

(B) is directly proportional to the mass of satellite m

(C) is inversely proportional to v^2

(D) is inversely proportional to v

23. Two masses m_1 and m_2 attached by a very thin string, they turns around the other end which is fixed at a point O (图 6) with the same period T.

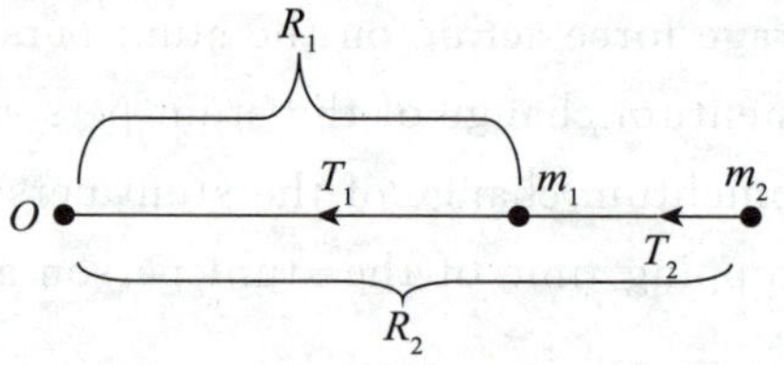

图 6

Which of the following statements are correct?

(A) The tension T_1 exerted at m_1 is greater than T_2 exerted at m_2

(B) The tension T_2 exerted at m_2 is greater than T_1 exerted at m_1

(C) The m_1 and m_2 have the same linear velocity

(D) The m_1 and m_2 have the same angular velocity

24. In rotational motion, the rotational inertia depends on

(A) the mass of object only

(B) the position of axis which the object turns around

(C) the mass distribution of the object

(D) the volume of the object only

25. A block is on a rotational plane. The first observer is standing in rest frame, and the second observer is standing on the center of this plane. In the following statements, choose two correct answer.

(A) The first observer says that the block is rotating with the plane, because there is a centrifugal force exerted on it

(B) The second observer says that the block is kept balance because there are two forces exerted on it. One is the centripetal force, the other is the centrifugal force

(C) The first observer also says the centrifugal force is provided by friction between the block and the plane

(D) The second observer also says that the centripetal force is provided by the friction between the block and plane

Answer of MSQ AP Physics 1

No. of Q	Choice
1	A, D
2	B, C
3	A, C
4	C, D
5	B, D
6	B, C
7	B, D
8	B, C
9	A, B
10	A, D
11	B, C
12	B, D
13	B, C
14	A, D
15	A, D
16	B, D
17	C, D
18	A, D
19	C, D
20	B, D
21	B, D
22	A, C
23	A, D
24	B, C
25	B, D

Multi-Select Questions of AP Physics 2

Directions: Each of the questions or incomplete statements below is followed by four suggested answers or completions. Select two that are best in each case and place the letter of your choice in the corresponding box on the student answer sheet.

Note: To simplify calculations, you may use $g=10\ \text{m/s}^2$ in all problems.

1. Water flows through the pipe shown in figure 1. At the larger end, the pipe has diameter D or radium R and the speed of the water is v_1. What is the speed of the water at the smaller end, where the pipe has diameter d or radium r?

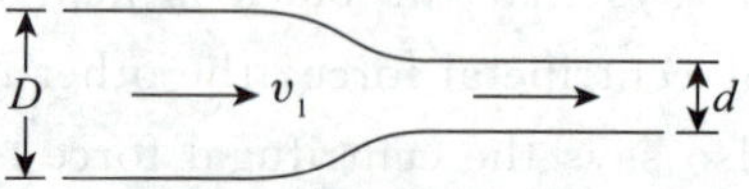

图 1

(A) $\left(\frac{d}{D}\right)v_1$ (B) $\left(\frac{R^2}{r^2}\right)v_1$ (C) $\left(\frac{d^2}{D^2}\right)v_1$ (D) $\left(\frac{D^2}{d^2}\right)v_1$

2. The temperature of an ideal gas is directly proportional to which of the following?

(A) Average translational kinetic energy of the molecules

(B) Average velocity of the molecules

(C) Average potential energy of the molecules

(D) The pressure of the gas

3. Conducting sphere X is initially uncharged. Conducting sphere Y has twice the diameter of sphere X and initially has charge q. If the spheres are connected by a long thin wire, which two of the followings are true once equilibrium has been reached?

(A) Sphere Y has half the potential of sphere X

(B) Spheres X and Y have the same potential

(C) Sphere Y has twice the potential of sphere X

(D) Sphere X has half the charge of sphere Y

4. If the separation between the plates of an isolated charged parallel-plate capacitor is increased slightly, which of the following statements are true?

(A) The capacitance of capacitor decreases

(B) The stored electrostatic energy increases

(C) The force of attraction between the plates increases

(D) The capacitance of capacitor increases

5. An electron of charge $-e$ and a long straight wire carrying a current I to the right are both in the plane of the page, as shown in figure 2. In the position shown, the electron is a distance R from the wire and is moving directly toward it with speed v.

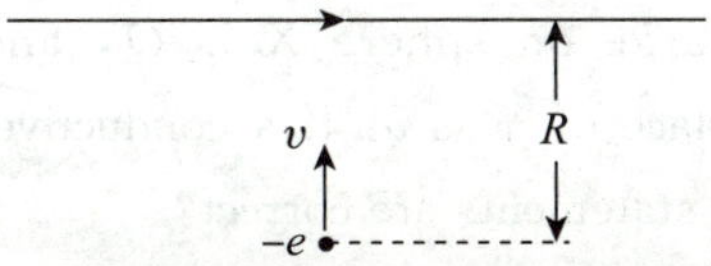

图 2

What is the relation between the direction of the magnetic force F on the electron and the magnetic field B and the velocity v when it is in this position?

(A) F is perpendicular to the velocity

(B) F is to the left and perpendicular to the magnetic field B

(C) F is to the right and perpendicular to the magnetic field B

(D) F, v and B are perpendicular each other

6. A uniform magnetic field B is directed out of the page, as shown in figure 3. A loop of wire of area is in the plane of the page. What changes can produce emf in the loop?

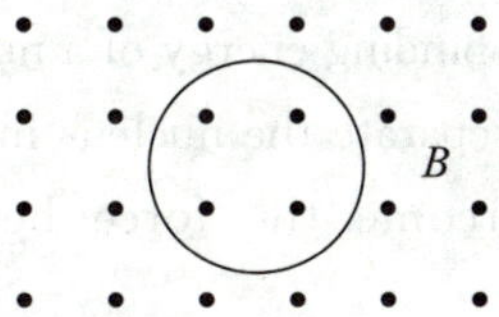

图 3

(A) Changing the strength of the magnetic field

(B) The loop is rotated around the center of the loop in the page

(C) The loop is rotated around the axis which is inside the plane and going through the center of the loop

(D) The loop moves to right with constant speed

7. In a Young's double-slit demonstration of the interference of light, in order to increase the width between fringes, what quantities should be change?

(A) Increase the frequency of the light

(B) Increase the wavelength of the light

(C) Decrease the distance between the slits and the screen

(D) Decrease the distance between the two slits

8. In photoelectric effect experiment, in order to increase the kinetic energy of photoelectrons, we should

(A) increase the frequency of incident light

(B) change the material which has higher work function

(C) change the material which has lower work function

(D) increase the wavelength of incident light

9. The two charged metal spheres X and Y shown in figure 4 are far apart, and each is

isolated from all other charges. The radius of sphere X is greater than that of sphere Y, and the magnitudes of the charge on sphere X is Q, and the magnitude of charge on the surface of Y is q. If a conductive wire connects the two spheres, which statements are correct?

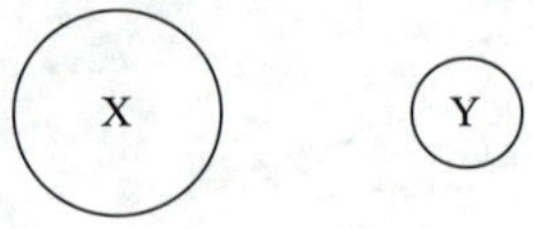

图 4

(A) The electric potential of the two spheres is the same

(B) The charge on both spheres is equal to $(Q+q)/2$

(C) The electric potential at X is larger than that at Y

(D) The charge on sphere X is more than that on sphere Y

10. An ion with a charge $+Q$ moves with constant speed v in a circular path of radius R in a uniform magnetic field of magnitude B. In order to increase the radius R, which statements are true?

(A) Increase the speed of the charge

(B) Increase the strength of magnetic field B

(C) Decrease the strength of magnetic field B

(D) Decrease the speed of the charge

11. Correct statements about the binding energy of a nucleus include which of the followings?

(A) It is the energy needed to separate the nucleus into its individual protons and neutrons

(B) It is the energy to overcome the force between protons and neutrons in a nuclear region

(C) It is the energy equivalent of the apparent loss of mass of its nucleon constituents

(D) It is the energy needed to divide the nucleus into two parts

12. Two long, straight, parallel wires in the plane of the page carry currents I_1 and I_2 in the same direction as shown in figure 5, and I_2 is larger than I_1. Which of the following correctly describe the forces acting on the wires?

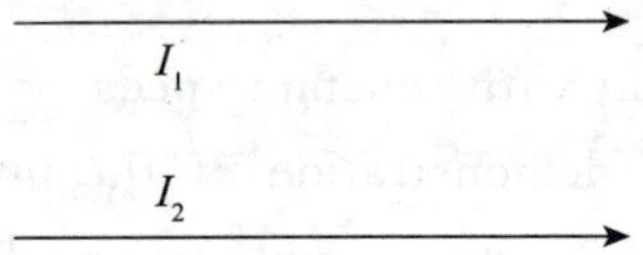

图 5

(A) The two wires attractive and the forces exerted on each wire are the same in magnitude

(B) The two wires attractive, and the force exerted on I_2 is larger than that exerted on I_1

(C) The two wires attractive, and the force exerted on I_1 is larger than that exerted on I_2

(D) The two wires attractive, and the forces are equal and directly proportional to the product of I_1 and I_2

13. An ideal gas may be taken from one state to another state with a different pressure, volume, and temperature along several different paths. Quantities that will always be the same for this process, regardless of which path is taken, include which of the following?

(A) The change in internal energy of the gas

(B) The heat exchanged between the gas and its surroundings

(C) The change of the temperature

(D) The work done by the gas

14. Two parallel wires, one carrying a current I_1, the other carrying a current I_2, the force between them depends on

(A) the direction of current I_1 and I_2

(B) the magnitude of I_1 only

(C) the magnitude of I_2 only

(D) the product of I_1 and I_2

15. The hollow metal sphere shown in figure 6 is positively charged. Point C is the center of the sphere and point P is any other point within the sphere. Which of the following are true of the electric field and the electric potential at these points?

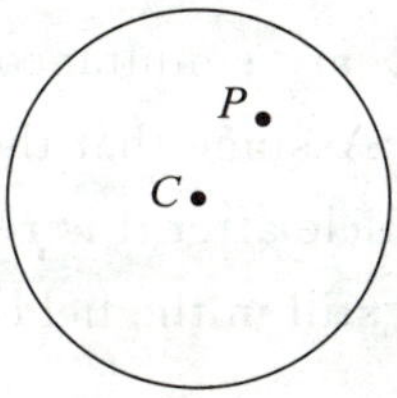

图 6

(A) The electric field is zero at C, but at P it is not zero and is directed inward

(B) The electric potential at C is higher than that at P

(C) The electric field is zero at both C and P

(D) The electric potential is the same at point C and P

16. The total capacitance of several capacitors in parallel is the sum of the individual capacitances for which of the following reasons?

(A) The charge on each capacitor depends on its capacitance, but the potential difference across each is the same

(B) The charge is the same on each capacitor, but the potential difference across each capacitor depends on its capacitance

(C) Equivalent capacitance is always greater than the largest capacitance

(D) The potential difference across each capacitor is the same, the ratio between the charge and the potential difference is the individual capacitance

17. If the temperature is kept as constant, the resistance of a metal wire will depends on

(A) the length of the wire only

(B) the cross section of the wire only

(C) both the length and cross section of the wire

(D) the material of the wire

18. Charges $-Q$ and $+Q$ are located on the x-and y-axes, respectively, each at a distance d from the origin O, as shown in figure 7. Where the electric potential is zero?

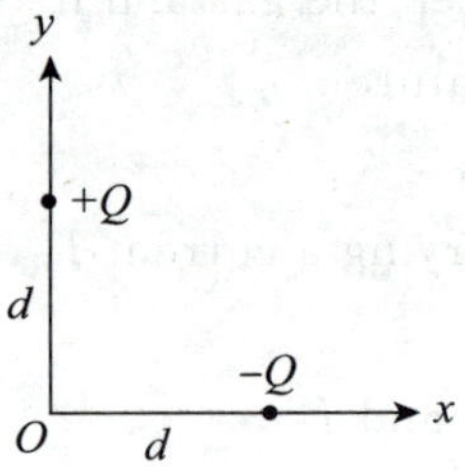

图 7

(A) At the origin point O

(B) At any points along the x axis

(C) At any points along the y axis

(D) At any points along a line which passes through origin O and has 45° with both x and y axis

19. An electron e and a proton p are simultaneously released from rest in a uniform electric field E, as shown in figure 8. Assume that the particles are sufficiently far apart so that the only force acting on each particle after it is released is that due to the electric field. At a later time when the particles are still in the field, the electron and proton will have

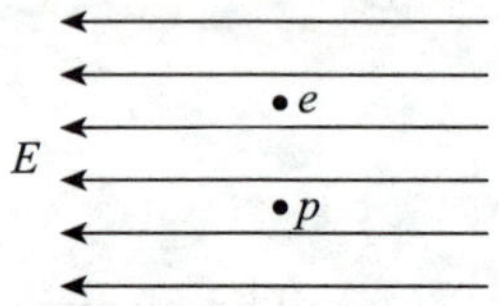

图 8

(A) the same direction of motion

(B) different magnitude of acceleration

(C) the same displacement

(D) the same magnitude of force acting on them

20. Which of the following will occur if the average speed of the gas molecules in a closed rigid container is increased?

(A) The temperature of the gas will increase

(B) The density of the gas will increase

(C) The pressure of the gas will increase

(D) The pressure of the gas will decrease

21. In photoelectric effect, in order to increase the kinetic energy of photoelectron, which of following quantities can be change?

(A) Choose the metal with higher work function

(B) Choose the metal with lower work function

(C) Increase the frequency of incident light

(D) Decrease the frequency of incident light

22. A T-shaped tube with a constriction is inserted in a vessel containing a liquid, as shown in figure 9. What happens if air is blown through the tube from the left, as shown by the arrow in the diagram?

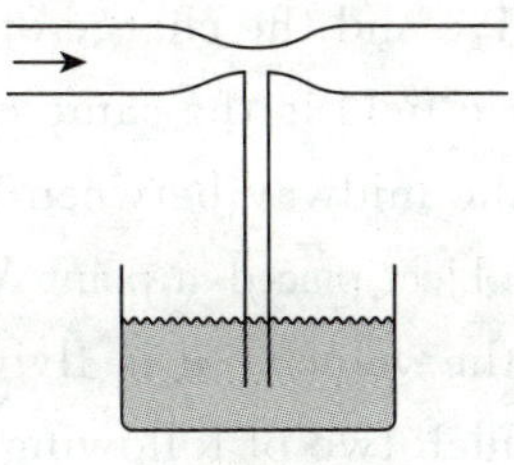

图 9

(A) The liquid level in the tube rises to a level above the surface of the liquid surrounding the tube

(B) The liquid level in the tube falls below the level of the surrounding liquid

(C) The pressure in the tube is lower than the atmosphere

(D) The pressure in the tube is higher than the atmosphere

23. In time t, an amount of heat Q flows through the solid door of area A and thickness d represented in figure 10. The temperatures on each side of the door are T_2 and T_1, respectively. Which of the following changes would be certain to decrease Q?

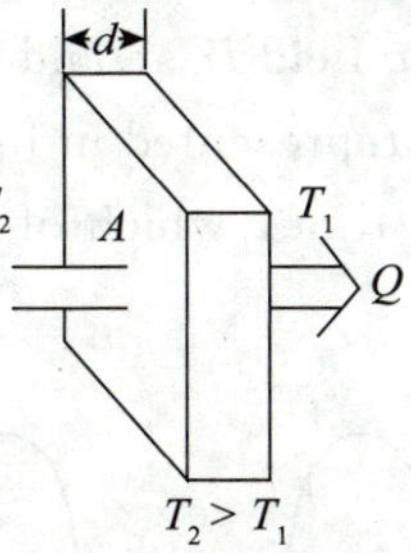

图 10

(A) Increasing A only (B) Decreasing d only

(C) Decreasing $T_2 - T_1$ (D) Decreasing A

24. The nuclide $^{214}_{82}$Pb emits an electron and becomes nuclide X. Which of the following give the mass number and atomic number of nuclide X?

(A) The mass number is 214 (B) The atomic number is 83

(C) The mass number is 213 (D) The atomic number is 81

25. The diagram (图 11) shows electric field lines in an isolated region of space containing two small charged spheres, Y and Z. Which of the following statements are true?

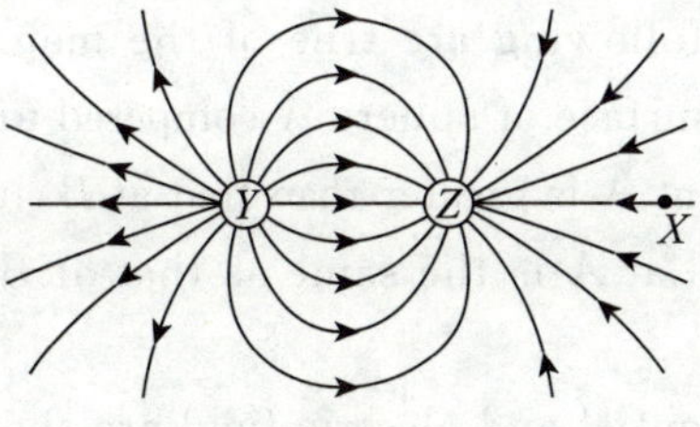

图 11

(A) The charge on Y is negative and the charge on Z is positive

(B) The strength of the electric field is the same everywhere

(C) The electric potential in the midway between Y and Z is zero

(D) A small negatively charged object placed at point X would tend to move toward the right

26. A beam of electrons with the velocity v is flying to the right and into a uniform electric field shown in figure 12. Which two of following ways have to be satisfied the beam of electrons can keep a straight line moving to the right?

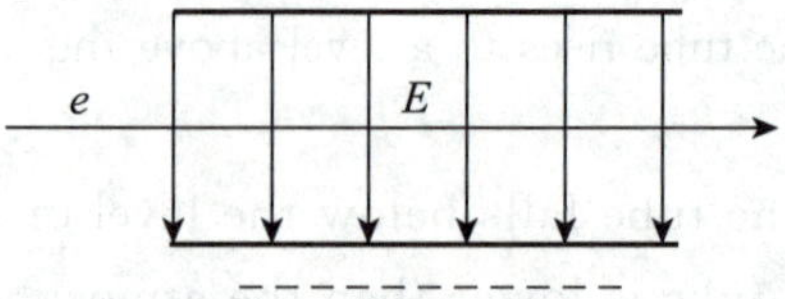

图 12

(A) Put a magnetic field B perpendicular into the page

(B) Put a magnetic field B perpendicular out of the page

(C) The magnitude of magnetic field B should be equal to E/v

(D) The magnitude of magnetic field B should be equal to v/E

27. The two circular wire loops represented in figure 13 are coaxial. If the current I in loop A is increasing with respect to time, which of the following are true of the induced current in loop B ?

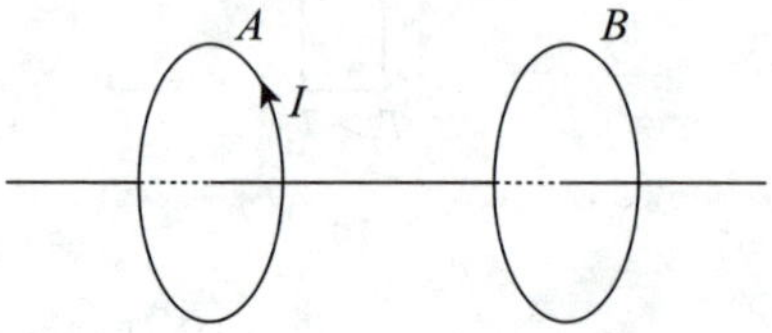

图 13

(A) It is in the opposite direction as that in loop A and the magnitude of the induced current depends on the rate of the change of current I

(B) It is in the same direction as that in loop A and also increasing with respect to time

(C) It is in the same direction as that in loop A and constant with respect to time

(D) It is in the opposite direction as that in loop A and constant with respect to time if the change of current I is linear

28. Two charged metal spheres are connected by a conducting wire. Sphere A is larger than sphere B. Which of the following are true of the magnitude of the electric potential and the electric field E at the surface of sphere A compared to that at the surface of sphere B ?

(A) The electric potential at A is greater than that at B, the electric field are the same

(B) The electric potential at A is the same as that at B, the electric field at B is less than that at A

(C) Both the electric potential and electric field are the same at A and B

(D) The electric potential at A is the same as that at B, the electric field at A is greater than that at B

29. Which of the following experiments review the wave property of the light? Choose two of them.

(A) The diffraction experiment of the light
(B) The reflection law of the light
(C) Yong's double-slits interference experiment
(D) Compton's effect of the light

30. The energy released from the nuclear reaction depends on

(A) the mass of the nucleus which join in the reaction
(B) the banding energy of the nucleons
(C) the mass defect in the reaction
(D) the difference of banding energy between the most stable nucleus and react nucleus

Answer of MSQ AP Physics 2

No. of Q	Choices
1	B, D
2	A, D
3	B, D
4	A, B
5	A, C
6	A, C
7	B, D
8	A, C
9	A, D
10	A, C
11	A, C
12	A, D
13	A, C
14	A, D
15	C, D
16	A, D
17	C, D
18	A, D
19	B, D
20	A, C
21	B, C
22	A, C
23	C, D
24	A, B
25	C, D
26	A, C
27	A, D
28	B, D
29	A, C
30	C, D

B Practice: Short Answer Questions

AP Physics 1: Short Answer Questions

1. (10 points)

A small and a large sphere, of mass M and $3M$ respectively, are arranged as shown on the left side of figure 1. The spheres are then simultaneously dropped from rest. When the large sphere strikes the floor, the spheres have fallen a height H. Assume air resistance is negligible. Express all answers in terms of M, H, and fundamental constants, as appropriate.

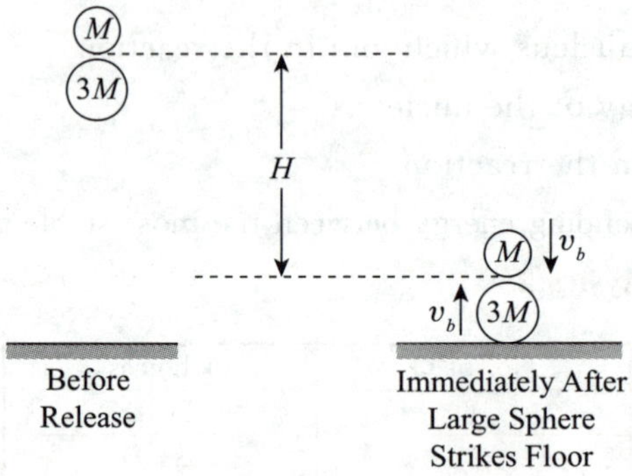

图 1

(a) Derive an expression for the speed v_b with which the large sphere strikes the floor.

Immediately after striking the floor, the large sphere is moving upward with speed v_b and collides head-on with the small sphere, which is moving downward with identical speed v_b at that instant. Immediately after the collision, the small sphere moves upward with speed v_S and the large sphere has speed v_L.

(b) Derive an equation that relates v_b, v_S, and v_L.

In this particular situation $v_L=0$.

(c) Use your relationship from part (b) to determine the speed of the small sphere in terms of v_b.

(d) Indicate whether the collision is elastic. Justify your answer using your results from parts (b) and (c).

(e) Determine the height h that the small sphere rises above its lowest position, in terms of the original height H.

2. (11 points)

Four light bulbs are connected in a circuit with a 24 V battery as shown in figure 2.

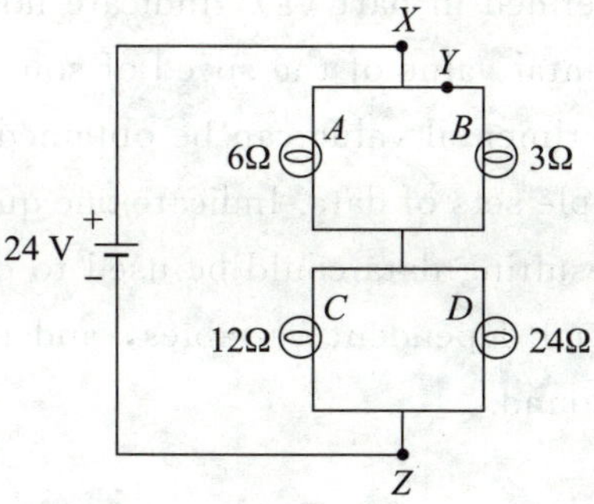

图 2

(a) i. Determine the average potential energy change of an electron as it moves from point Z to point X.

ii. Indicate whether the electron gains or loses potential energy as it moves from point Z to point X.

________Gains energy ________Loses energy

(b) Calculate the equivalent resistance of the circuit.

(c) i. Calculate the magnitude of the current through point Y.

ii. Indicate on the diagram the direction of the current through point Y.

3. (10 points)

You are given the apparatus represented in figure 3. A glass tube is fitted with a movable piston that allows the indicated length L to be adjusted. A sine-wave generator with an adjustable frequency is connected to a speaker near the open end of the tube. The output of a microphone at the open end is connected to a waveform display. You are to use this apparatus to measure the speed of sound in air.

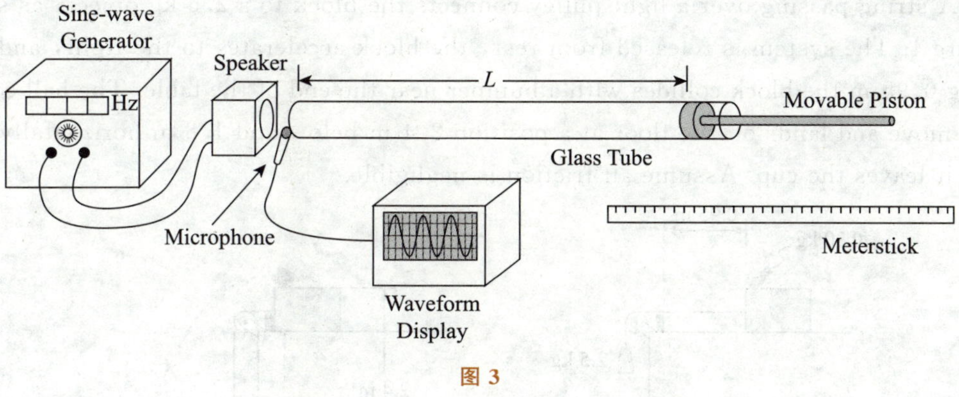

图 3

(a) Describe a procedure using the apparatus that would allow you to determine the speed of sound in air.

Clearly indicate what quantities you would measure and with what instrument each measurement would be made. Represent each measured quantity with a different symbol.

(b) Using the symbols defined in part (a), indicate how your measurements can be used to determine an experimental value of the speed of sound.

(c) A more accurate experimental value can be obtained by varying one of the measured quantities to obtain multiple sets of data. Indicate one quantity that can be varied, and describe how a graph of the resulting data could be used to determine the speed of sound. Clearly identify independent and dependent variables, and indicate how the slope of the graph relates to the speed of sound.

4. (11 points)

A 20 kg box on a horizontal frictionless surface is moving to the right at a speed of 4.0 m/s. The box hits and remains attached to one end of a spring of negligible mass whose other end is attached to a wall. As a result, the spring compresses a maximum distance of 0.50 m, and the box then oscillates back and forth.

(a)

i. The spring does work on the box from the moment the box first hits the spring to the moment the spring first reaches its maximum compression. Indicate whether the work done by the spring is positive, negative, or zero.

______ Positive ______ Negative ______ Zero

Justify your answer.

ii. Calculate the magnitude of the work described in part i.

(b) Calculate the spring constant of the spring.

(c) Calculate the magnitude of the maximum acceleration of the box.

(d) Calculate the frequency of the oscillation of the box.

5. (10 points)

A 0.30 kg ball is in a cup of negligible mass attached to a block of mass M that is on a table. A string passing over a light pulley connects the block to a 2.5 kg object, as shown in figure 4. The system is released from rest, the block accelerates to the right, and after moving 0.95 m the block collides with a bumper near the end of the table. The ball continues to move and lands on the floor at a position 2.4 m below and 1.8 m horizontally from where it leaves the cup. Assume all friction is negligible.

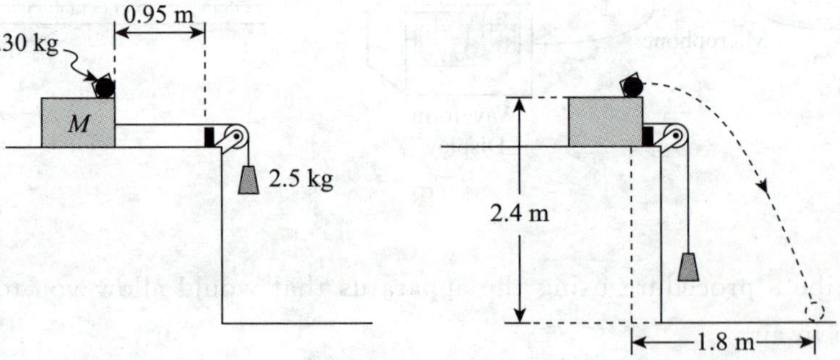

图 4

(a) Calculate the speed of the ball just after the block hits the bumper and the ball leaves the cup.

(b) Calculate the magnitude of the acceleration of the block as it moves across the table.

(c) Calculate the mass M of the block.

(d) If the mass of the ball is increased, the horizontal distance it travels before hitting the floor will decrease.

Explain why this will happen.

6. (9 points)

A 0.40 kg object moves in a straight line under the action of a net force. The graph (图 5) shows the velocity as a function of time for the object during a 25 s interval. At time $t=0$, the object is at the position $x=0$.

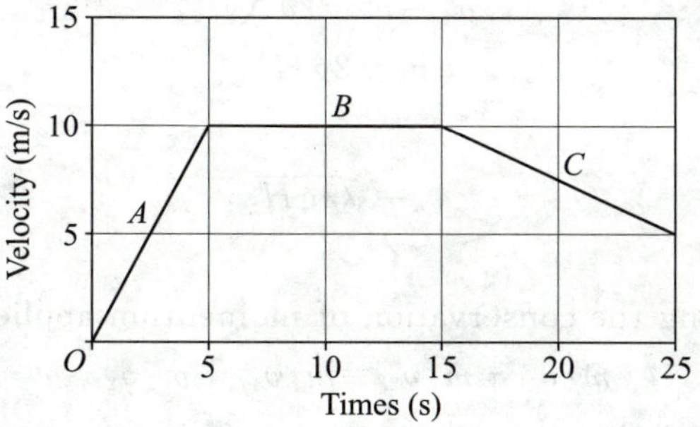

图 5

(a) On the grid (图 6), sketch a graph of the acceleration as a function of time for the object. Label the scale for the acceleration.

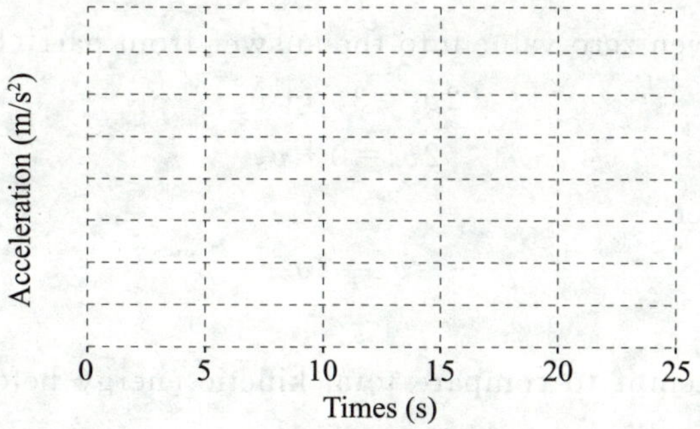

图 6

(b) Calculate the position of the object at $t=5.0$ s.

(c) On which segment of the graph is the net force acting on the object zero?

________ A ________ B ________ C

Justify your answer.

(d) Calculate the net force on the object during the first 3.0 s of the motion.

Scoring guide line

Question 1

10 points total **distribution of points**

(a) 2 points

For a statement that shows the conservation of energy for the large sphere 1 point

$$\Delta U_{3M}=\Delta K_{3M}$$

$$3MgH=\frac{1}{2}(3M)v_b^2$$

For a correct answer (or equivalent expression for v_b) 1 point

$$v_b=\sqrt{2gH}$$

Alternate solution Alternate points

For using a proper kinematic approach 1 point

$$v_f^2=v_0^2+2a\Delta y$$

$$v_b^2=2gH$$

For a correct answer 1 point

$$v_b=\sqrt{2gH}$$

(b) 2 points

For stating or showing the conservation of momentum applied to the collision 1 point

$$m_1v_{1i}+m_2v_{2i}=m_1v_{1f}+m_2v_{2f}$$

For stating or showing that the spheres are initially traveling in opposite directions 1 point

$$3Mv_b+M(-v_b)=3Mv_L+Mv_S$$

$$2v_b=3v_L+v_S$$

(c) 1 point

Substituting the given zero value into the answer from part (b)

$$2v_b=3v_L+v_S$$

$$2v_b=0+v_S$$

For a correct answer 1 point

$$v_S=2v_b$$

(d) 3 points

For any correct attempt to compare total kinetic energy before the collision to total kinetic energy after the collision 1 point

$$K_i=K_{3Mi}+K_{Mi} \quad \text{and} \quad K_f=K_{3Mf}+K_{Mf}$$

For correct substitutions of the expression from part (c), and the correct masses in the kinetic energy terms 1 point

$$K_i=\frac{1}{2}(3M)(v_b)^2+\frac{1}{2}(M)(-v_b)^2=2Mv_b^2$$

$$K_f=\frac{1}{2}(M)(2v_b)^2=2Mv_b^2$$

For correctly stating that the collision is elastic 1 point

(e) 2 points

For a statement of conservation of energy for ball M as it rises to the new height h

1 point

$$U_{gf}=K_0$$

$$Mgh=\frac{1}{2}Mv_S^2$$

$$h=\frac{(2v_b)^2}{2g}=\frac{4v_b^2}{2g}=\frac{2\ (\sqrt{2gH}\,)^2}{g}$$

For a correct answer consistent with the expression for v_b obtained in part (a)

1 point

$$h=4H$$

Alternate solution Alternate points

For using a correct kinematic approach to solve for maximum height of ball M

1 point

$$v_f^2=v_0^2+2a\Delta y$$

$$v_0=v_S=2v_b=2\sqrt{2gH}$$

$$0=[2(\sqrt{2gH}\,)]^2-2gh$$

For a correct answer consistent with the expression for v_b obtained in part (a)

1 point

$$h=4H$$

Question 2

11 points total **Distribution of points**

(a) 3 points

(i) 2 points

For substituting the electron charge into the equation for potential energy 1 point

$$\Delta U=-W=-q\Delta V=-(-1.6\times10^{-19}\text{ C})\ (0-24\text{ V})$$

For the correct answer, with units 1 point

$$\Delta U=-3.8\times10^{-18}\text{ J}=-24\text{ eV}$$

Note: Full credit is given for the correct answer, with units, with no supporting calculations.

(ii) 1 point

For selecting "Loses energy" 1 point

(b) 4 points

For recognizing that the equivalent resistance is a series sum of two parallel combinations 1 point

For the correct calculation of R_{AB} 1 point

$$\frac{1}{R_{AB}}=\frac{1}{R_A}+\frac{1}{R_B}=\frac{1}{6\ \Omega}+\frac{1}{3\ \Omega}$$

$$R_{AB}=2\Omega$$

For the correct calculation of R_{CD} 1 point

$$\frac{1}{R_{CD}}=\frac{1}{R_C}+\frac{1}{R_D}=\frac{1}{12\ \Omega}+\frac{1}{24\ \Omega}$$

$$R_{CD}=8\ \Omega$$

For the correct calculation of R_T 1 point

$$R_T=R_{AB}+R_{CD}=2\ \Omega+8\ \Omega$$

$$R_T=10\ \Omega$$

(c) 4 points

(i) 3 points

For calculation of the total current, with calculations consistent with the value of R_T found in part (b) 1 point

$$I_T=\frac{V_T}{R_T}=\frac{24\ \text{V}}{10\ \Omega}$$

$$I_T=2.4\ \text{A}$$

For the use of $V_{AB}=I_T R_{AB}$ to find V_{AB} 1 point

$$V_{AB}=I_T R_{AB}=(2.4\ \text{A})(2\ \Omega)$$

$$V_{AB}=4.8\ \text{V}$$

For the use of $I_Y=V_{AB}/R_B$ to find I_Y 1 point

$$I_Y=\frac{V_{AB}}{R_B}=\frac{4.8\ \text{V}}{3\ \Omega}$$

$$I_Y=1.6\ \text{A}$$

(ii) 1 point

For an arrow drawn at point Y pointing to the right 1 point

Question 3

10 points total **Distribution of points**

(a) and (b) 6 points

These two parts are closely linked; therefore they are scored as a unit.

For indicating the use of the sine-wave generator to send a sound wave of a given frequency into the glass tube 1 point

For indicating adjustment of the movable piston until the sound picked up by the microphone and shown on the waveform display indicates that resonance occurs (maximum amplitude of standing wave) or until resonance is heard by ear 1 point

For indicating the use of the meterstick to measure the distance L from the piston to the left-hand end of the tube at resonance 1 point

For a statement indicating that L is proportional, but not equal, to the wavelength 1 point

For defining variables for frequency and wavelength 1 point

For indicating that $v=\lambda f$ should be used with the measurements to determine an experimental value of the speed of sound 1 point

Example

Send a sound wave of frequency f into the glass tube using the sine-wave generator and speaker. Move the piston all the way to the left end of the tube. Pull the piston to the right until the sound picked up by the microphone and shown on the waveform display indicates that resonance occurs. Use the meterstick to measure the distance L between the piston and the left-hand end of the tube. For a tube closed at one end, the wavelength λ is equal to $4L/n$, with the first resonance at $n=1$. Using the above measurements, an experimental value of the speed of sound can be determined using the equation $v=\lambda f$.

(c) 4 points

For indicating an appropriate variable that can be varied to obtain multiple sets of data 1 point

For correctly identifying appropriate independent and dependent variables to be graphed 1 point

For indicating an appropriate plot that will produce a linear graph 1 point

For stating how the slope of this graph can be used to determine the speed of sound v 1 point

Examples

One of the measured variables that could be varied in order to obtain multiple sets of data is the frequency f. If f is varied, this means it is the independent variable, and the dependent variable is the wavelength λ. A plot of λ versus $1/f$ will produce a linear graph, the slope of which is the speed of sound.

One of the measured variables that could be varied in order to obtain multiple sets of data is the wavelength λ. If λ is varied, this means it is the independent variable, and the dependent variable is the frequency f. A plot of f versus $1/\lambda$ will produce a linear graph, the slope of which is the speed of sound.

Question 4

11 points total **Distribution of points**

(a) 5 points

i. 2 points

For selecting "Negative" 1 point

For a correct justification 1 point

Examples

The spring pushes on the box in a direction opposite to box's motion, therefore the work done is negative.

The box slows down. So its kinetic energy decreases. Therefore the work done is negative.

ii. 3 points

For a correct expression relating work to change in kinetic energy 1 point

$$W=\Delta K$$

$$W=\frac{1}{2}m(v_2^2-v_1^2) = -\frac{1}{2}mv_1^2$$

For substituting the correct values 1 point

$$W=-\left(\frac{1}{2}\right)(20\ \text{kg})\ (4.0\ \text{m/s})^2$$

For a correct answer 1 point

$W=160$ J (either positive or negative value acceptable)

(b) 2 points

For a correct expression of conservation of energy 1 point

$$K_1=U_2$$

$$\frac{1}{2}mv_1^2=\frac{1}{2}kx_2^2$$

For correct substitution of values consistent with the answer from part (a) ii 1 point

$$160\ \text{J}=\left(\frac{1}{2}\right)k(0.50\ \text{m})^2$$

$$k=1\ 280\ \text{N/m}$$

(c) 2 points

For a correct expression of the maximum acceleration 1 point

$$a=F_{\max}/m=kx_{\max}/m$$

For correct substitution of values consistent with the answer from part (b) 1 point

$$a=\ (1\ 280\ \text{N/m})\ (0.50\ \text{m})/(20\ \text{kg})$$

$$a=32\ \text{m/s}^2$$

(d) 2 points

For a correct expression of the frequency 1 point

$$f=\frac{1}{T}=\frac{1}{2\pi}\sqrt{\frac{k}{m}}$$

For correct substitution of values consistent with the answer from part (b) 1 point

$$f=\left(\frac{1}{2\pi}\right)\sqrt{\frac{1\ 280\ \text{N/m}}{20\ \text{kg}}}$$

$$f=1.27\ \text{Hz}$$

Question 5

10 point total **Distribution of points**

(a) 3 points

For using a correct expression for the vertical motion to determine time t 1 point

$$\Delta y=v_{yi}t+\frac{1}{2}at^2$$

$$v_{yi}=0,\ \text{so}\ t=\sqrt{\frac{2\ \Delta y}{a}}$$

For substituting the correct height and acceleration into the vertical equation 1 point

$$t=\sqrt{\frac{2\times 2.4\text{m}}{9.8\ \text{m/s}^2}}=0.70\ \text{s}$$

For substituting the correct values into the constant horizontal velocity equation 1 point

$$v_x=\frac{\Delta x}{t}$$

$$v_x=\frac{1.8\ \text{m}}{0.70\ \text{s}}=2.6\ \text{m/s}$$

(b) 2 points

For using a correct expression to determine the acceleration of the block 1 point

$$v_x^2=v_0^2+2ad$$

$$v_0=0$$

$$a=\frac{v_x^2}{2d}$$

For correct substitutions consistent with part (a) 1 point

$$a=\frac{(2.6\ \text{m/s})^2}{2\times 0.95\ \text{m}}=3.5\ \text{m/s}$$

(c) 3 points

For using Newton's second law

$$\sum F=ma$$

$$m_{obj}g=(m_{obj}+M+m_{ball})a$$

For a correct value for the total force on the system 1 point

$$m_{obj}g=2.5\ \text{kg}\times 9.8\ \text{m/s}^2=25\ \text{N}$$

For correct values in the expression for the total mass being accelerated 1 point

$$(m_{obj}+M+m_{ball})a=(2.5\ \text{kg}+0.3\ \text{kg}+M)a=(2.8\ \text{kg}+M)a$$

For the correct substitution of the acceleration of the block from part (b) 1 point

$$25\ \text{N}=(2.8\ \text{kg}+M)(3.5\ \text{m/s}^2)$$

$$M=4.24\ \text{kg}$$

(d) 2 points

For stating that with the same force, the horizontal acceleration of the ball will be less as a result of the increased mass of the system 1 point

For stating the launch velocity will be lower 1 point

Example

A larger combined mass of the system will result in a smaller acceleration, so the ball has a smaller speed upon launch. However, the time to hit the floor will be the same since it's falling the same vertical distance with the same vertical acceleration in freefall. So the horizontal distance ($v_{launch}\times t_{fall}$) will be less.

One earned point was deducted for any incorrect statement in the reasoning not included in the points shown.

Question 6

9 total points **Distribution of points**

(a) 3 points

For using a linear scale on the vertical axis 1 point

For showing horizontal lines in the three appropriate time frames 1 point

For having the correct acceleration values for segments B and C 1 point

$a=\Delta v/\Delta t$ (or the slope of the velocity versus time curve)

For segment A: $a=(10\text{ m/s}-0\text{ m/s})/(5\text{ s}-0\text{ s})=2\text{ m/s}^2$

For segment B: slope is zero, so the acceleration is zero

For segment C: $a=(5\text{ m/s}-10\text{ m/s})/(25\text{ s}-15\text{ s})=-0.5\text{ m/s}^2$ (图 7)

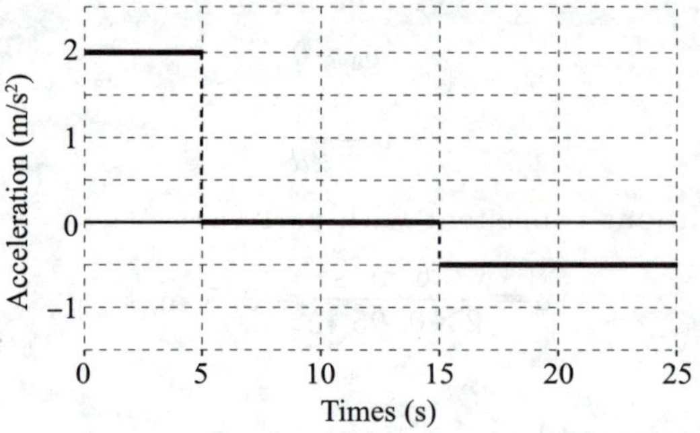

图 7

(b) 2 points

$x=x_0+v_0t+\frac{1}{2}at^2$, where $x_0=0$

For substituting $v_0=0$ and $t=5$ s into the equation above 1 point

For substituting the value on the graph of the constant acceleration between 0 and 5 s or an explicit calculation of 2 m/s^2 1 point

$$x=(1/2)(2\text{ m/s}^2)(5\text{ s})^2$$

$$x=25\text{ m}$$

Alternate solution Alternate points

Position is the area under the velocity versus time curve

For correctly substituting into the expression for the area under the curve 1 point

$$x=(1/2)(10\text{ m/s})(5\text{ s})$$

For the correct answer 1 point

$$x=25\text{ m}$$

(c) 2 points

For selecting segment B 1 point

For an appropriate justification 1 point

Example

$F_{net}=ma=m(\Delta v/\Delta t)$, the velocity is constant for this segment, so $\Delta v=0$; therefore, $F_{net}=0$.

(d) 2 points

For using Newton's second law 1 point

$$F_{net}=ma$$

$$F_{net}=(0.40\ \text{kg})(2\ \text{m}/\text{s}^2)$$

For the correct answer 1 point

$$F_{net}=0.80\ \text{N}$$

AP Physics 2: Short Answer Questions

1. (10 points)

A glass U-tube with a uniform diameter of 0.850 cm is used to determine the density of an oil. As shown in figure 1, a 24.5 cm column of water balances a 27.2 cm column of the oil so that interfaces A and B of the mercury with the other liquids are at the same height. The density of water is 1.00×10^3 kg/m^3.

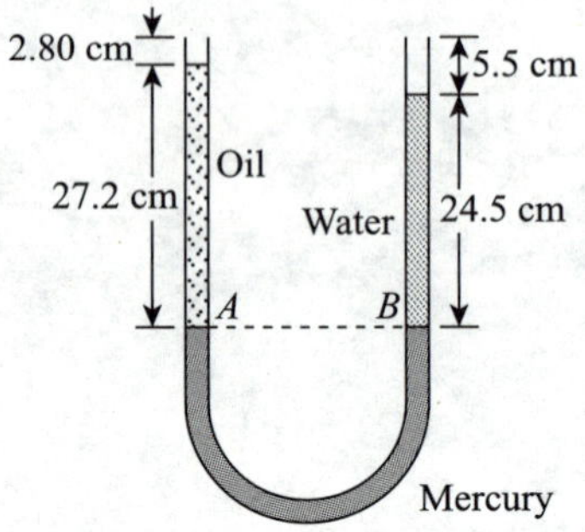

图 1

(a) Calculate the density of the oil.

(b) Calculate the absolute pressure at B, the interface between the water and the mercury.

A new tube, identical to the U-tube except for a cone shape on the left, as shown in figure 2, is filled with the same volume of mercury that was in the U-tube. The mercury is at the same height on both sides of the new tube as it was in the U-tube, as shown by the dashed line. The same volumes of oil and water that were in the U-tube are now poured into the new tube, on the left and right respectively.

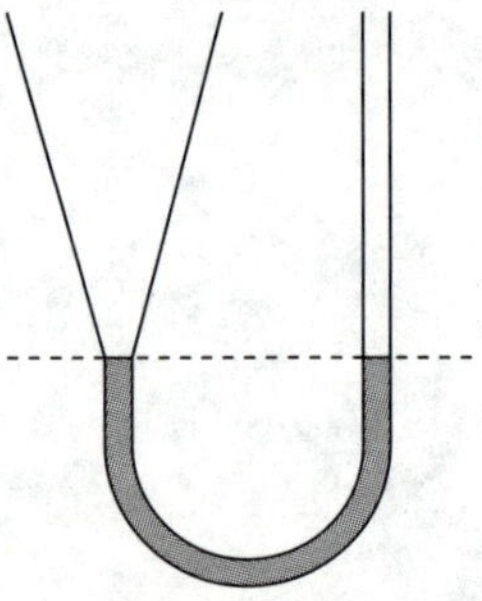

图 2

(c) Indicate the new position of B relative to A.

______ Above A ______ Below A ______ At the same height as A

Justify your answer.

(d) A small piece of wood with density less than that of the oil is placed so that it floats in the left side of the tube. Indicate whether the pressure at the bottom of the tube increases, decreases, or remains the same.

______ Increases ______ Decreases ______ Remains the same

2. (10 points)

A cylindrical container is fitted with a frictionless piston that is initially locked in place. The cylinder contains a fixed amount of an ideal gas that is initially at room temperature and atmospheric pressure.

(a) The cylinder is placed in a hot-water bath. On the axes below (图 3), sketch a graph of pressure versus temperature for the process the gas undergoes as a result, and indicate the direction of the process on the graph.

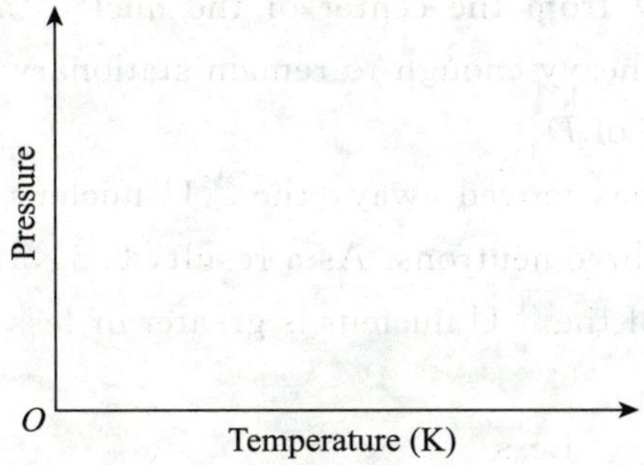

图 3

(b) The cylinder is removed from the hot-water bath. After equilibrium is reached, the lock is removed so the piston is free to move. Indicate whether the piston moves up, moves down, or remains stationary.

______ Moves up ______ Moves down ______ Remains stationary

Justify your answer.

(c) When the system is again at equilibrium, the piston is pushed down very slowly. On the axes below (图 4), sketch a graph of pressure versus volume for the process the gas undergoes as a result, and indicate the direction of the process on the graph. Label this process "C."

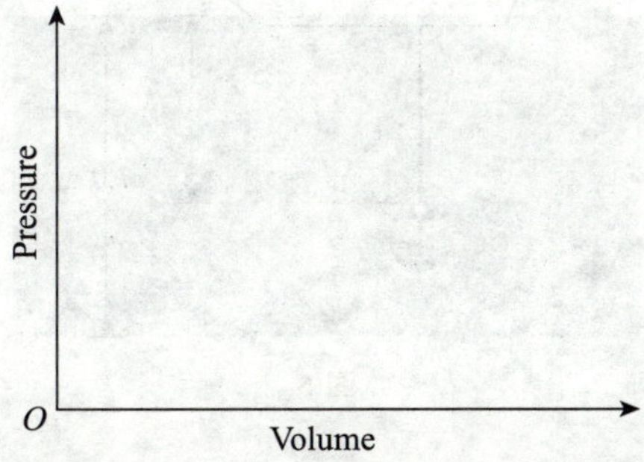

图 4

(d) Now the piston is pulled up quickly, so no heat is added to or removed from the gas during the process. On the axes above, sketch a graph of pressure versus volume for the process the gas undergoes as a result, and indicate the direction of the process on the graph. Label this process "*D*."

3. (10 points)

The momentum of a particular proton is 5.5×10^{-20} kg m/s. Relativistic effects can be ignored throughout this question.

(a) Calculate the de Broglie wavelength of the proton.

(b) Calculate the kinetic energy of the proton.

The proton is directed toward a very distant stationary uranium nucleus, $^{235}_{92}U$. The proton reaches a distance D from the center of the nucleus and then reverses direction. Assume that the nucleus is heavy enough to remain stationary during the interaction.

(c) Calculate the value of D.

(d) After the proton has moved away, the $^{235}_{92}U$ nucleus spontaneously fissions into $^{148}_{57}La$ and $^{84}_{35}Br$, along with three neutrons. As a result, 2.5×10^{-11} J of energy is released. Indicate whether the mass of the $^{235}_{92}U$ nucleus is greater or less than the mass of the fission products.

______ Greater ______ Less

Calculate the mass difference.

4. (10 points)

A sailboat at rest on a calm lake has its anchor dropped a distance of 4.0 m below the surface of the water (图 5). The anchor is suspended by a rope of negligible mass and volume. The mass of the anchor is 50 kg, and its volume is 6.25×10^{-3} m³. The density of water is 1 000 kg/m³.

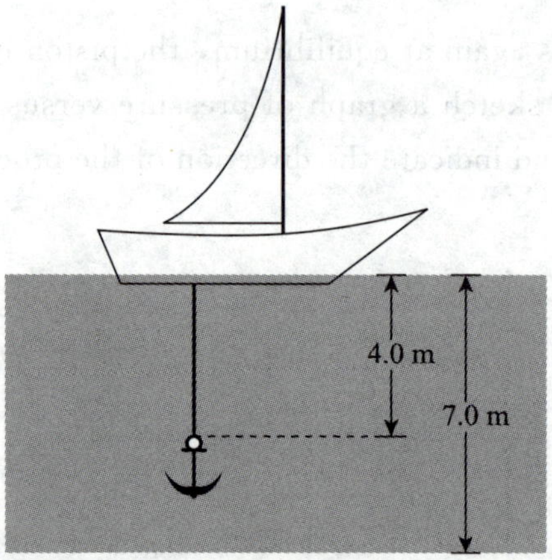

图 5

(a) On the dot that represents the anchor, draw and label the forces (not components) that act on the anchor.

(b) Calculate the magnitude of the buoyant force acting on the anchor. If you need to draw anything other than what you have shown in part (a) to assist in your solution, use the space below. DO NOT add anything to the figure in part (a).

(c) Calculate the tension in the rope. If you need to draw anything other than what you have shown in part (a) to assist in your solution, use the space below. DO NOT add anything to the figure in part (a).

(d) The bottom of the boat is at a depth d below the surface of the water. Suppose the anchor is lifted back into the boat so that the bottom of the boat is at a new depth d_1 below the surface of the water. How does d_1 compare to d ?

______ $d_1<d$ ______ $d_1=d$ ______ $d_1>d$

Justify your answer.

5. (10 points)

A student is asked to experimentally determine the index of refraction of the semicircular block of transparent plastic shown in figure 6. The student aims a red laser beam of wavelength $\lambda = 632$ nm at the center of the flat side of the block, as shown. The ray is refracted from air into the plastic and strikes the semicircular side of the block perpendicularly. The student uses a protractor to aim the laser at several different angles of incidence θ_i between 0° and 90° and to measure the angles of refraction θ_r. The student's data are given in table 1.

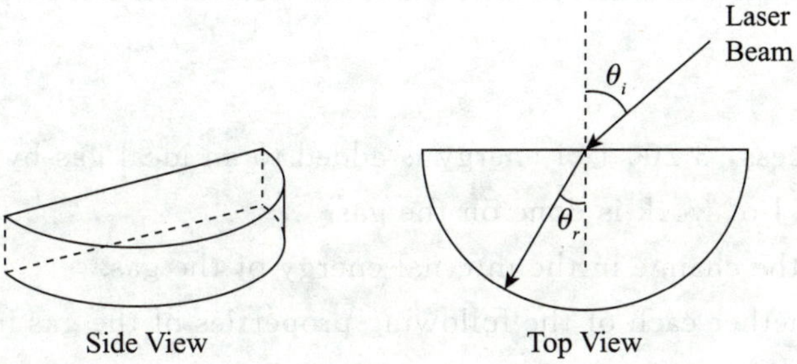

图 6

表 1

θ_i	0°	15°	30°	45°	60°	75°
θ_r	0°	10°	21°	30°	37°	44°
$\sin\theta_i$	0	0.26	0.50	0.71	0.87	0.97
$\sin\theta_r$	0	0.17	0.36	0.50	0.60	0.70

(a) On the grid below (图 7), plot data that will allow the index of refraction of the plastic to be calculated from a straight line that represents the data. Clearly label the axes, including the scales.

图 7

(b) On your graph, draw a straight line that best represents the data. Use the slope of the line to determine the index of refraction of the plastic.

(c) The student now wants to confirm the result obtained in part (b) by using the critical angle for the plastic. Describe one experimental method the student can use to measure the critical angle. Indicate how the index of refraction can be determined from this measurement.

6. (10 points)

In a certain process, 3 200 J of energy is added to an ideal gas by heating. During the same process, 2 100 J of work is done on the gas.

(a) Determine the change in the internal energy of the gas.

(b) Indicate whether each of the following properties of the gas increases, decreases, or remains the same during the process.

i. Volume

______ Increases ______ Decreases ______ Remains the same

Justify your answer.

ii. Temperature

______ Increases ______ Decreases ______ Remains the same

Justify your answer.

iii. Pressure

______ Increases ______ Decreases ______ Remains the same

Justify your answer.

Suppose that in a different process 1 800 joules of work is done on the ideal gas at a constant temperature.

(c) Determine the change in internal energy of the gas during the process.

(d) Which of the following correctly describes the energy transfer by heating, if any, between the gas and its surroundings?

_____ Energy is transferred into the gas. _____ Energy is transferred out of the gas.

_____ There is no energy transfer by heating.

Justify your answer.

7. (11 points)

Two long, straight horizontal wires are near each other and parallel, with one directly above the other as shown in figure 8. Wire X is fixed in place and connected to a battery (not shown) so that it carries a current of 65 A. Wire Y, which is part of a second circuit, is free to move vertically and is suspended at rest by the magnetic force between the wires. The mass per length of wire Y is 5.6×10^{-3} kg/m. Neglect effects from the parts of the circuits that are not shown.

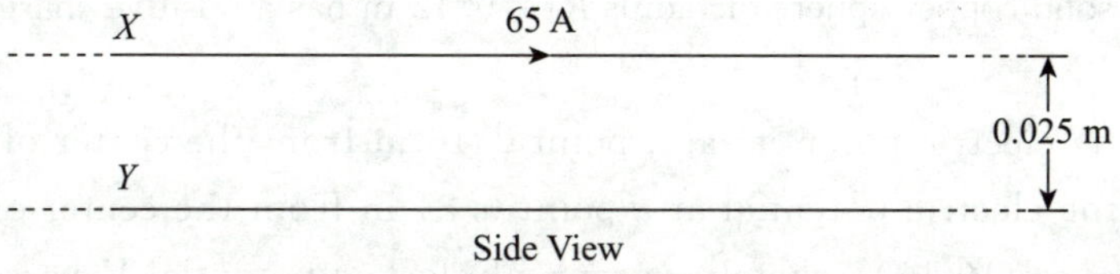

图 8

(a) Calculate the magnitude of the magnetic field produced by wire X at the position of wire Y.

(b)

i. Calculate the magnitude of the current in wire Y.

ii. Indicate the direction of the current in wire Y.

_____ To the left _____ To the right _____ Neither left nor right, since there is no current

(c) Now wire Y is moved to a new position that is closer to wire X, but wire Y is still below wire X and is still carrying the same current as determined in part (b). Wire Y is released from rest. Describe the initial motion of wire Y. Justify your answer.

(d) Suppose wire Y is moved to a position 0.025 m above wire X. What changes in current, if any, must occur to maintain equilibrium?

8. (10 points)

The energy-level diagram for an isolated hypothetical atom is shown in figure 9.

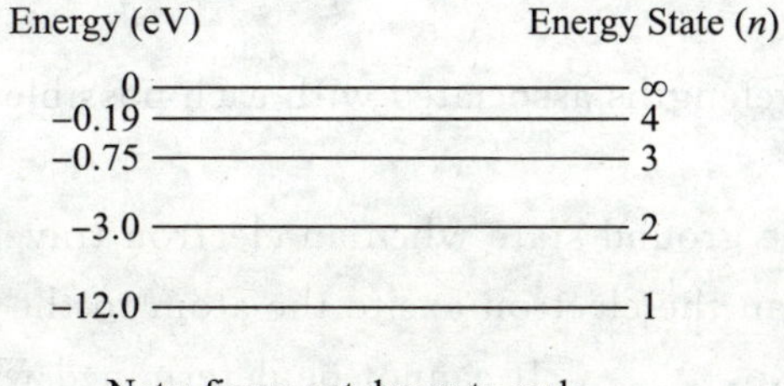

图 9

(a) A collection of such atoms with electrons in the $n=3$ state undergo transitions in which the atoms only emit photons, and the electrons eventually end in the $n=1$ state. On the diagram, draw arrows to indicate all possible transitions, given the starting and ending states for the electrons.

(b) Calculate the longest wavelength of photons that the atom can emit during the transitions identified in part (a).

(c) What is the ionization energy of an atom in the ground state?

(d) Photons of energy 11.0 eV are incident on the atom. What effect can this have on an electron in the $n=1$ state? Justify your answer.

(e) Photons of energy 14.0 eV are incident on the atom. What effect can this have on an electron in the $n = 1$ state? Justify your answer.

9. (12 points)

An isolated, solid copper sphere of radius $R_1=0.12$ m has a positive charge of 6.4×10^{-9} C.

(a)

i. Calculate the electric potential at a point 0.10 m from the center of the sphere.

ii. Calculate the electric potential at a point 0.24 m from the center of the sphere.

(b) On the axes (图 10), sketch a graph of electric potential V versus radius r from the center of the sphere. Label the value at $r=0$ on the vertical axis.

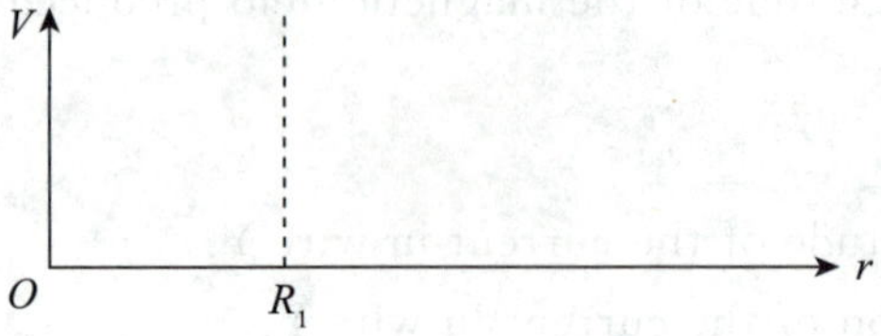

图 10

(c)

i. Determine the magnitude of the electric field at a point 0.10 m from the center of the sphere.

ii. Determine the magnitude of the electric field at a point 0.24 m from the center of the sphere.

10. (10 points)

The allowed energies of a simple hypothetical atom are -6.0 eV, -3.0 eV, and -1.0 eV.

(a) Draw the atom's energy-level diagram. Label each level with the energy and the principal quantum number.

(b) Calculate the wavelengths associated with each possible transition between energy levels for the atom.

(c) The atom is in the ground state when an electron traveling with a speed of 1.3×10^6 m/s collides with it. Can the electron excite the atom to the $n=2$ state?

______ Yes, ______ No, ______ It cannot be determined with the information given.

Justify your answer.

(d) Another electron excites the atom from the ground state to the $n=2$ state. The atom then decays back to the ground state by emitting a photon.

i. Calculate the energy of the emitted photon in joules.

ii. In what region of the electromagnetic spectrum is the radiation?

______ Radio, ______ X-rays, ______ Visible light

______ It cannot be determined with the information given.

Scoring guide line

Question 1

10 points total **Distribution of points**

(a) 3 points

For a statement or equation that the pressures at interfaces A and B are equal 1 point

$$P_{atm}+\rho_o g h_o=P_{atm}+\rho_w g h_w$$

For a substitution of the correct density and the correct heights 1 point

$$\rho_o=\rho_w h_w/h_o=(1\ 000\ \text{kg/m}^3)(24.5\ \text{cm})/(27.2\ \text{cm})$$

For the correct answer 1 point

$$\rho_o=901\ \text{kg/m}^3$$

(b) 2 points

Use equation for absolute pressure

$$P=P_0+\rho g h$$

For using atmospheric pressure for P_0 1 point

For using the correct height (in meters) with the correct density and a correct acceleration owing to gravity 1 point

$$P=P_0+\rho g h=(1.0\times10^5\ \text{Pa})+(1\ 000\ \text{kg/m}^3)(9.8\ \text{m/s}^2)(0.245\ \text{m})$$

$$P=1.02\times10^5\ \text{Pa}$$

Note: The use of 1.013×10^5 Pa and 10 m/s^2 is acceptable.

(c) 3 points

For selecting "Above A" 1 point

For a statement that the height of the oil above the mercury is now lower 1 point

For a statement that the pressure is lower at interface A owing to the lower height 1 point

(d) 1 point

For selecting "Increases" 1 point

Units: 1 point

For correct units in both numerical answers of parts (a) and (b) 1 point

Question 2

10 points total **Distribution of points**

(a) 3 points

For a graph showing pressure proportional to temperature (i. e., a straight line segment that, if extended, would pass through the origin) 1 point

For showing that the initial pressure and initial temperature are not zero 1 point

For a final state that is at a higher pressure and temperature than the initial state (regardless of the shape of the path) (图 11) 1 point

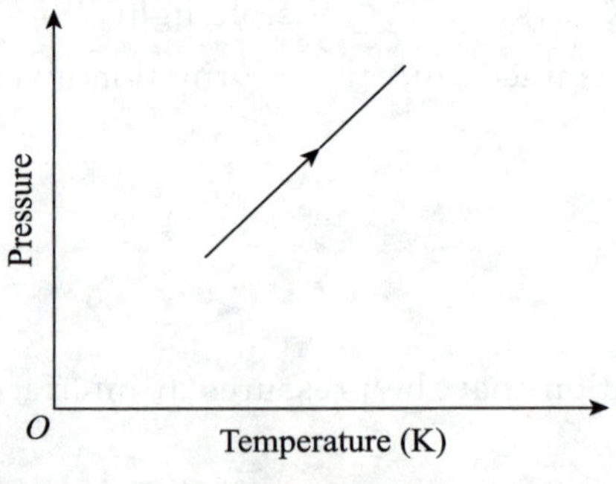

图 11

(b) 2 points

For selecting "Moves down" 1 point

For recognition of the piston's mass, which is pulled down by the force of gravity 1 point

Note: One point could be earned for selecting "Remains stationary" with a clear explanation that the internal and external pressures are equal or that the system returned to the original pressure P_0

(c) and (d) 5 points

These two parts are closely linked; therefore they are scored as a unit.

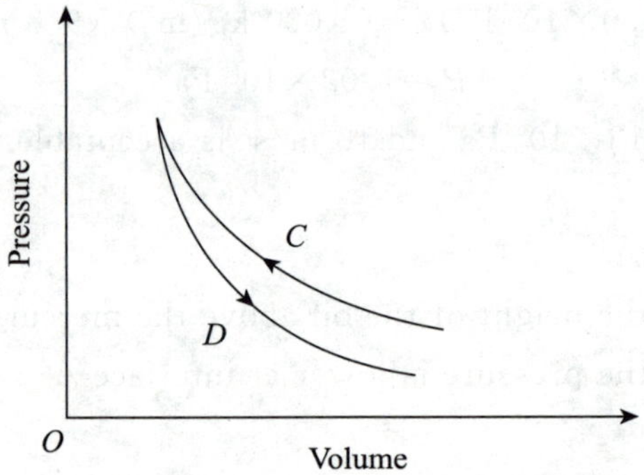

图 12

For drawing curve C as concave up, with a negative slope 1 point

For drawing curve D as concave up, with a negative slope 1 point

For drawing the final state of curve C and the initial state of curve D as the only point where the two curves intersect 1 point

For drawing curve C above curve D 1 point

For correct labels and directions of arrows on both processes（图 12） 1 point

Question 3

10 points total **Distribution of points**

(a) 2 points

For correct substitution of the momentum value into the de Broglie wavelength relationship 1 point

$$\lambda=\frac{h}{p}=\frac{6.63\times10^{-34}\ \text{J}\cdot\text{s}}{5.5\times10^{-20}\ \text{kg}\cdot\text{m/s}}$$

For a correct answer, with units 1 point

$$\lambda=1.2\times10^{-14}\ \text{m}$$

(b) 2 points

For correct substitution of the momentum into an equation to compute the speed of the proton and substituting the speed into the equation for kinetic energy 1 point

$$v=\frac{p}{m_p}=\frac{5.5\times10^{-20}\ \text{kg}\cdot\text{m/s}}{1.67\times10^{-27}\ \text{kg}}=3.3\times10^{7}\ \text{m/s}$$

$$K=\frac{1}{2}mv^2=\frac{1}{2}(1.67\times10^{-27}\ \text{kg})\ (3.3\times10^{7}\ \text{m/s})^2$$

For a correct answer, with units 1 point

$$K=9.1\times10^{-13}\ \text{J}$$

Alternate solution Alternate points

Derive formula for kinetic energy

$$K=\frac{1}{2}mv^2=\frac{1}{2}m\left(\frac{p}{m}\right)^2=\frac{p^2}{2m}$$

For using $K=\frac{p^2}{2m}$ to find the kinetic energy 1 point

$$K=\frac{(5.5\times10^{-20}\ \text{kg}\cdot\text{m/s})^2}{2\times1.67\times10^{-27}\ \text{kg}}$$

For a correct answer, with units 1 point

$$K=9.1\times10^{-13}\ \text{J}$$

(c) 3 points

For an attempt to apply conservation of energy to the system 1 point

$$U_1+K_1=U_2+K_2$$

U_1 is approximately zero (the proton is initially far away from the uranium nucleus)

$K_2=0$ (the proton is instantaneously at rest)

Therefore $K_1=U_2$

For any correct expression that shows the electrostatic potential energy of the system at the proton's closest approach equal to the kinetic energy determined in part (b), using either symbols or values from the problem 1 point

$$K_1=U_2=\frac{kq_1q_2}{r}=\frac{k(92e)(e)}{D}$$

$$K_1 = \frac{92ke^2}{D}$$

$$D = \frac{92ke^2}{K} = \frac{(92)(9\times10^9\ \text{N}\cdot\text{m}^2/\text{C}^2)(1.6\times10^{-19}\ \text{C})^2}{9.0\times10^{-13}\ \text{J}}$$

For a correct answer, with units 1 point

$$D = 2.3\times10^{-14}\ \text{m}$$

(d) 3 points

For selecting "Greater" 1 point

For using the mass-energy relationship $E=mc^2$ to solve for the mass defect of the uranium decaying into the daughter particles plus excess energy 1 point

$$E = \Delta mc^2$$

$$\Delta m = \frac{E}{c^2} = \frac{2.5\times10^{-11}\ \text{J}}{(3.0\times10^8\ \text{m/s})^2}$$

For an answer with any proper units of mass 1 point

$$\Delta m = 2.8\times10^{-28}\ \text{kg or } \Delta m = 0.17\ \text{amu}$$

Question 4

10 points total **Distribution of points**

(a) 3 points

For showing the buoyant force in the correct direction and labeling it 1 point

For showing the tension in the correct direction and labeling it 1 point

For showing the gravitational force in the correct direction and labeling it (图 13) 1 point

图 13

One earned point was deducted for any extraneous forces.

(b) 2 points

For use of the correct expression for the buoyant force 1 point

$$F_B = \rho Vg$$

Substitute correct values

$$F_B = (1\,000\ \text{kg/m}^3)(6.25\times10^{-3}\ \text{m}^3)(9.8\ \text{m/s}^2)$$

For a correct answer, with units 1 point

$$F_B = 61.3\ \text{N}\ (62.5\ \text{N if using } g = 10\ \text{m/s}^2)$$

(c) 3 points

For a correct expression of Newton's second law when the anchor is at equilibrium 1 point

$$F_T + F_B = mg$$

For substituting a value consistent with the buoyant force from part (b) 1 point

For substituting correct values for calculation of the gravitational force 1 point

$$F_T = (50\ \text{kg})(9.8\ \text{m/s}^2) - (61.3\ \text{N})$$

$$F_T = 429\ \text{N}\ (438\ \text{N if using } g = 10\ \text{m/s}^2)$$

(d) 2 points

For selecting $d_1 > d$ 1 point

For a correct justification 1 point

Example

When the anchor is lifted onto the boat, the buoyant force on the boat must now support the weight of both the boat and the anchor. This will increase the buoyant force, which requires a greater volume of water, which means the boat will reach a greater depth into the water.

Question 5

10 points total **Distribution points**

(a) 3 points

For labeling both axes with variables that will yield a straight-line graph 1 point

For correctly scaling both axes 1 point

For correctly plotting the points (图 14) 1 point

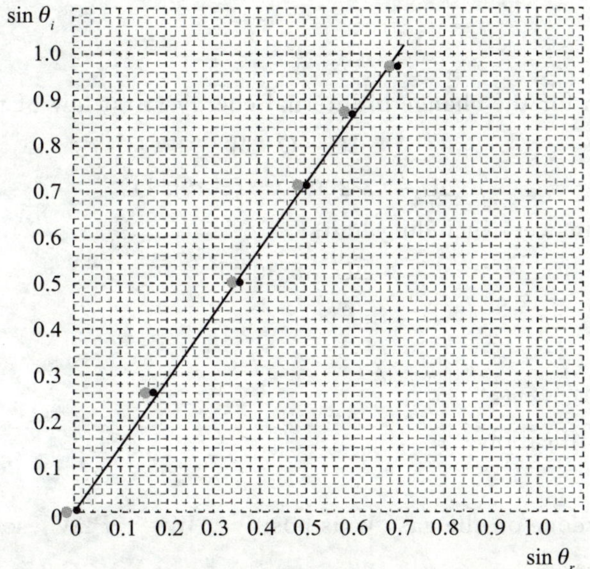

图 14

(b) 4 points

For a straight line that best represents the data 1 point

For relating the slope to the index of refraction using Snell's law 1 point

$n=$slope (or 1/slope if plotting $\sin\theta_r$, as a function of $\sin\theta_i$)

For substituting two points on the line 1 point

For the example graph shown, $n=\dfrac{1.00-0.20}{0.70-0.14}$

For a calculated index of refraction in the range 1.2～1.6 1 point

$$n=1.43$$

(c) 3 points

For a new experimental method in which the student aims the laser at the curved side of the block 1 point

For stating that the critical angle is the minimum angle at which no light is refracted or total internal reflection is reached 1 point

For relating the index of refraction to the critical angle and stating that $n_2=1$ (the second medium is air) 1 point

Example

Aim the laser through the curved side of the block at normal incidence, so that it emerges from the center of the flat side. Pivot the block about the center of the flat side so that the emerging beam just grazes the flat side (or the beam begins to disappear).

Measure the angle between the incident beam and the normal to the flat side.

This is the critical angle θ_c. The index of refraction is given by $n=1/\sin\theta_c$.

Question 6

10 total points **Distribution of points**

(a) 1 point

Use the equation for the change in internal energy (first law of thermodynamics)

$$\Delta U=Q+W$$

$$\Delta U=(3\ 200\ \text{J})+(2\ 100\ \text{J})$$

For the correct answer, including units 1 point

$$\Delta U=5\ 300\ \text{J}$$

(b) 6 point

i. 2 points

For selecting "Decreases" 1 point

For a correct justification 1 point

Examples

Because work is done on the gas, W is positive. $W=-P\Delta V$, so ΔV must be negative and the volume decreases.

Because work is done on the gas, W is positive and the gas is compressed. If the gas is compressed, the volume decreases.

ii. 2 points

For selecting "Increases" 1 point

For a correct justification 1 point

Example

The internal energy increases (as shown in part (a)) and the temperature of an ideal gas increases as the internal energy increases.

iii. 2 points

For selecting "Increases" 1 point

For a correct justification 1 point

Example

From the ideal gas law, PV/T is constant. So if V decreases and T increases, P must increase.

(c) 1 point

For stating the internal energy does not change 1 point

$$\Delta U = 0$$

(d) 2 points

For selecting "Energy is transferred out of the gas" 1 point

For a correct justification 1 point

Example

$\Delta U = Q + W$ is zero for an ideal gas at constant temperature and W is positive since work is done on the gas. Therefore Q is negative, meaning energy is transferred out of the gas.

Question 7

11 points total **Distribution of points**

(a) 2 points

For the correct substitution of the values of μ_0, I, and r into the correct expression for the magnetic filed around a wire 1 point

$$B = \frac{\mu_0 I}{2\pi r} = \frac{(4\pi \times 10^{-7}\ \text{T} \cdot \text{m/A})(65\ \text{A})}{(2\pi)(0.025\ \text{m})}$$

For the correct answer 1 point

$$B = 5.2 \times 10^{-4}\ \text{T}$$

(b) 5 points

i. 4 points

For an expression of equilibrium between the magnetic and gravitational forces on the wire 1 point

$$mg = F_B$$

For correct expressions for the magnetic and gravitational forces 1 point

$$mg = BIL$$

For a correct expressions for the current in terms of the linear density 1 point

$$I=\left(\frac{m}{L}\right)\frac{g}{B}$$

For correctly substituting the value from part (a) into the above equation to solve for the current 1 point

$$I=(5.6\times10^{-3}\ \text{kg/m})\frac{9.8\ \text{m/s}^2}{5.2\times10^{-4}\ \text{T}}$$

$$I=106\ \text{A}\ (\text{or } 108\ \text{A using } g=10\ \text{m/s}^2)$$

ii. 1 point

For selecting "To the right" 1 point

(c) 2 points

For stating that the wire would move upward 1point

For stating that the wire would accelerate with a correct justification 1 point

Example

As wire Y is moved closer to X, the magnetic field from X will increase. This will create an upward net force on Y that will accelerate Y upward toward wire X.

(d) 2 point

For stating that the direction of the current in one of the wires must change 2 point

Examples

The current must be reversed in one wire, but not in the other.

Reverse the current in wire X.

Change the direction of current in wire Y.

Question 8

10 points total **Distribution of points**

(a) 3 points

For an arrow from the $n=3$ to the $n=2$ energy state 1 point

For an arrow from the $n=2$ to the $n=1$ energy state 1 point

For an arrow from the $n=3$ to the $n=1$ energy state (图 15) 1 point

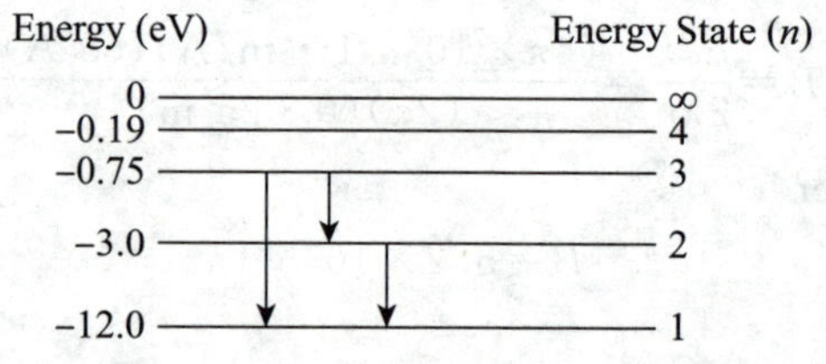

图 15

One earned point was deducted if there were any incorrect or extra lines on the diagram.

(b) 2 points

Use an expression relating energy to wavelength

$$E=hf=hc/\lambda$$

For substitution using the correct 3 to 2 transition energy 1 point

$$\lambda = hc/E = (1\ 240\ \text{eV} \cdot \text{nm})/[(3.0-0.75)\ \text{eV}]$$

For an answer consistent with whichever transition was chosen 1 point

$$\lambda = 551\ \text{nm or } 5.51\times10^{-7}\ \text{m}$$

(138 nm for the 2 to 1 transition, and 110 nm for the 3 to 1 transition)

(c) 1 point

For the correct, positive value for the ionization energy 1 point

$$E = 12\ \text{eV or } 1.9\times10^{-18}\ \text{J}$$

(d) 2 points

For correctly stating that a 11.0 eV photon has no effect 1 point

For a correct explanation 1 point

Example

Since there is no energy level at −1.0 eV, an 11.0 eV photon will have no effect on electrons in the −12.0 eV ($n=1$) energy state.

(e) 2 points

For correctly stating that a 14.0 eV photon will remove the electron from (or ionize) the atom 1 point

For a correct explanation 1 point

Example

A 14.0 eV photon has sufficient energy to ionize the atom, since only 12.0 eV is required for ionization.

Question 9

12 total points **Distribution of points**

(a) 5 points

i. 3 points

For writing the appropriate expression for the electric potential, given the spherical symmetry of the situation 1 point

$$V = \frac{kq}{r}$$

The electric field inside a conducting sphere in equilibrium is zero, so the potential inside the sphere is constant, uniform, and equal to the potential at the surface.

$$V = \frac{kq}{R_1}$$

For correct substitutions 1 point

$$V = \frac{(9.0\times10^9\ \text{N}\cdot\text{m}^2/\text{C}^2)(6.4\times10^{-9}\ \text{C})}{0.12\ \text{m}}$$

For the correct answer 1 point

$$V = 480\ \text{V}$$

ii. 2 points

For writing the appropriate expression for the electric potential 1 point

$$V=\frac{kq}{r}$$

$$V=\frac{(9.0\times10^{9}\ \mathrm{N\cdot m^2/C^2})\ (6.4\times10^{-9}\ \mathrm{C})}{0.24\ \mathrm{m}}$$

For the correct answer calculated 1 point

$$V=240\ \mathrm{V}$$

(b) 3 points

For labeling the potential at $r=0$ with the value from part (a) i 1 point

For drawing a horizontal line from $r=0$ to $r=R_1$ that is continuous with the other part of the graph (图 16) 1 point

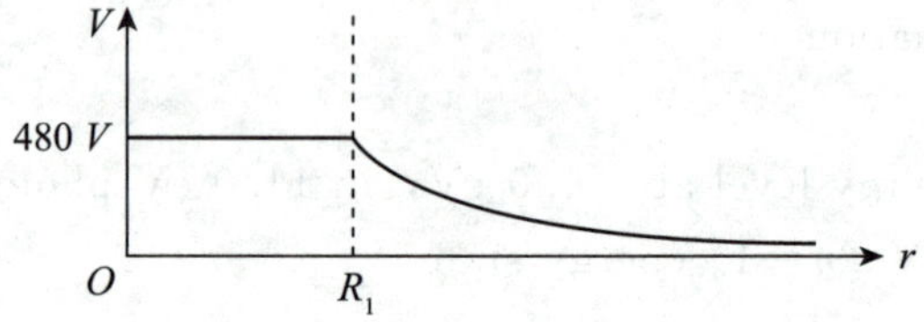

图 16

For drawing a reasonable curve for the $1/r$ dependence in the region beyond R_1 1 point

(c) 4 points

i. 2 points

$$r=0.10\ \mathrm{m}<0.12\ \mathrm{m}=R_1$$

There is no electric field inside a conducting sphere in equilibrium.

For the correct answer 2 points

$$E=0\ \mathrm{N/C}$$

ii. 2 points

$r=0.24\ \mathrm{m}>0.12\ \mathrm{m}=R_1$, so the charge can be treated as a point charge at the center of the sphere.

For writing the appropriate expression for the electric field 1 point

$$E=\frac{kq}{r^2}$$

$$E=\frac{(9.0\times10^{9}\ \mathrm{N\cdot m^2/C^2})(6.4\times10^{-9}\ \mathrm{C})}{(0.24\ \mathrm{m})^2}$$

For the correct answer 1 point

$$E=1\ 000\ \mathrm{N/C}$$

Question 10

10 points total **Distribution of points**

(a) 2 points

For a correct energy-level diagram (图 17) with horizontal lines and -6 eV at the bottom 1 point

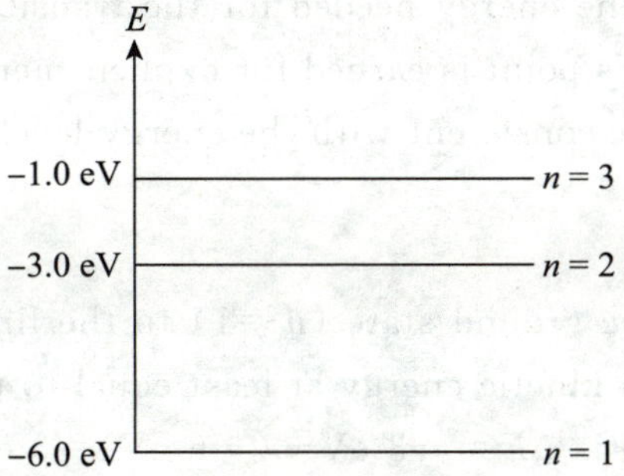

Note: figure not to scale

图 17

Note: The energy-level diagram does not need to be drawn to scale in order to earn this point.

For correct labeling of energy levels and quantum numbers (e.g., $n=1$ correlated with -6 eV) and no other incorrect energy levels shown. 1 point

(b) 3 points

For correctly relating photon energy to wavelength 1 point

$$E=hc/\lambda \text{ or } E=hf \text{ and } c=\lambda f \text{ or } E=pc \text{ and } p=h/f$$

The energy in these equations is equal to the change in energy for each transition, so the wavelength of a photon for each transition becomes $\lambda=hc/\Delta E$, where the value of hc is given on the equation sheet.

For the indication that there are three possible transitions 1 point

For the substitution of correct values of ΔE in all attempted transitions 1 point

$$n=3 \rightarrow n=1$$

$$\Delta E=-1\text{ eV}-(-6\text{ eV})=5\text{ eV}$$

$$\lambda_{n=3\rightarrow n=1}=\frac{1.24\times10^{3}\text{ eV}\cdot\text{nm}}{5.0\text{ eV}}=248\text{ nm}$$

$$n=3 \rightarrow n=2$$

$$\Delta E=-1\text{ eV}-(-3\text{ eV})=2\text{ eV}$$

$$\lambda_{n=3\rightarrow n=2}=\frac{1.24\times10^{3}\text{ eV}\cdot\text{nm}}{2.0\text{ eV}}=620\text{ nm}$$

$$n=2 \rightarrow n=1$$

$$\Delta E=-3\text{ eV}-(-6\text{ eV})=3\text{ eV}$$

$$\lambda_{n=2\rightarrow n=1}=\frac{1.24\times10^{3}\text{ eV}\cdot\text{nm}}{3.0\text{ eV}}=413\text{ nm}$$

(c) 3 points

For selecting "Yes," along with a statement that the kinetic energy of the electron is greater than the transition energy. (Note: The selection of "yes" with no justification does not earn this point.) 1 point

Points are earned for an appropriate justification as follows:

For a correct calculation of the electron's kinetic energy 1 point

For an identification of the energy needed for the transition from ground state to the first excited state (Note: This point is earned for explicit mention of the needed excitation energy, either 3 eV or a value consistent with the energy-level diagram shown in part (a).)

1 point

Example

To excite the atom in the ground state ($n=1$) to the first excited state ($n=2$), the incident electron must have a kinetic energy at least equal to the $n=1 \rightarrow n=2$

transition energy, which is $\Delta E = -3\ \text{eV} - (-6\ \text{eV}) = 3\ \text{eV}$.

The kinetic energy of the incident electron is

$$K = \frac{1}{2} m_e v^2$$

$$K = \frac{1}{2}(9.11 \times 10^{-31}\ \text{kg})(1.3 \times 10^6\ \text{m/s})^2 = 7.7 \times 10^{-19}\ \text{J} = 4.8\ \text{eV}$$

Since the kinetic energy of the electron is greater than the transition energy, the electron can excite the atom to the $n=2$ state.

Note: Credit is earned for a "No" response with complete justification only if the justification is consistent with any errors in the energy-level diagram or in the calculation of kinetic energy.

Note: Full credit can also be earned for an answer that (1) calculates the speed of an electron having a kinetic energy equal to that of the energy required to raise the atom from the ground state to the first excited state and (2) then shows that that speed is less than the speed of the electron given in the question, thus concluding that the electron can excite the atom from the ground state to the first excited state.

(d) 2 points

i. 1 point

The energy of the emitted photon for the $n=2 \rightarrow n=1$ transition is calculated in part (b).

$$\Delta E = -3\ \text{eV} - (-6\ \text{eV}) = 3\ \text{eV}$$

Converting this value to joules:

$$\Delta E = 3\ \text{eV}\left(\frac{1.60 \times 10^{-19}\ \text{J}}{1\ \text{eV}}\right)$$

For the correct answer

$$\Delta E = 4.8 \times 10^{-19}\ \text{J}$$ 1 point

Note: This point is also awarded for an answer consistent with the energy-level diagram shown in part (a).

ii. 1 point

For selecting "Visible light" only 1 point

The wavelength for the $n=2 \rightarrow n=1$ transition is found in part (b).

$$\lambda_{n=2 \rightarrow n=1} = 413\ \text{nm}$$

A wavelength of 413 nm is the wavelength for violet in the visible light region of the electromagnetic spectrum.

C Practice: Mock Examination（模拟试题）

AP Physics 1 Practice Exam（1）

Section Ⅰ: Multiple Choice

DO NOT OPEN THIS BOOKLET UNTIL YOU ARE TOLD TO DO SO.

Instructions

Section Ⅰ of this exam contains 50 multiple-choice questions. Pages containing equations and other information are also printed in this booklet. Calculators, rulers, and straightedges may be used in this section.

Indicate all of your answers to the multiple-choice questions on the answer sheet.

No credit will be given for anything written in this exam booklet, but you may use the booklet for notes or scratch work.

For questions 1 through 45, select the single best answer choice for each question.

After you have decided which of the choices is best, fill in the appropriate letter in the corresponding space on the answer sheet.

For questions 46 through 50, select the two best answer choices for each question.

After you have decided which two of the choices are best, enter both letters in the corresponding space on the answer sheet.

Use your time effectively, working as quickly as you can without losing accuracy. Do not spend too much time on any one question. Go on to other questions and come back to the ones you have not answered if you have time. It is not expected that everyone will know the answers to all of the multiple-choice questions.

Your total score on Section Ⅰ is based only on the number of questions answered correctly.

Points are not deducted for incorrect answers or unanswered questions.

Total Time

90 minutes

Number of Questions

50

Percent of Total Score

50%

Writing Instrument

Pencil required

Electronic Device

Calculator allowed

AP Physics 1 Practice (1)

Section Ⅰ

50 Questions

Time—90 minutes

Note: To simplify calculations, you may use $g = 10\ \text{m/s}^2$ in all problems.

Directions: Each of the questions or incomplete statements below is followed by 4 suggested answers or completions. Select the one that is the best in each case and then enter the appropriate letter in the corresponding space on the answer sheet.

1. An object is moving to the right with speed v_i when a force of magnitude F is exerted on it. In which of the following situations is the object's direction of motion changing and kinetic energy decreasing at the instant shown?

(A)

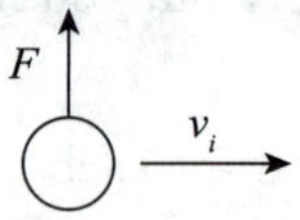

(B)

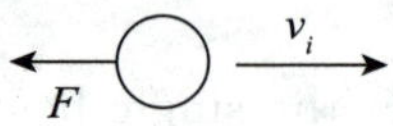

(C)

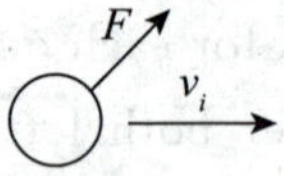

(D)

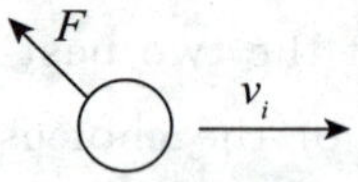

2. A ball is suspended by a lightweight string, as shown in figure 1. The ball is displaced to position 1 and released. The four labeled positions are spaced along the arc of the ball's motion. Between which adjacent pairs of positions is the change in kinetic energy of the ball greatest?

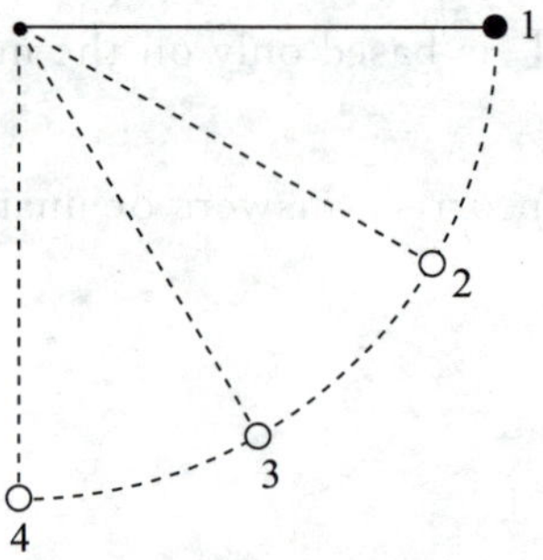

图 1

(A) 1 and 2

(B) 2 and 3

(C) 3 and 4

(D) The change is the same for all adjacent pairs

3. A newly discovered planet is found to have (2/3) ρ_E density and radius $2R_E$, where ρ_E and R_E are the density and radius of the Earth, respectively. The surface gravitational

field of the planet is most nearly

(A) 1.7 N/kg (B) 3.3 N/kg (C) 6.7 N/kg (D) 13 N/kg

4. A bus is initially traveling north at a constant speed, as shown in figure 2. As the bus starts to make a left turn without changing speed, a passenger notices that a box on the floor starts sliding toward the right side of the bus. Which of the following top views of the box, when correctly labeled, would best represent all of the horizontal forces exerted on the box as it starts sliding?

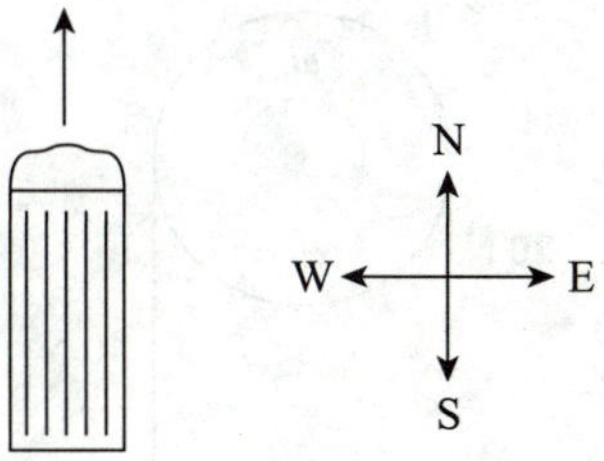

图 2

(A) 

(B) 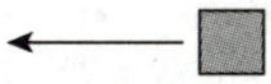

(C) 

(D) 

Questions 5—6 refer to the following material.

Three identical blocks each take a different path from a height h to the ground. Block A is released from rest and falls vertically. Block B is released from rest and slides down a frictionless incline. Block C is projected horizontally with an initial speed v (图 3).

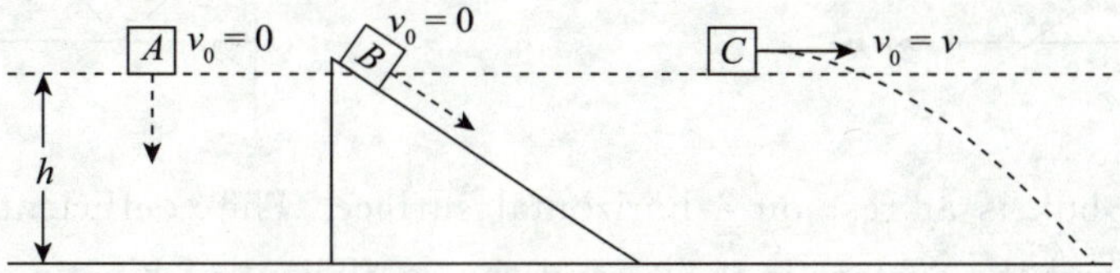

图 3

5. Which block takes the longest time to reach the ground?

(A) A (B) B (C) C

(D) The blocks take the same time to reach the ground

6. Which block has the greatest speed just before hitting the ground?

(A) A (B) B (C) C

(D) The blocks reach the ground with the same speed

7. A ball is dropped from rest and falls to the floor. The initial gravitational potential energy of the ball-Earth-floor system is 10 J. The ball then bounces back up to a height

where the gravitational potential energy is 7 J. What was the mechanical energy of the ball-Earth-floor system at the instant the ball left the floor?

(A) 0 J (B) 3 J (C) 7 J (D) 10 J

8. A disk is initially rotating counterclockwise around a fixed axis with angular speed ω_0. At time $t=0$, the two forces shown in figure 4 are exerted on the disk. If counterclockwise is positive, which of the following could show the angular velocity of the disk as a function of time?

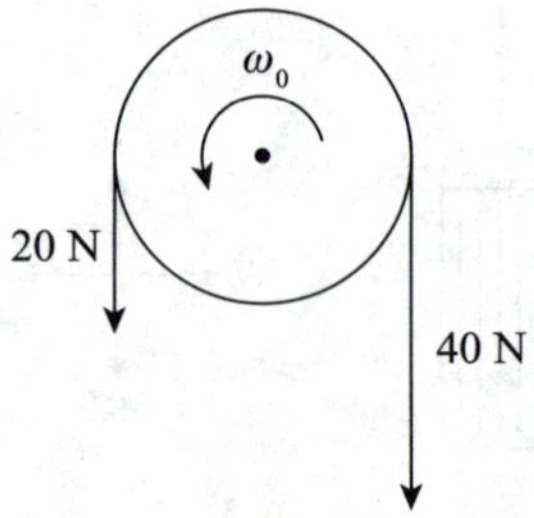

图 4

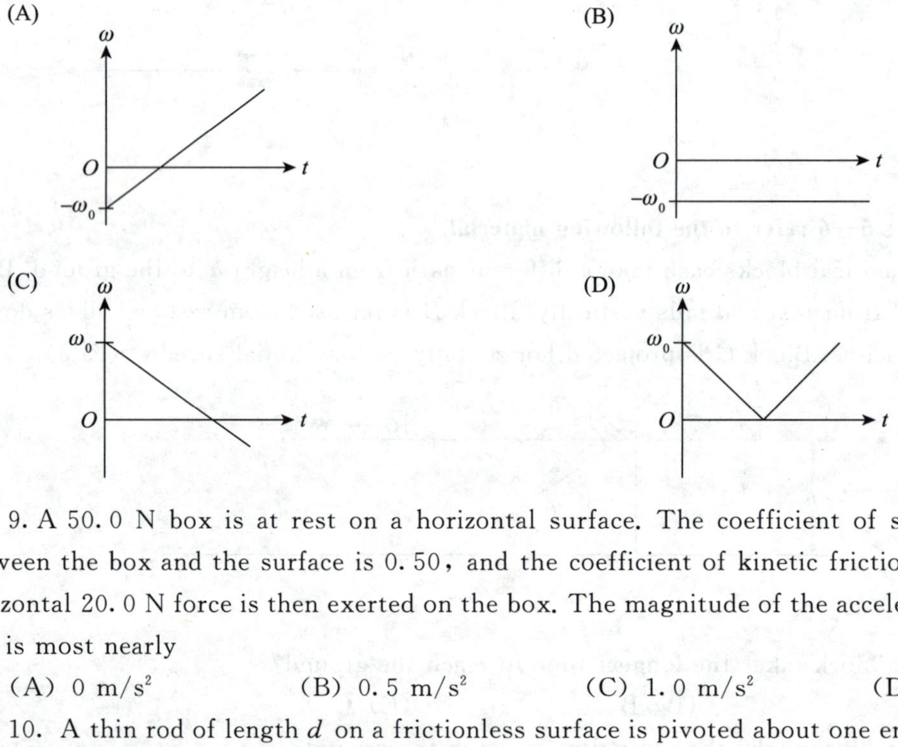

9. A 50.0 N box is at rest on a horizontal surface. The coefficient of static friction between the box and the surface is 0.50, and the coefficient of kinetic friction is 0.30. A horizontal 20.0 N force is then exerted on the box. The magnitude of the acceleration of the box is most nearly

(A) 0 m/s^2 (B) 0.5 m/s^2 (C) 1.0 m/s^2 (D) 4.0 m/s^2

10. A thin rod of length d on a frictionless surface is pivoted about one end, as shown in figure 5, and can rotate freely. The rod is at rest when it is struck by a sphere with linear momentum of magnitude p_i perpendicular to the rod. The sphere rebounds along its original line of motion with momentum of magnitude p_f. What is the magnitude of the angular momentum of the rod immediately after the collision?

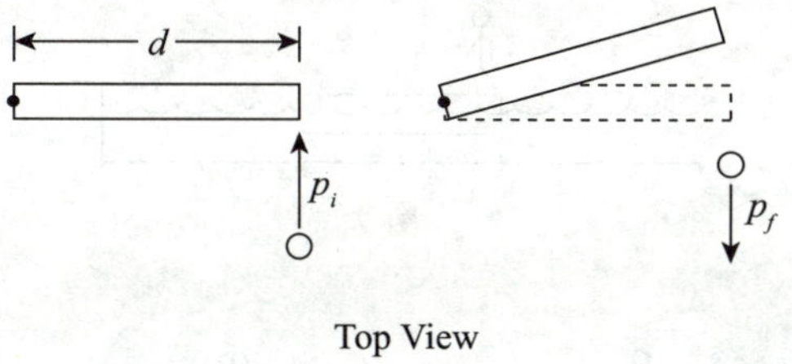

图 5

(A) $p_f - p_i$ (B) $p_f + p_i$

(C) $(p_f - p_i)d$ (D) $(p_f + p_i)d$

11. A spacecraft is placed in a circular orbit around a planet with mass 6.4×10^{23} kg. The spacecraft orbits at a height of 4.5×10^{7} m above the planet's surface. What additional information is needed to calculate the speed of the spacecraft in the orbit?

(A) No additional information

(B) The planet's radius only

(C) The spacecraft's mass only

(D) Both the planet's radius and the spacecraft's mass

12. Two carts, of mass $2m$ and m, approach each other head-on with the same speed v, as shown in figure 6. When the carts collide, they hook together. Assuming positive momentum is to the right, which of the following best represents the momentum of the cart of mass m as a function of time before and after the collision?

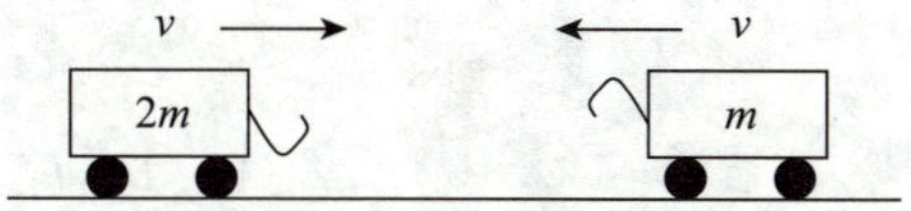

图 6

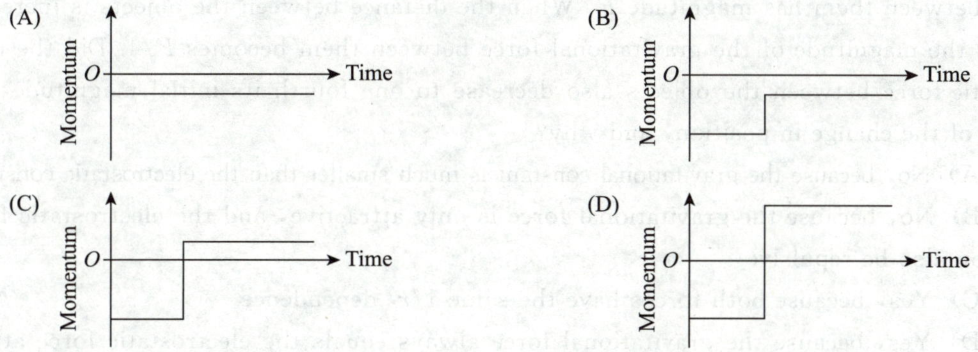

13. A student stands at one end of a raft floating in a pool with equally spaced marks along the bottom, as shown in figure 7. The student and the raft have the same mass. The student walks to the opposite end of the raft. Which of the following best shows the final locations of the raft and student relative to the marks at the bottom of the pool? Assume that there is no drag force between the raft and the water.

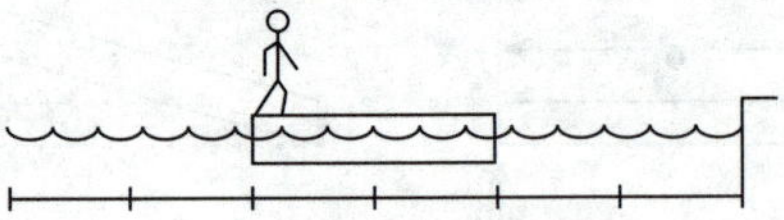

图 7

(A)

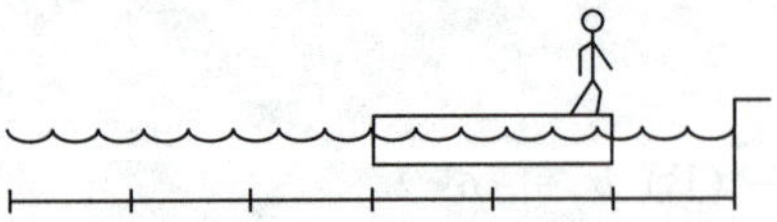

(B)

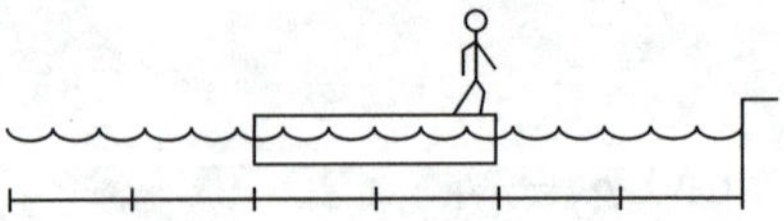

(C)

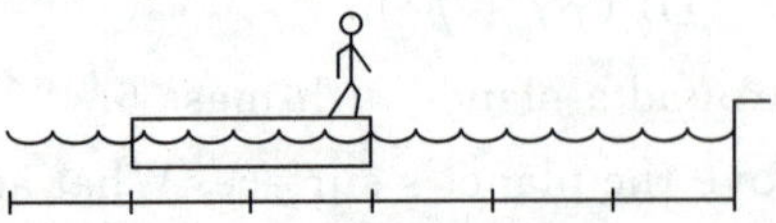

(D)

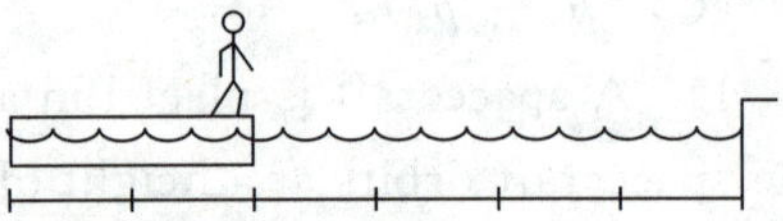

14. A circuit is designed with two resistors, $R_1=200\ \Omega$ and $R_2=400\Omega$, and a battery with internal resistance $R=10\Omega$, as shown in figure 8. What is the relationship between the three labeled currents?

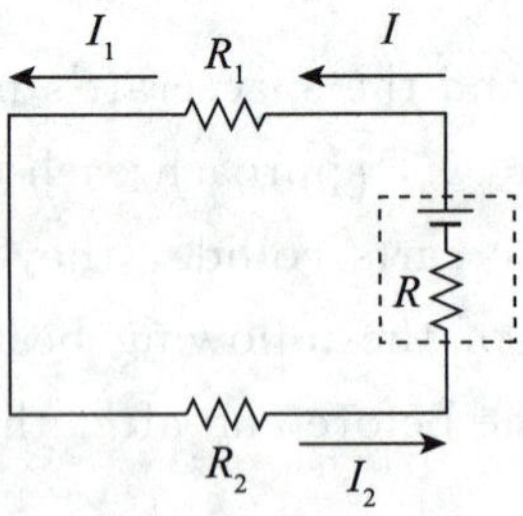

图 8

(A) $I_1+I_2=I$ (B) $I>I_1>I_2$

(C) $I_2>I_1>I$ (D) $I_1=I_2=I$

15. When two charged, massive objects are placed a distance r apart, the gravitational force between them has magnitude F. When the distance between the objects is increased to $2r$, the magnitude of the gravitational force between them becomes $F/4$. Did the electrostatic force between the objects also decrease to one fourth its initial magnitude as a result of the change in position, and why?

(A) No, because the gravitational constant is much smaller than the electrostatic constant

(B) No, because the gravitational force is only attractive, and the electrostatic force can also be repulsive

(C) Yes, because both forces have the same $1/r^2$ dependence

(D) Yes, because the gravitational force always equals the electrostatic force at any given distance

16. Cart A moves horizontally on a frictionless track. It collides and sticks to identical cart B, which is at rest. The speed of cart A before the collision is measured to be 3.0 m/s, with an uncertainty of ±0.2 m/s. Students are asked to predict the speed of the carts after the collision. Given the uncertainty of the measurement for the speed of cart A, which of the

following is the best prediction?

(A) 1.1 to 1.9 m/s (B) 1.2 to 1.8 m/s

(C) 1.3 to 1.7 m/s (D) 1.4 to 1.6 m/s

17. A lion is running at constant speed toward a gazelle that is standing still, as shown in figure 9. After several seconds, the gazelle notices the lion and accelerates directly toward him, hoping to pass the lion and force him to reverse direction. As the gazelle accelerates toward and past the lion, the lion changes direction and accelerates in pursuit of the gazelle. The lion and the gazelle eventually each reach constant but different speeds.

图 9

Which of the following sets of graphs shows a reasonable representation of the velocities of the lion and the gazelle as functions of time ?

(A) lion gazelle

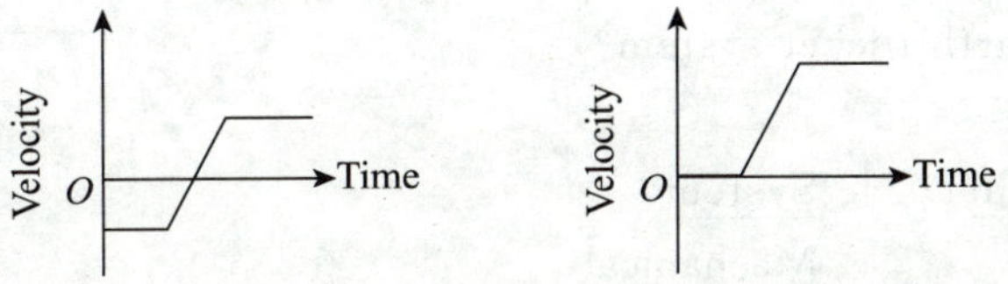

(B) lion gazelle

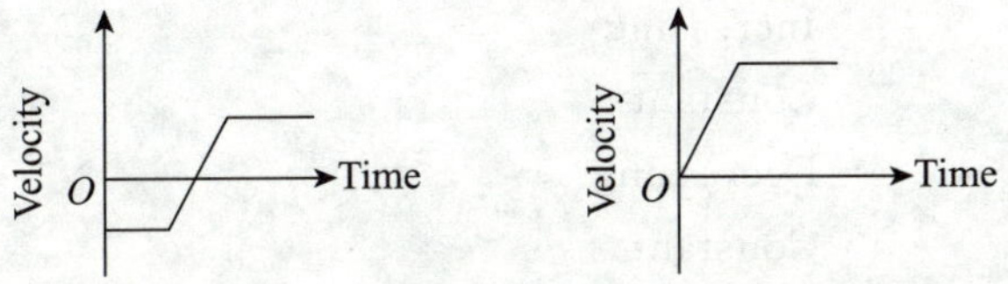

(C) lion gazelle

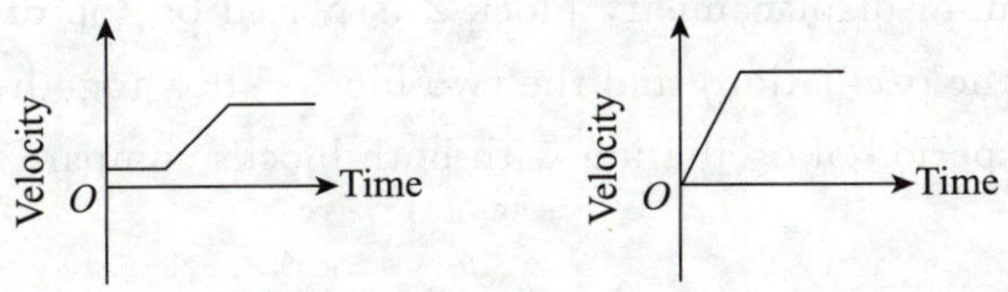

(D) lion gazelle

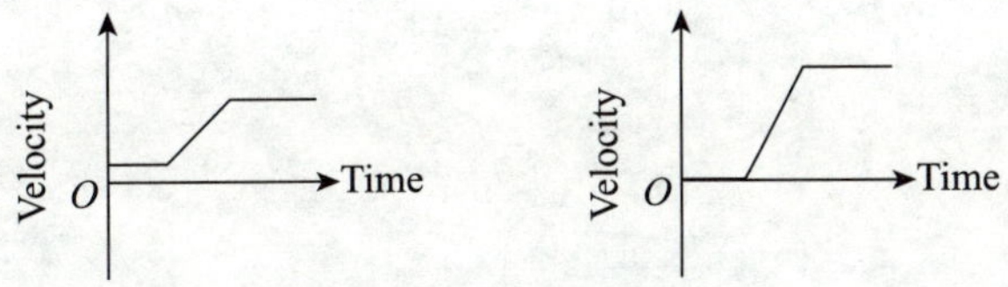

18. A spacecraft of mass 4 000 kg is traveling in a straight line in the positive direction. Engines can be fired so that the force exerted on the spacecraft is in the positive or negative direction.

Graph 10 shows data for the force during one interval. Which of the following is the best estimate of the net change in the speed of the spacecraft from time $t=0$ to time $t=4$ s?

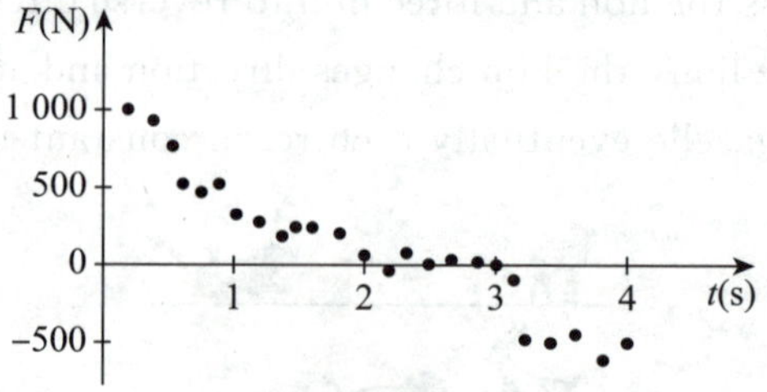

图 10

(A) $+0.4$ m/ s　　(B) $+0.1$ m/ s

(C) -0.1 m/ s　　(D) -0.4 m/ s

19. A rocket is continuously firing its engines as it accelerates away from the Earth. For the first kilometer of its ascent, the mass of fuel ejected is small compared to the mass of the rocket. For this distance, which of the following indicates the changes, if any, in the kinetic energy of the rocket, the gravitational potential energy of the Earth-rocket system, and the mechanical energy of the Earth-rocket system?

	Rocket Kinetic Energy	System Gravitational Potential Energy	System Mechanical Energy
(A)	Increasing	Increasing	Increasing
(B)	Increasing	Increasing	Constant
(C)	Increasing	Decreasing	Decreasing
(D)	Decreasing	Increasing	Constant

20. Block 1 is attached to a spring and oscillates on a horizontal frictionless surface. When block 1 is at a point of maximum displacement, block 2 is placed on top of it from directly above without interrupting the oscillation, and the two blocks stick together. How do the maximum kinetic energy and period of oscillation with both blocks compare to those of block 1 alone?

	Maximum Kinetic Energy	Period
(A)	Smaller	Smaller
(B)	Smaller	Greater
(C)	The same	Smaller
(D)	The same	Greater

21. A child slides from rest down slides A and B shown in figure 11. The slides are the same height, and the coefficient of friction between the slides and the child is the same. Which of the following compares the change ΔK in the kinetic energy of the child and the change ΔU in the potential energy of the child-Earth system for the two slides?

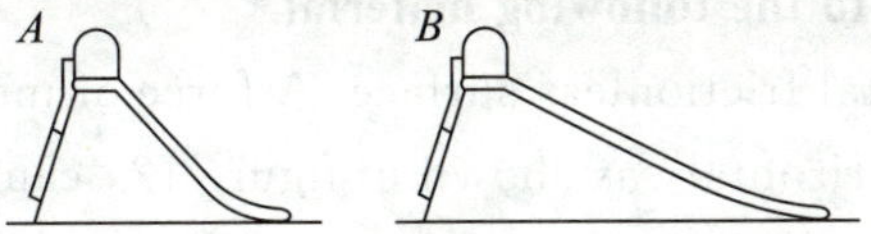

图 11

(A) $\Delta K_A=\Delta K_B$; $\Delta U_A=\Delta U_B$
(B) $\Delta K_A<\Delta K_B$; $\Delta U_A>\Delta U_B$
(C) $\Delta K_A>\Delta K_B$; $\Delta U_A=\Delta U_B$
(D) $\Delta K_A>\Delta K_B$; $\Delta U_A>\Delta U_B$

Questions 22—24 refer to the following material.

A student is observing an object of unknown mass that is oscillating horizontally at the end of an ideal spring. The student measures the object's period of oscillation with a stopwatch.

22. The student wishes to determine the spring constant of the spring using the measurements of the period of oscillation. Which of the following pieces of equipment would provide another measured quantity that is sufficient information to complete the determination of the spring constant?

(A) Meterstick (B) Motion sensor (C) Balance (D) Photogate

23. Using a number of measurements, the student determines the following (表 1).

表 1

Spring constant	85 N/m
Mass of object	0.50 kg
Amplitude of oscillation	0.30 m
Maximum speed of object	3.9 m/s

The total energy of the object-spring system is most nearly

(A) 0.98 J (B) 3.8 J (C) 7.6 J (D) 12.8 J

24. While the object is continuously oscillating, the student determines the maximum speed of the object during two oscillations. The first speed is 3.5 m/s and the second speed is 2.7 m/s. Which of the following could account for the decrease in the object's maximum kinetic energy?

(A) Energy was transferred from the object to the spring, which increased the maximum potential energy of the spring
(B) Energy was transferred from the spring to the object, which decreased the maxi-

mum potential energy of the spring

(C) As energy was transferred back and forth between the object and the spring, a greater average share of the energy became potential energy of the spring

(D) The object-spring system lost energy to its surroundings

Questions 25—26 refer to the following material.

A crate is on a horizontal frictionless surface. A force of magnitude F is exerted on the crate at an angle θ to the horizontal, as shown in figure 12, causing the crate to slide to the right. The surface exerts a normal force of magnitude F_N on the crate. As the crate slides a distance d, it gains an amount of kinetic energy ΔK. While F is kept constant, the angle is now doubled but is still less than 90°. Assume the crate remains in contact with the surface.

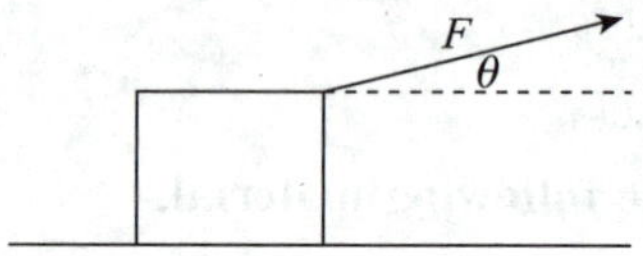

图 12

25. How does the new normal force exerted on the crate compare to F_N?

(A) The new normal force is greater than F_N

(B) The new normal force is less than F_N

(C) The new normal force is equal to F_N

(D) The new normal force is greater or less than F_N depending on the value of θ

26. As the crate slides a distance d, how does the new gain in kinetic energy compare to ΔK ?

(A) The new gain is greater than ΔK

(B) The new gain is less than ΔK

(C) The new gain is equal to ΔK

(D) The new gain is greater or less than ΔK depending on the value of θ

27. A block on a level surface is attached to one end of a spring, as shown in figure 13. The other end of the spring is attached to a wall. There is friction between the block and the surface. A person displaces the block from its equilibrium position and releases it. Which of the following shows the mechanical energy E as a function of time t for the system that includes only the block and the system that includes the block and spring?

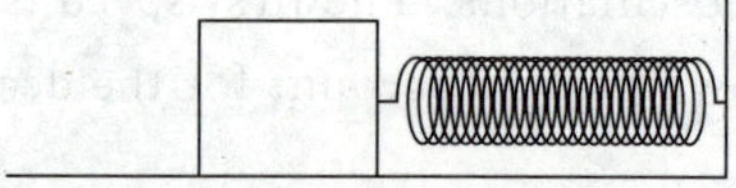

图 13

(A)

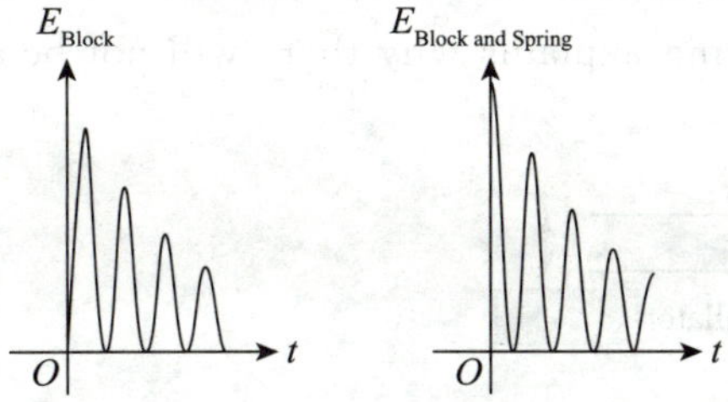

(B)

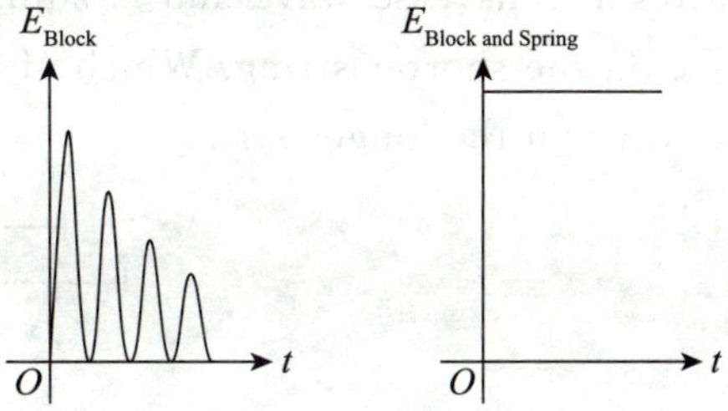

(C)

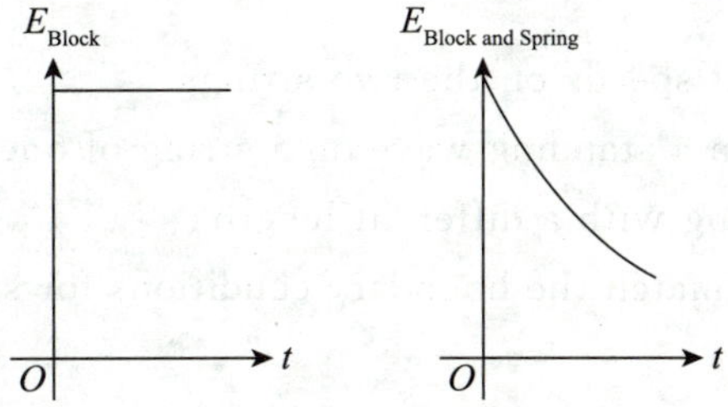

(D)

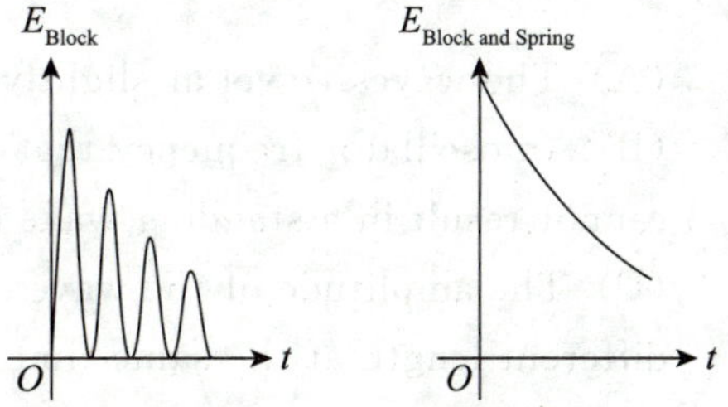

28. A transverse wave is traveling on a string. The graph（图 14）shows position as a function of time for a point on the string. If the frequency of the wave is doubled, what is the new average speed of the point?

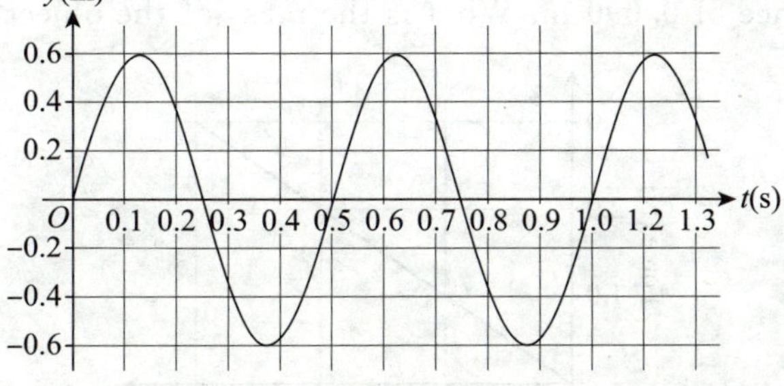

图 14

(A) 0 m/s　　(B) 2.4 m/s　　(C) 4.8 m/s　　(D) 9.6 m/s

29. Alan and Beverly are on opposite sides of a large room when Alan says something to Beverly. Beverly does not hear him, so Alan repeats the message louder and Beverly now hears it. Which of the following could be different about the second sound wave compared to the first that allows Beverly to hear it?

(A) The second wave travels to Beverly more quickly, so less energy is dissipated by the time the wave reaches her

(B) The second wave reflects more off the walls of the room

(C) The air molecules disturbed by the second wave undergo a greater displacement from their equilibrium positions

(D) The air molecules disturbed by the second wave are, on average, more closely spaced

30. Two strings differing only in length are attached to the same oscillator, as shown in figure 15. Both are fixed at the other end and are under the same tension. The oscillator

creates a transverse wave and is adjusted to the lowest frequency that creates a standing wave on the shorter string. Which of the following explains why there will not be a standing wave on the longer string?

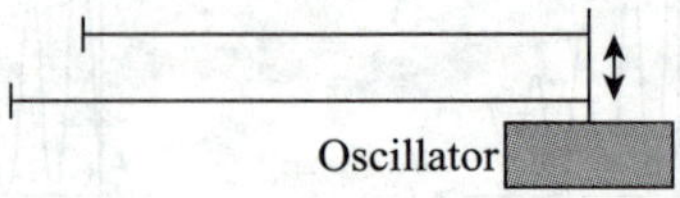

图 15

(A) The waves travel at slightly different speeds on the two strings

(B) An oscillator frequency that results in a standing wave on a string of one length cannot result in a standing wave on a string with a different length

(C) The amplitude of the wave does not match the boundary conditions for strings of different length at the same time

(D) The wavelength associated with the frequency does not match the boundary conditions set by the length of the longer string

31. Figure 16 shows the net force exerted on an object as a function of the position of the object. The object starts from rest at position $x=0$ m and acquires a speed of 3.0 m/s after traveling a distance of 0.090 m. What is the mass of the object?

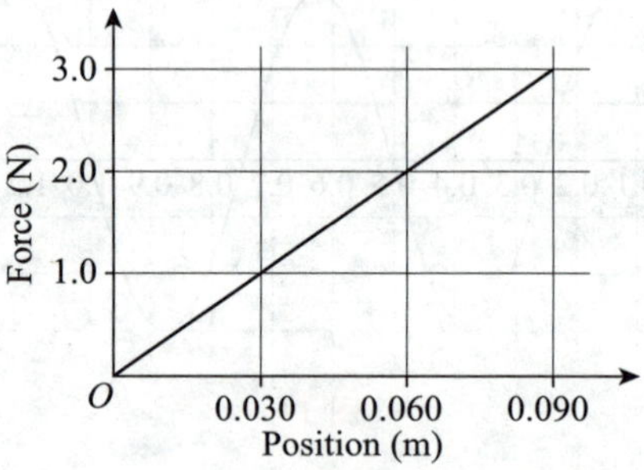

图 16

(A) 0.015 kg　　(B) 0.030 kg

(C) 0.045 kg　　(D) 0.060 kg

32. Two identical carts are free to move along a straight frictionless track. At time t_1, cart X is moving at 2.0 m/s when it collides with and sticks to cart Y, which is initially at rest. Which of the following graphs best shows the velocity of cart X before and after the collision?

(A)

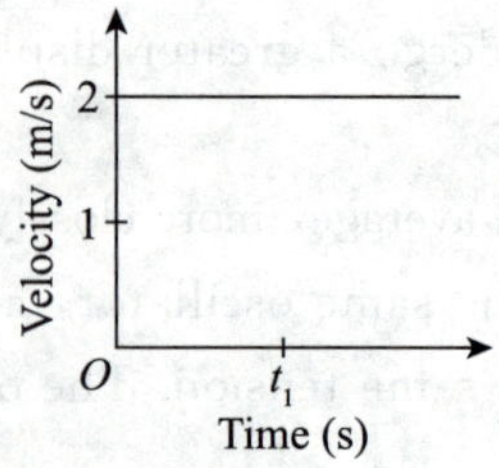

(B)

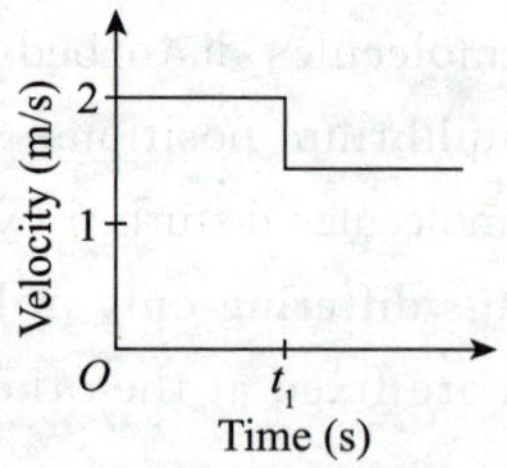

(C)

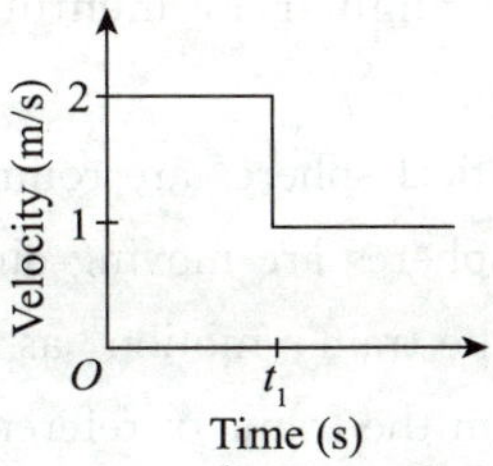

(D)

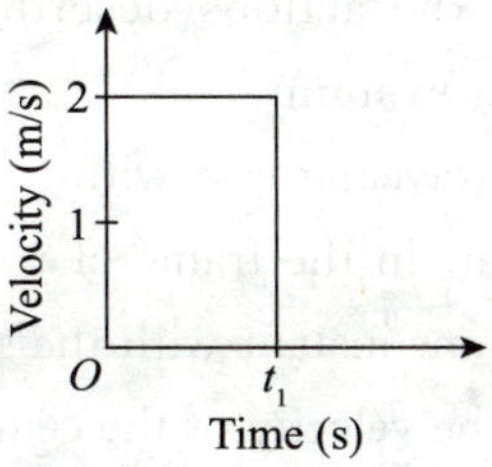

33. A spaceship and its shuttle pod are traveling to the right in a straight line with speed v, as shown in the top figure (图 17). The mass of the pod is m, and the mass of the spaceship is $6m$. The pod is launched, and afterward the pod is moving to the right with speed v_p and the spaceship is moving to the right with speed v_f, where $v_f > v$, as shown in the bottom figure. Which of the following is true of the speed v_c of the center of mass of the system after the pod is launched?

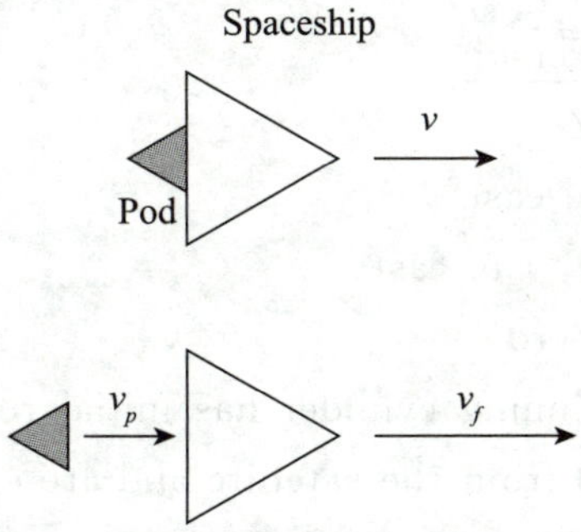

图 17

(A) $v_c = v_f$ (B) $v < v_c < v_f$ (C) $v_c < v$ (D) $v_c = v$

34. The diagram (图 18) shows a top view of a child of mass M on a circular platform of mass $5M$ that is rotating counterclockwise. Assume the platform rotates without friction. Which of the following describes an action by the child that will result in an increase in the total angular momentum of the child-platform system?

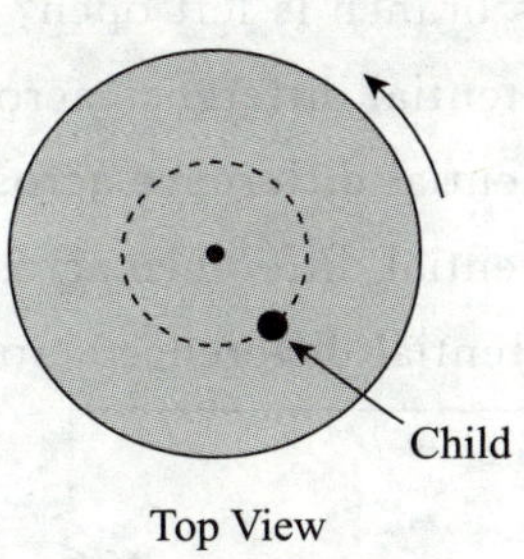

图 18

(A) The child moves toward the center of the platform

(B) The child moves away from the center of the platform

(C) The child moves along a circle concentric with the platform (dashed line shown) opposite the direction of the platform's rotation

(D) None of the actions described will change the total angular momentum of the child platform system

35. A train is traveling east with constant speed v_t. Two identical spheres are rolling on the floor of one train car. In the frame of reference of the train, the spheres are moving directly toward each other at one instant with the same speed v_p parallel to the train's motion, as shown in figure 19. What is the velocity of the center of mass of the spheres in the frame of reference of the train and in the frame of reference of a person standing at rest alongside the train?

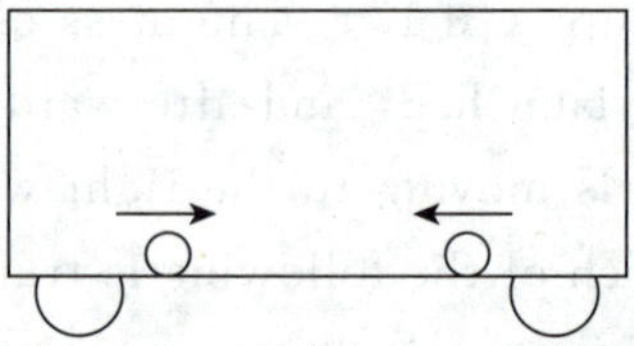

图 19

	Train	Person
(A)	Zero	Zero
(B)	Zero	v_t east
(C)	v_t east	v_p+v_t east
(D)	v_p+v_t east	Zero

36. A satellite that is a spinning cylinder has initial rotational inertia I_0 and angular velocity ω_0. Solar panels unfold from the satellite and are extended outward. The satellite then has rotational inertia $I_f=aI_0$ and angular velocity $\omega_f=b\omega_0$, where a and b are constants. Which of the following is true about the constants a and b ?

(A) $a=1$ and $b=1$
(B) $a>1$ and $b<1$
(C) $a>1$ and $b=1$
(D) $a<1$ and $b<1$

37. When the circuit shown in figure 20 is set up, the potential difference across the battery is 3.0 V. By how much will the magnitude of the potential difference across R_2 change when R_3 is removed and its branch is left open?

(A) The magnitude of the potential difference across R_2 does not change
(B) The magnitude of the potential difference across R_2 decreases by 0.5 V
(C) The magnitude of the potential difference across R_2 increases by 0.5 V
(D) The magnitude of the potential difference across R_2 increases by 1.0 V

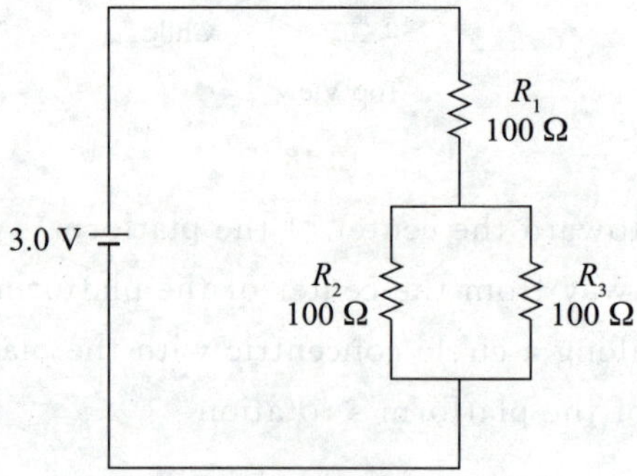

图 20

38. Two wave pulses are propagating along a straight line toward each other as shown in figure 21. Which of the following is the resultant when the centers of the pulses align?

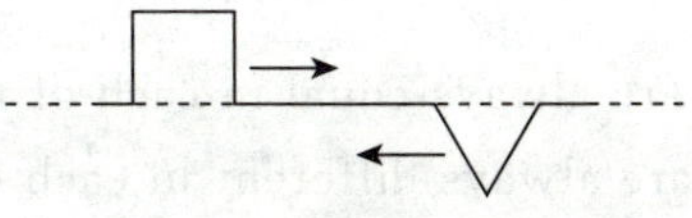

图 21

(A)
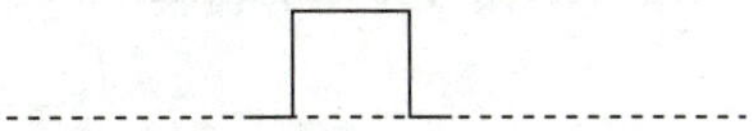

(B)
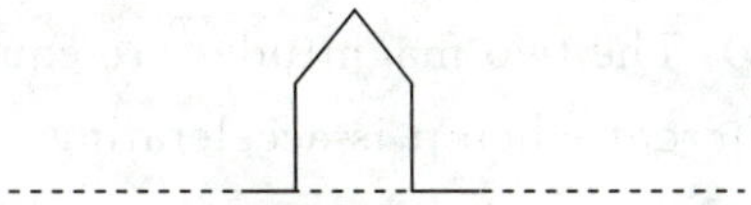

(C)

(D)

39. A person holds a 4.0 kg block at position A shown in figure 22, in contact with an uncompressed vertical spring with a spring constant of 500 N/m. The person gently lowers the block from rest at position A to rest at position B. Which of the following describes the change in the energy of the block-spring-Earth system as a result of the block being lowered?

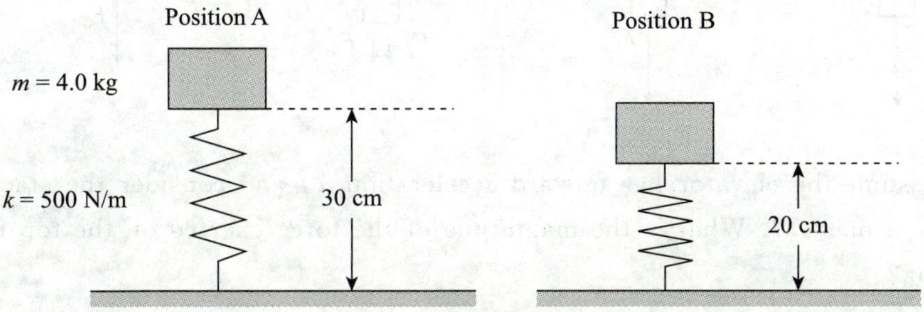

图 22

(A) The energy decreases by approximately 1.5 J
(B) The energy decreases by approximately 2.5 J
(C) The energy increases by approximately 4.0 J
(D) The energy of the system does not change

Questions 40—42 refer to the following material.

The stacks of boxes shown in figure 23 are inside an elevator that is moving upward. The masses of the boxes are given in terms of the mass M of the lightest box.

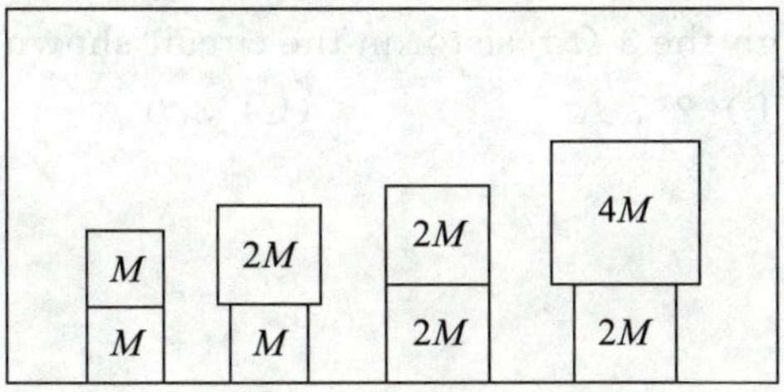

图 23

40. How does the magnitude of the force exerted by the top box on the bottom box compare with the magnitude of the force exerted by the bottom box on the top box for each of the stacks?

(A) The two magnitudes are always equal in each of the stacks

(B) The two magnitudes are always different in each of the stacks

(C) The two magnitudes are equal when the boxes have equal mass and different when the boxes have different masses

(D) The two magnitudes are equal when the elevator is moving at constant speed and different when it is accelerating

41. Assume the elevator is moving at constant speed, and consider the bottom box in the stack that has two boxes of mass $2M$. Let F_{floor} be the force exerted by the floor on the box, F_g be the force exerted by gravity on the box, and F_{box} be the force exerted by the top box on the bottom box.

Which of the following best represents the forces exerted on the bottom box?

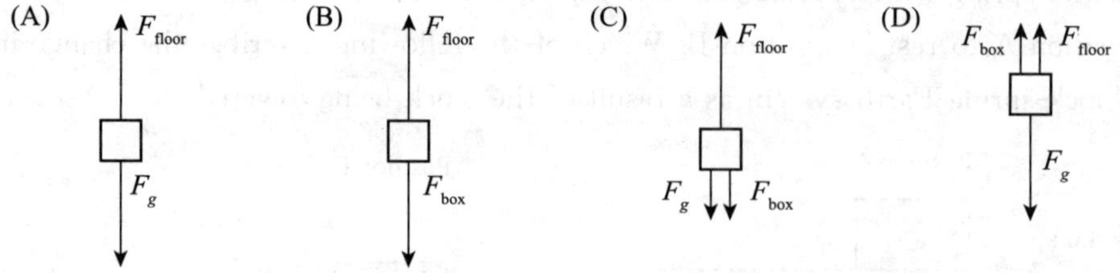

42. Assume the elevator has upward acceleration a, and consider the stack that has two boxes of mass M. What is the magnitude of the force exerted on the top box by the bottom box?

(A) Mg (B) Ma

(C) $M(a-g)$ (D) $M(a+g)$

43. A researcher is analyzing data from a high-energy particle collision. The result of the analysis gives a value of 8.8×10^{-19} C$\pm0.1\times10^{-19}$ C for the charge of one of the emitted particles. Should the researcher accept the value?

(A) Yes, it is a perfectly acceptable value

(B) No, because the value is much bigger than the elementary charge

(C) No, because the value is not an integer multiple of the elementary charge

(D) No, because the uncertainty is more than 1% of the value

44. The current through the 3 Ω resistor in the circuit shown in figure 24 is most nearly

(A) 1.3 A (B) 2.7 A (C) 3.0 A (D) 4.0 A

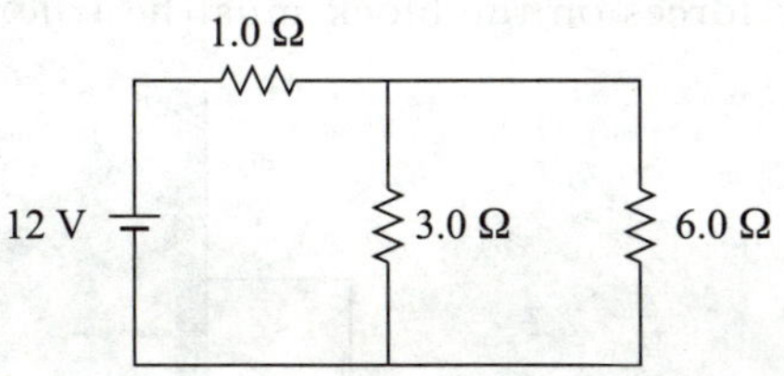

图 24

45. A student wants to determine the speed of sound at an elevation of one mile. To do this the student performs an experiment to determine the resonance frequencies of a tube that is closed at one end. The student takes measurements every day for a week and gets different results on different days. Which of the following experiments would help the student determine the reason for the different results?

(A) Repeating the experiment on several 10°C days and several 20°C days

(B) Repeating the experiment using a longer tube

(C) Repeating the experiment using a wider range of frequencies of sound

(D) Repeating the original experiment for an additional week

Directions: For each of the questions or incomplete statements below, two of the suggested answers will be correct. For each of these questions, you must select both correct choices to earn credit. No partial credit will be earned if only one correct choice is selected. Select the two that are best in each case and then enter both of the appropriate letters in the corresponding space on the answer sheet.

46. A uniform plank is placed with a pivot at its center. A block is placed on the plank to the left of the pivot, as shown in figure 25. A student is asked to place a second block of greater mass on the plank so it will balance when horizontal. Which of the following quantities are needed to determine where the second block should be placed? Select two answers.

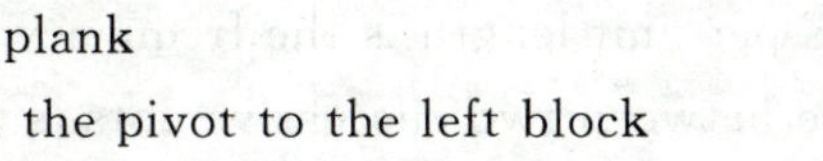
图 25

(A) The mass of the plank

(B) The mass of each block

(C) The length of the plank

(D) The distance from the pivot to the left block

47. A block is held at rest against a wall by a force of magnitude F exerted at an angle θ from the horizontal, as shown in figure 26. Let F_g be the gravitational force exerted by the Earth on the block, F_N be the normal force exerted by the wall on the block, and F_f be the frictional force exerted by the wall on the block. Which of the following statements

about the magnitudes of the forces on the block must be true? Select two answers.

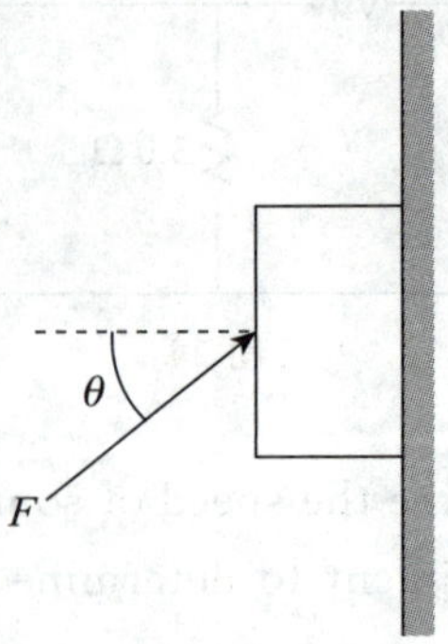

图 26

(A) $F = F_g \sin\theta$ (B) $F\cos\theta = F_N$

(C) $F\sin\theta = F_g \pm F_f$ (D) $F = F_g + F_N \pm F_f$

48. Some students have determined the gravitational mass of an object and want to compare it to the object's inertial mass. Procedures that would allow them to accomplish this include which of the following? Select two answers.

(A) Hanging the object vertically from a spring scale and recording the scale reading

(B) Placing the object on one side of a double pan balance, adding objects of known mass to the other side until the masses are balanced, and recording the amount of mass added

(C) Attaching the object to a spring of known spring constant, allowing it to oscillate horizontally on a nearly frictionless surface, and measuring the period

(D) Attaching the object to a force sensor, using the sensor to pull the object across a nearly frictionless horizontal surface, and measuring the acceleration

49. Figure 27 shows a representation of a wave traveling in a uniform medium at a particular instant.

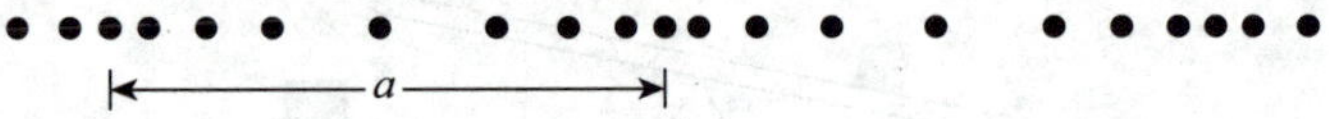

图 27

Correct statements about the wave include which of the following? Select two answers.

(A) It is a longitudinal wave

(B) Distance a is the wavelength

(C) The number of dots per unit length is the frequency

(D) The largest distance between two successive dots is the amplitude

50. A geologist is using a sensor to record waves in a layer of rock and is viewing them on a monitor screen. At one moment the monitor shows the signal in figure 28. If the signal is created by just two traveling waves, what can be concluded from the signal about the waves when they reach the sensor? Select two answers.

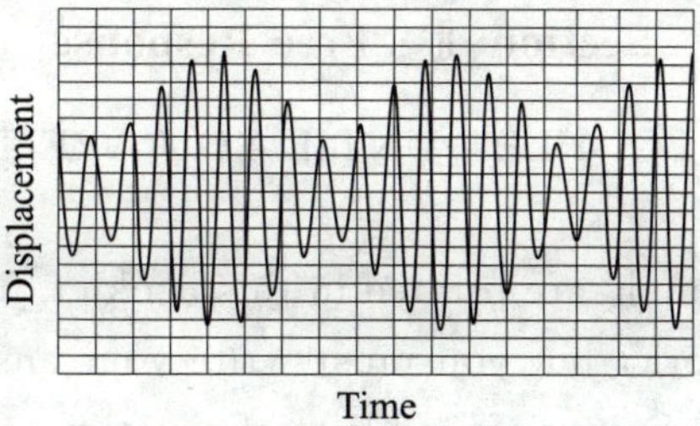

图 28

(A) The waves have different frequencies
(B) The waves have different amplitudes
(C) The waves are traveling in opposite directions
(D) The waves are traveling at different speeds

Section Ⅱ: Free Response

DO NOT OPEN THIS BOOKLET UNTIL YOU ARE TOLD TO DO SO.

Instructions

The questions for Section Ⅱ are printed in this booklet. You may use any blank space in the booklet for scratch work, but you must write your answers in the spaces provided for each answer. A table of information and lists of equations that may be helpful are in the booklet. Calculators, rulers, and straightedges may be used in this section.

All final numerical answers should include appropriate units. Credit for your work depends on demonstrating that you know which physical principles would be appropriate to apply in a particular situation. Therefore, you should show your work for each part in the space provided after that part. If you need more space, be sure to clearly indicate where you continue your work. Credit will be awarded only for work that is clearly designated as the solution to a specific part of a question. Credit also depends on the quality of your solutions and explanations, so you should show your work.

Write clearly and legibly. Cross out any errors you make; erased or crossed-out work will not be scored. You may lose credit for incorrect work that is not crossed out.

Manage your time carefully. You may proceed freely from one question to the next.

You may review your responses if you finish before the end of the exam is announced.

Total Time

90 minutes

Number of Questions

5

Percent of Total Score

50%

Writing Instrument

Either pencil or pen with

black or dark blue ink

Electronic Device

Calculator allowed

Suggested Time

Approximately 25 minutes each for questions 2 and 3 and 13 minutes each for questions 1, 4, and 5

Weight

Approximate weights:

Questions 2 and 3:

26% each

Questions 1, 4, and 5:

16% each

AP Physics Practice 1 (1)
Section Ⅱ
5 Questions
Time—90 minutes

Directions: Questions 2 and 3 are long free-response questions that require about 25 minutes each to answer and are worth 12 points each. Questions 1, 4, and 5 are short free-response questions that require about 13 minutes each to answer and are worth 7 points each. Show your work for each part in the space provided after that part.

1. (7 points, suggested time about 13 minutes)

Figure 29 represents two carts, with magnets attached, that make up a system. The mass of one cart and magnet is 1 kg and the mass of the other is 5 kg. The carts are initially at rest on a frictionless track. They are released from rest and exert a repulsive force on each other. The track is not quite horizontal, with the right side slightly lower than the left side.

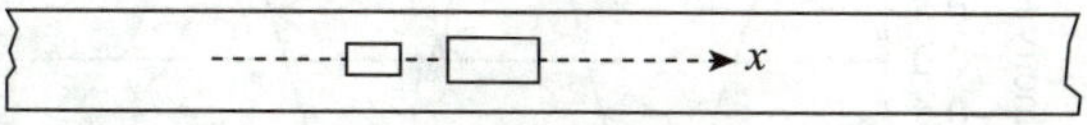

图 29

The speeds of the carts are measured over a 10 s interval. Graph 30 shows the momentum of the two carts as a function of time for this interval as they move along the x-axis.

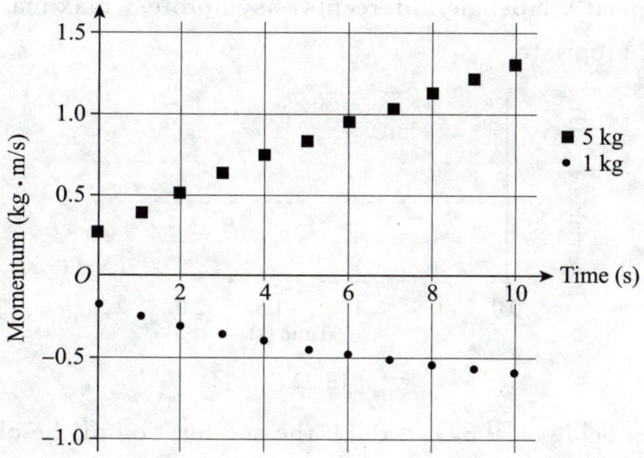

图 30

(a) Based on the graph, were the measurements started at the instant the carts were released?

Justify your answer.

(b) Calculate the magnitude of the external force exerted on the system.

(c) Suppose the experiment is repeated with different carts, so that the masses of cart plus magnet are 2 kg and 4 kg. Would your answer to part (b) be different with the new masses? Justify your answer.

2. (12 points, suggested about time 25 minutes)

A student hangs a 0.125 kg object on a spring, sets it into oscillation, and obtains the data for the position and velocity of the object as a function of time shown in figure 31.

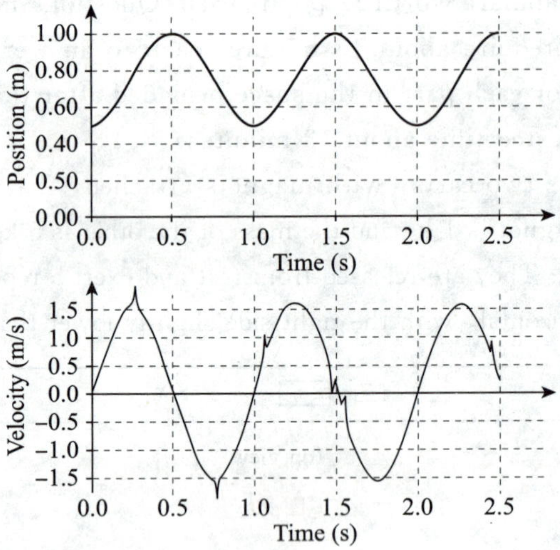

图 31

(a) On grid 32, sketch the potential energy of the object-spring-Earth system as a function of time. Explicitly label any intercepts, asymptotes, maxima, or minima with numerical values, as appropriate.

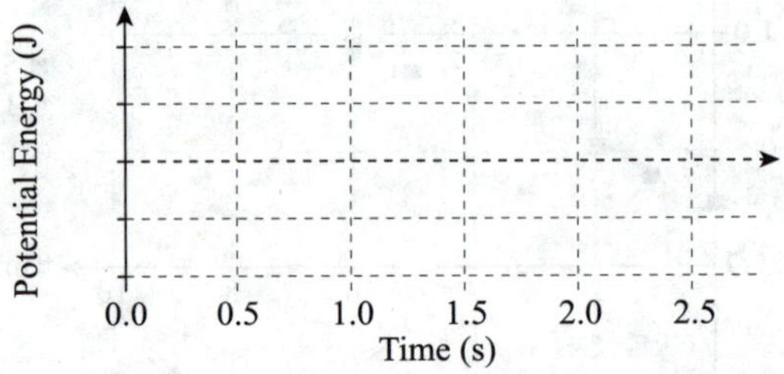

图 32

(b) On the axes below (图 33), sketch the acceleration of the object as a function of time. Explicitly label any intercepts, asymptotes, maxima, or minima with numerical values, as appropriate.

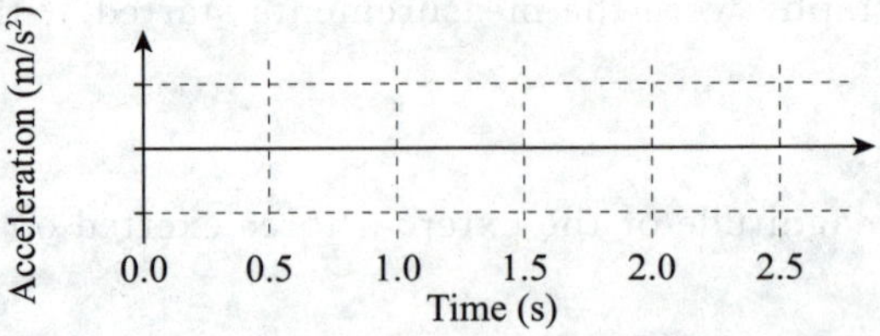

图 33

Next the student is given a rubber band and asked to determine whether the relationship between the restoring force exerted by the rubber band and the amount it is stretched is the same as that of an ideal spring.

(c) Describe an experimental procedure that the student could use to collect the necessary data, including all the equipment the student would need.

(d) How should the student analyze the data to determine whether the relationship between the restoring force exerted by the rubber band and the amount it is stretched is the same as that of an ideal spring? What evidence from the analysis would be used to make the determination?

3. (12 points, suggested time about 25 minutes)

In a physics lab, students will compare the two circuits shown in figure 34. In one circuit, lightbulbs A and B will be connected to a battery in series. In the other circuit, lightbulbs C and D will be connected to an identical battery in parallel. All four lightbulbs are identical.

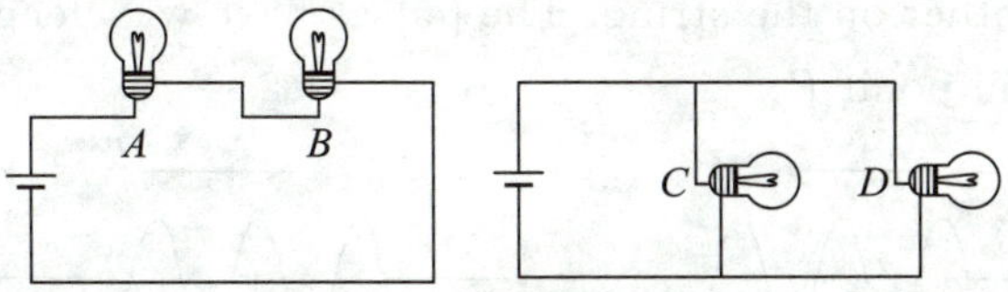

图 34

Student 1 predicts that the bulbs will be brighter in the series arrangement, while Student 2 disagrees and predicts that the bulbs will be brighter in the parallel arrangement. Their arguments are as follows.

Student 1: "Bulbs A and B will be brighter than C and D because in the series circuit, each bulb gets all the current coming out of the battery. But in the parallel circuit each bulb gets only half the current."

Student 2: "But you're not taking into account that bulb A uses up some of the current before it reaches bulb B, making the bulbs dimmer in the series circuit. And in the parallel circuit, the full voltage of the battery is across each bulb, so it's as if the other bulb weren't even there. So bulbs C and D will be brighter."

(a) For part (a), ignore whether the students' predictions are correct or incorrect. Do not simply repeat the students' arguments as your answers.

i. Which aspects of Student 1's reasoning, if any, are correct? Explain your answer.

ii. Which aspects of Student 1's reasoning, if any, are incorrect? Explain your answer.

iii. Which aspects of Student 2's reasoning, if any, are correct? Explain your answer.

iv. Which aspects of Student 2's reasoning, if any, are incorrect? Explain your answer.

Together the students write the following equations to describe the two circuits.

Series Arrangement

Equation 1: $I_S = I_A = I_B$

Equation 2: $\Delta V - I_A R_A - I_B R_B = 0$

Parallel Arrangement

Equation 3: $I_P = I_C + I_D$

Equation 4: $\Delta V - I_C R_C = \Delta V - I_D R_D = 0$

(b)

i. Indicate all of the equations above that support Student 1's reasoning, and explain how each equation provides support.

ii. Indicate all of the equations above that support Student 2's reasoning, and explain how each equation provides support.

(c) Resolve the two lines of reasoning about the two circuits to conclude which arrangement will have brighter bulbs. Use the equations above and any other equations that you need, and explain how the equations support your conclusion.

4. (7 points, suggested time about 13 minutes)

A string is held taut, with each of its ends fixed to an oscillator that creates wave pulses on the string. Figure 35 shows identical pulses at a particular instant when they are traveling toward each other on the string. The pulses have wavelength l and amplitude A, and are equidistant from point P.

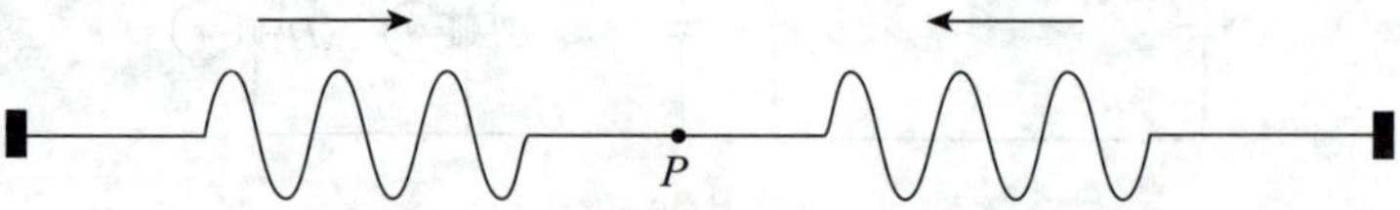

图 35

(a) On figure 36, the dashed line at zero displacement represents the string with no pulses. On the figure, draw the shape of the string at the instant when the pulses overlap by 3 of the length of each pulse.

图 36

(b) On the axes in figure 37, sketch a graph of displacement as a function of time for point P, from time $t=0$ when the pulses just begin to overlap until $t=t_1$ when they once again do not overlap.

图 37

(c) Next the oscillators are adjusted to create continuous waves of wavelength l, instead of the wave pulses shown. A standing wave is created on the string. What can be inferred about the length of the string? Justify your answer. Equations may be part of your reasoning, but equations alone are not sufficient justification.

5. (7 points, suggested time about 13 minutes)

Figure 38 shows part of a system consisting of a block at the top of an inclined plane that rests on a table, which is located on the Earth. The block and plane are at rest when the block is released. In trial 1 there is no friction between the block and the plane or between the plane and the table. In trial 2 the plane is fixed to the table so it cannot move, but there is still no friction between the block and the plane.

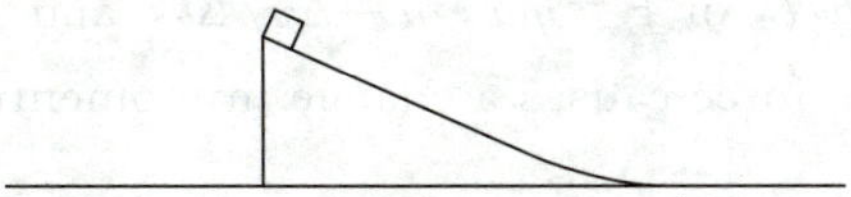

图 38

Indicate whether the speed of the block relative to the table when the block reaches the bottom of the plane is greater in trial 1 or trial 2. Justify your answer in a clear, coherent, paragraph-length explanation.

STOP

END OF EXAM

IF YOU FINISH BEFORE TIME IS CALLED,

YOU MAY CHECK YOUR WORK ON THIS SECTION.

Answers for AP Physics 1 Practice (1)

1—D	14—D	27—D	40—A
2—A	15—C	28—D	41—C
3—D	16—D	29—C	42—D
4—B	17—A	30—D	43—C
5—B	18—B	31—B	44—B
6—C	19—A	32—C	45—A
7—C	20—D	33—D	46—B, D
8—C	21—C	34—D	47—B, C
9—A	22—C	35—B	48—C, D
10—D	23—B	36—B	49—A, B
11—B	24—D	37—C	50—A, B
12—C	25—B	38—C	
13—C	26—B	39—A	

Question 1

7 points total **Distribution of points**

(a) 1 point

For a correct, justified response about the state of motion at time $t=0$. 1 point

For example:

No. The carts have a non-zero momentum at $t=0$, so this cannot be the instant they were released.

(b) 4 points

For using the relationship between force and momentum, or Newton's second law and the definitions of acceleration and momentum 1 point

$$F=\Delta p/t\text{, or }F=ma\text{, }a=\Delta v/\Delta t\text{, and }p=mv$$

For recognizing that the force causes a change in momentum or a change in velocity

1 point

For using reasonable values of momenta from the graph to calculate the force of the carts 1 point

$$F_5=\Delta p/\Delta t=(1.25\text{ kg}\cdot\text{m/s}-0.25\text{ kg}\cdot\text{m/s})/10\text{ s}=0.10\text{ N}$$

$$F_1=\Delta p/\Delta t=(-0.6\text{ kg}\cdot\text{m/s}+0.2\text{ kg}\cdot\text{m/s})/10\text{ s}=-0.04\text{ N}$$

For recognizing that the external force is the sum of the two forces on the cart

1 point

$$F_{ext}=F_5+F_1=0.10\text{ N}-0.04\text{ N}=0.06\text{ N}$$

Calculating only the force on one of the carts earns a maximum of 1 point.

(c) 2 points 2 points

For a correct, justified response about the effect of changing the masses of the carts

For example:

No. The calculations in part (b) account for the effect of the external force on the motion of the entire system. The changes in individual masses do not change the total mass of the system, so the net effect on the system remains the same.

In the absence of other correct reasoning, recognizing that the mass is the same for both situations earned 1 point.

Question 2

12 points total **Distribution of points**

(a) 3 points

Zero potential energy can be defined at any point. For convenience define it to be at the equilibrium point of the hanging object. The graph above uses this definition.

$$\text{Then }E=U+K=\frac{1}{2}mv_{max}^2\text{, so}$$

$$U=\frac{1}{2}mv_{max}^2-K=\frac{1}{2}mv_{max}^2-\frac{1}{2}m[v(t)]^2.$$

The amplitude of the velocity graph is approximately 1.6 m/s, and the period is 1 s.

So $v(t) = (1.6\ \text{m/s})\ \sin(2\pi t/1\ \text{s})$.

$$U = \frac{1}{2}(0.125\ \text{kg})(1.6\ \text{m/s})^2[1-\sin^2(2\pi t/1\ \text{s})]$$
$$=(0.16\ \text{J})[1-\sin^2(2\pi t/1\ \text{s})]$$

For a graph with a reasonable sine squared shape（图 39） 1 point

For having the maxima of the graph at the labeled times 1 point

For having maxima and minima with a potential energy difference of about 0.16 J 1 point

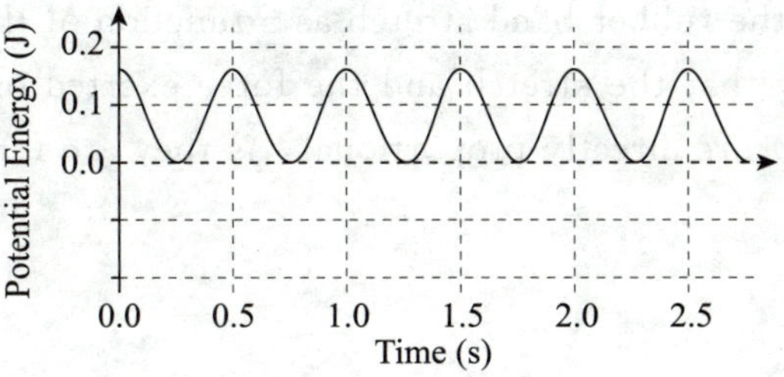

图 39

(b) 3 points

For a graph that is either a sine or cosine curve（图 40） 1 point

For a graph that has the appropriate phase relationship to the given position and velocity graphs 1 point

$$a_{max} = F_{max}/m = kx_{max}/m$$

The spring constant k can be determined using the period.

$$T = 2\pi\sqrt{m/k}\text{, so } k = 4\pi^2 m/T^2 = 4\pi^2(0.125\ \text{kg})/(1\ \text{s})^2 = 0.5\pi^2\ \text{kg/s}^2$$
$$a = (0.5\pi^2\ \text{kg/s}^2)(0.25\ \text{m})/0.125\ \text{kg} = 9.87\ \text{m/s}^2$$

For an amplitude of approximately 10 m/s^2 1 point

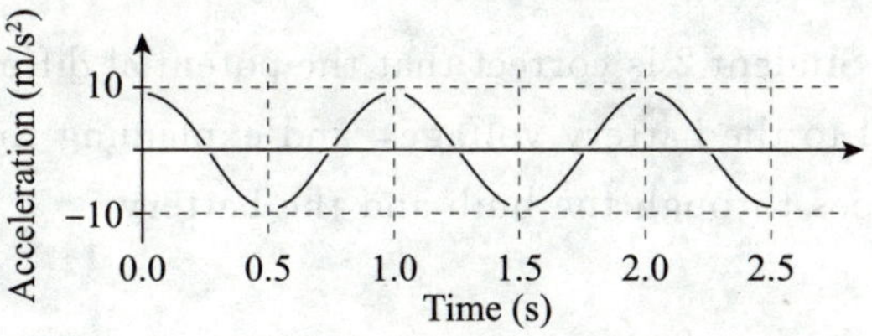

图 40

(c) 3 points

For a reasonable experimental setup that would allow useful measurements 1 point

For indicating all measurements needed and no extraneous ones 1 point

For evidence that multiple trials are performed

For example: Hook the rubber band over a horizontal rod, and measure its length. Hang an object of known mass from it and measure the new length. Repeat for a number of different objects. 1 point

(d) 3 points

For an appropriate analysis that would check for a linear relationship between force and rubber band stretch 1 point

For indicating an appropriate characteristic from the analysis that signifies a linear relationship 1 point

For indicating that a linear graph means that the rubber band exhibits $F \propto x$ like an ideal spring 1 point

For example: Plot the rubber band stretch as a function of the mass of the objects. A linear graph would show that the stretch and the force exerted on the rubber band (i. e. the weight of the object) are directly proportional, as they are for a spring.

Question 3

12 points total **Distribution of points**

(a) 5 points

(i)

For indicating that Student 1 is correct that bulbs A and B get all the current coming out of the battery, and explaining that it is because there is only one loop in the circuit 1 point

For indicating that bulbs C and D each get half the current, and explaining that it is because of the junction rule 1 point

(ii)

For indicating that while bulbs C and D do get half the current in their circuit, Student 1 is incorrect in assuming that this means they must have less current than A and B, and explaining that it is because the total current is not the same in both circuits 1 point

(iii)

For indicating that Student 2 is correct that the potential difference across bulbs C and D is the same and equal to the battery voltage, and explaining that for each bulb you can trace a loop that only goes through the bulb and the battery 1 point

(iv)

For indicating that Student 2 is incorrect in saying that current is "used up" by bulb A before it reaches bulb B, and explaining that a current moves an equal amount of charge through all the circuit elements carrying it 1 point

(b) 3 points

(i)

Equations 1 and 3 support Student 1's reasoning.

For indicating that equation 1 shows that series bulbs A and B each carry the total current in the circuit 1 point

For indicating that equation 3 shows that parallel bulbs C and D share the total current in the circuit 1 point

(ii)

Equation 4 directly supports Student 2's reasoning. Equation 2 can also be considered as providing support for the implied contrast between circuits.

For indicating that equation 4 shows that the potential difference across bulbs C and D is the same 1 point

No penalty for also indicating that equation 2 shows that the voltage is shared in the series circuit.

(c) 4 points

For indicating a valid quantity that can be used to compare bulb brightness across circuits (power, current, or voltage) 1 point

For correctly manipulating equations in an attempt to determine some quantity for the circuits with a valid assumption 1 point

For stating a conclusion that follows from the work for the two points above (even if that work is incorrect) 1 point

For explaining how the calculations support the conclusion (connecting ideas by using the chosen valid quantity to determine how it affects power which determines brightness) 1 point

For example, calculating currents and comparing power:

Substituting I_S into equation 2 and solving gives $I_S = \Delta V/(R_A + R_B)$. Solving each part of equation 4 gives $I_C = \Delta V/R_C$ and $I_D = \Delta V/R_D$

Since the resistors are identical, $I_S = I_C/2 = I_D/2$

Power is given by $P = I^2R$, and for identical resistors depends on the current. Therefore the power and thus the brightness of bulbs A and B is less.

Question 4

7 points total **Distribution of points**

(a) 3 points

For overlap of one wavelength showing something other than amplitude A 1 point

For showing amplitude $2A$ in the overlapping segment 1 point

For showing two wavelengths of amplitude A on both sides (图 41) 1 point

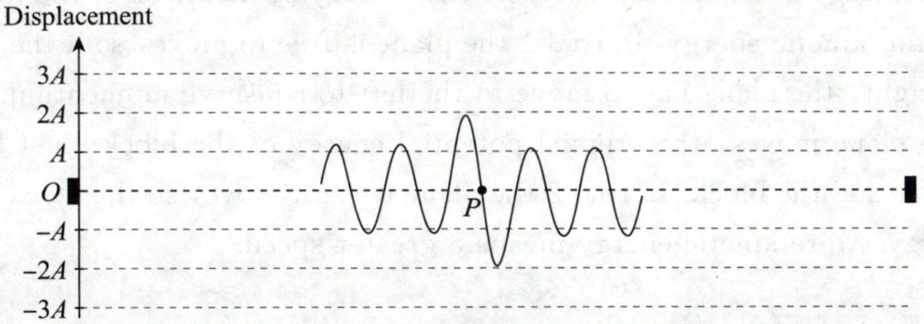

图 41

(b) 2 points

For a straight line at zero displacement (图 42) 2 points

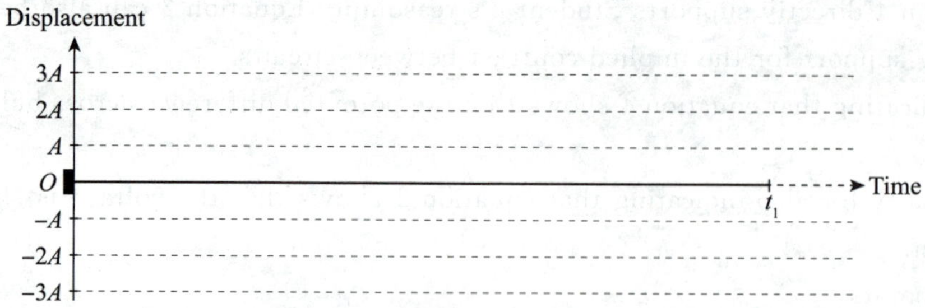

图 42

(c) 2 points

For a complete, correct response with justification 2 points

For example: The length is either a multiple of λ or $\frac{\lambda}{2}$. The ends are nodes, so the pattern has zero amplitude at the ends. That can only happen if the "middle" or "end" of a wavelength is at each end.

Question 5

7 points total **Distribution of points**

For any mention of energy conversion 1 point

For indicating that the same amount of potential energy is converted to kinetic energy in each trial 1 point

For indicating that the kinetic energy is shared between the block and plane in trial 1 1 point

For indicating that the block has all the kinetic energy in trial 2 1 point

For a description of the motion of the plane in each trial 1 point

For using momentum or forces to explain the motion in each trial 1 point

For a coherent argument that leads to a correct conclusion 1 point

For example:

The speed of the block is greater in trial 2. Since gravity is the only force doing work in both trials, energy is conserved. The potential energy of the block at the top of the plane is converted into kinetic energy. In trial 1 the plane is free to move, so if the block ends up going to the right, the plane has to move to the left to conserve momentum.

Since the plane moves, the original potential energy of the block must be divided between the plane and the block. In trial 2 the plane doesn't move, so the block gets all of the available energy. More kinetic energy means a greater speed.

AP Physics 1 Practice Exam (2)

Section Ⅰ: Multiple Choice

DO NOT OPEN THIS BOOKLET UNTIL YOU ARE TOLD TO DO SO.

Instructions

Section Ⅰ of this exam contains 50 multiple-choice questions. Pages containing equations and other information are also printed in this booklet. Calculators, rulers, and straightedges may be used in this section.

Indicate all of your answers to the multiple-choice questions on the answer sheet.

No credit will be given for anything written in this exam booklet, but you may use the booklet for notes or scratch work.

For questions 1 through 45, select the single best answer choice for each question.

After you have decided which of the choices is best, fill in the appropriate letter in the corresponding space on the answer sheet.

For questions 46 through 50, select the two best answer choices for each question.

After you have decided which two of the choices are best, enter both letters in the corresponding space on the answer sheet.

Use your time effectively, working as quickly as you can without losing accuracy. Do not spend too much time on any one question. Go on to other questions and come back to the ones you have not answered if you have time. It is not expected that everyone will know the answers to all of the multiple-choice questions.

Your total score on Section Ⅰ is based only on the number of questions answered correctly.

Points are not deducted for incorrect answers or unanswered questions.

Total Time

90 minutes

Number of Questions

50

Percent of Total Score

50%

Writing Instrument

Pencil required

Electronic Device

Calculator allowed

AP Physics 1 Practice (2)

Section Ⅰ

50 Questions

Time—90 minutes

Note: To simplify calculations, you may use $g=10\ m/s^2$ in all problems.

Direction: Each of the questions or imcomplete statements below is followed by 4 suggested answers or completions. Select the one that is the best in each case and then enter the appropriate letter in the corresponding space on the answer sheet.

1. A person throws a marble straight up into the air, releasing it a short height above the ground and catching it at that same height. If the air resistance is negligible, which of the following graphs of position versus time t is correct for the motion of the marble as it goes up and then comes down?

(A)

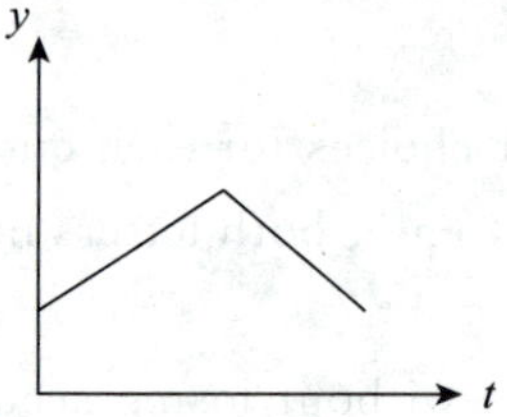

(B)

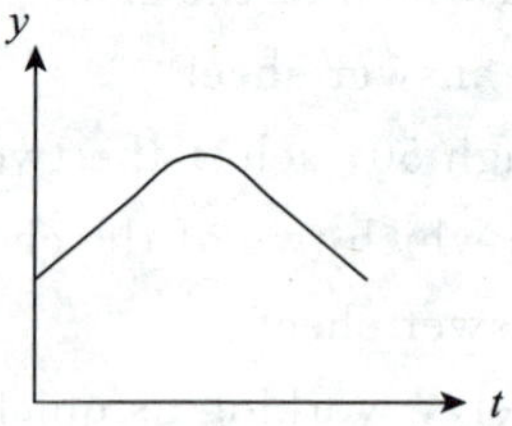

(C)

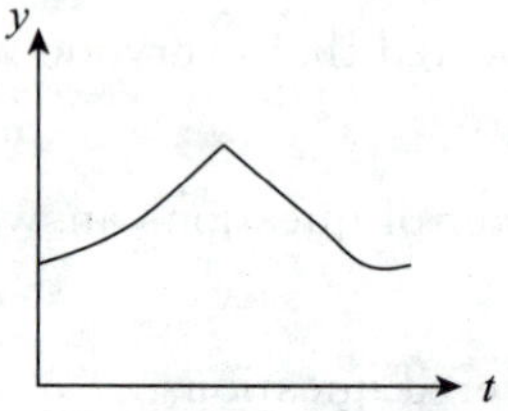

(D)

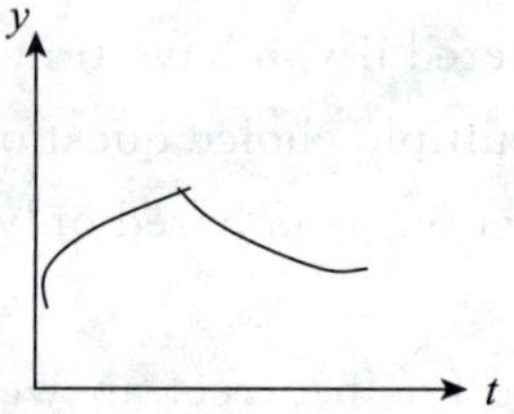

2. A rocket leaves a playground from the surface of the Earth. The radius of the Earth is R, and the weight of the payload on the surface of the Earth is W. The force of the Earth's gravity on the payload is $0.5\ W$ when the rocket's distance from the center of the Earth is

(A) R (B) $\sqrt{2}R$ (C) $2R$ (D) $2\sqrt{2}R$

Questions 3—4

An object is thrown with an initial speed v near the surface of the Earth. Assume that air resistance is negligible and the gravitational field is constant.

3. If the object is thrown vertically upward, the direction and magnitude of its acceleration while it is in the air is

(A) upward and decreasing (B) upward and constant

(C) downward and constant (D) downward and decreasing

4. If the object is thrown horizontally, the direction and magnitude of its acceleration while it is in the air is

(A) upward and decreasing (B) upward and constant

(C) downward and decreasing (D) downward and constant

5. An object weighing 120 N is set on a rigid beam of negligible mass at the distance of 3 m from a pivot, as shown in figure 1. A vertical force is to be applied to the other end of the beam at the distance of 4 m from the pivot to keep the beam at rest and horizontal. What is the magnitude F of the force required?

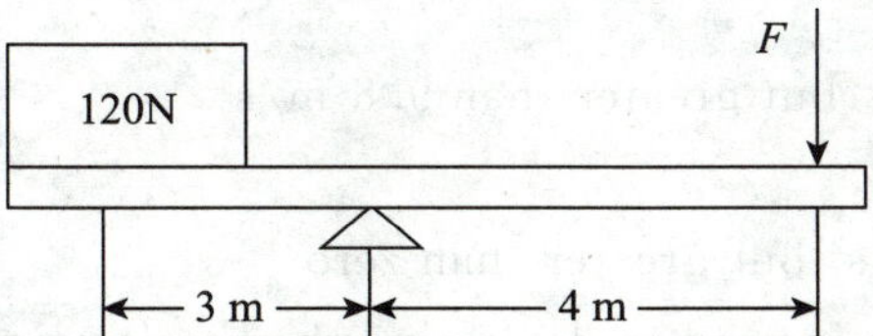

图 1

(A) 10 N (B) 30 N (C) 90 N (D) 120 N

6. Inertia can be best described as the

(A) force that keeps an object in motion with constant velocity

(B) force that keeps an object at rest

(C) force that overcomes friction

(D) property responsible for an object's resistance to changes in motion

Questions 7—9

A block released from rest at position A slides with negligible friction down an inclined track, around a vertical loop, and then along a horizontal track as shown in figure 2. The block never leaves the track.

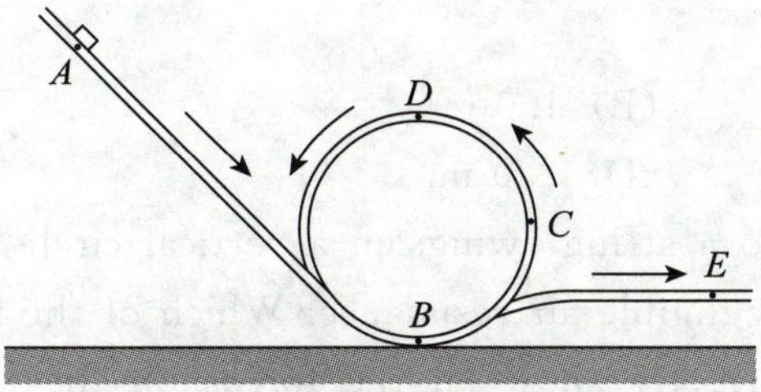

图 2

7. After the block is released, in which of the following sequences of positions is the speed of the block ordered from fastest to slowest?

(A) $B\ C\ D\ E$ (B) $B\ E\ C\ D$

(C) $D\ C\ E\ B$ (D) $E\ B\ C\ D$

8. Which of the following is true of the net force on the block when it is at position C?

(A) It is directed vertically downward only

(B) It is directed vertically upward only

(C) It is directed to the left only

(D) It has components both to the left and vertically downward

9. The gravitational potential energy and kinetic energy of the block are most nearly equal at which position ? (Consider the potential energy to be zero at position B)

(A) A (B) B (C) C (D) D

10. An artificial satellite orbits the Earth at a speed of 7 800 m/s and a height of 200 km above the Earth's surface. The satellite experiences an acceleration due to gravity

(A) 39 m/s^2

(B) less than 39 m/s^2 but greater than 9. 8 m/s^2

(C) 9. 8 m/s^2

(D) less than 9. 8 m/s^2 but greater than zero

11. A simple pendulum is used to determine the acceleration due to gravity at the surface of a planet. The pendulum has a length of 2 m and its period is measured to be 2 s. The value of g obtained in this investigation is most nearly

(A) 1 m/s^2 (B) 20m/s^2 (C) 5 m/s^2 (D) 10 m/s^2

12. Two blocks of masses 1. 0 kg and 2. 0 kg, respectively, are pushed by a constant applied force F across a horizontal frictionless table with constant acceleration such that the blocks remain in contact with each other, as shown in figure 3. The 1. 0 kg block pushes the 2. 0 kg block with a force of 2. 0 N. The acceleration of the two blocks is

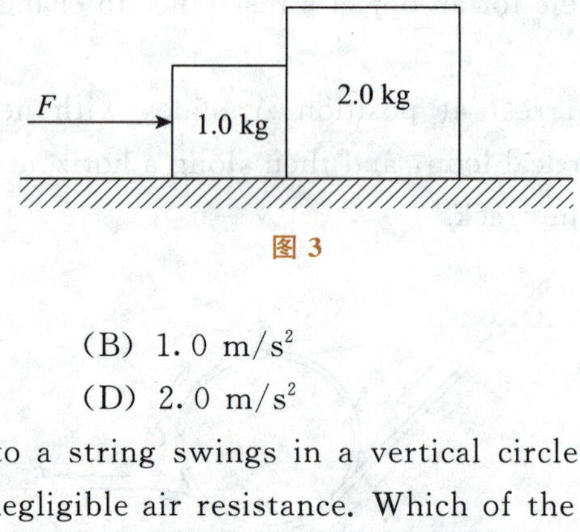

图 3

(A) 0 (B) 1. 0 m/s^2

(C) 1. 5 m/s^2 (D) 2. 0 m/s^2

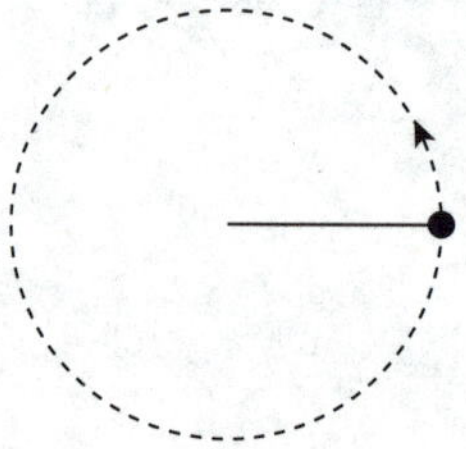

图 4

13. A rock attached to a string swings in a vertical circle, as shown in figure 4, with negligible air resistance. Which of the following diagrams could correctly show all the forces on the rock when the string is in the position above?

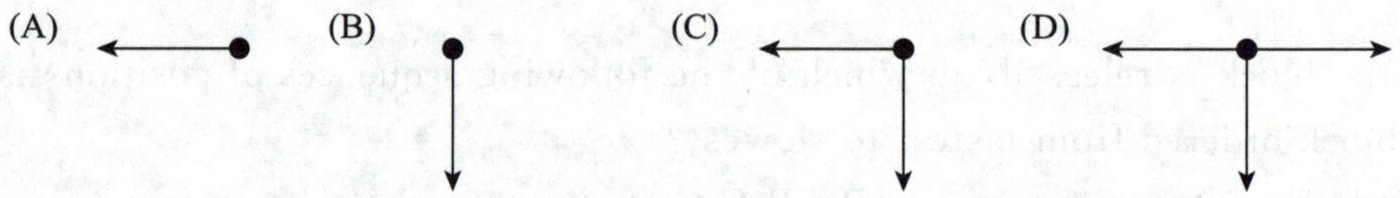

14. A conducting spherical shell of radius R has charge q uniformly distributed on its outer surface. Graph 5 represents electric field strength E versus distance r from the center of the spherical shell. Which of following graphs best represents E versus r when the charge on the spherical shell is doubled to $2q$?

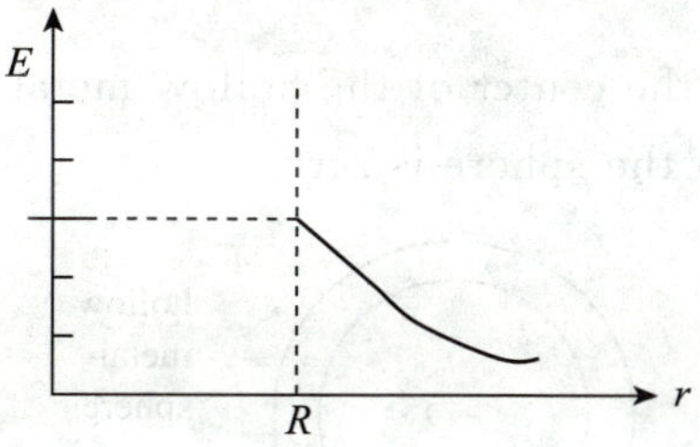

图 5

(A)

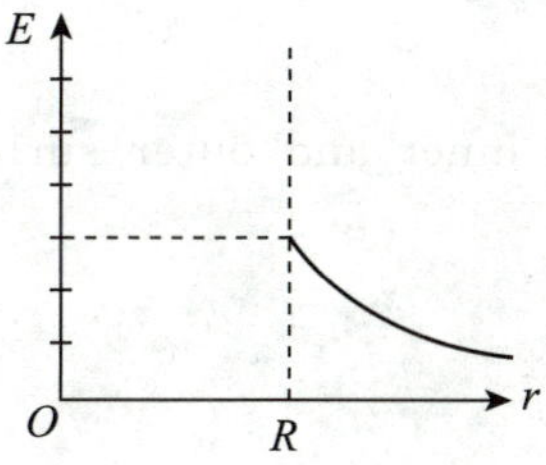

(B)

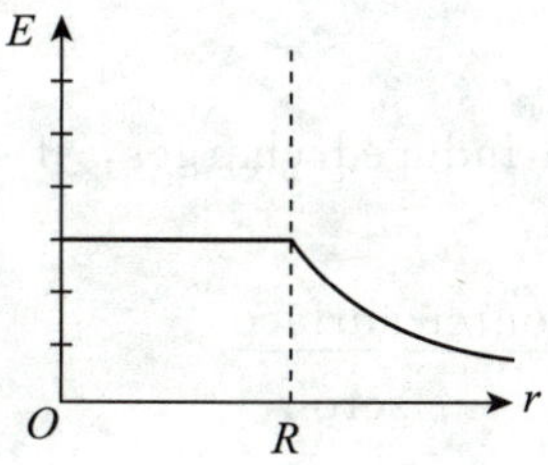

(C)

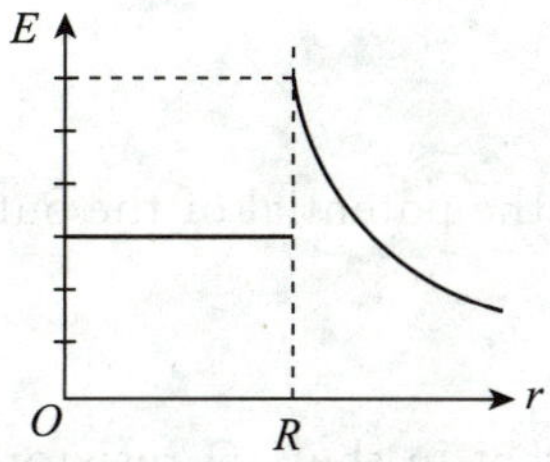

(D)

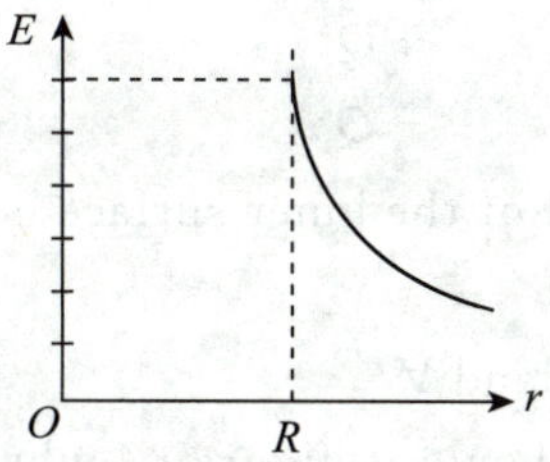

15. Which of the following is constant for an object in uniform circular motion?

(A) kinetic energy (B) velocity

(C) acceleration (D) displacement

16. Two point charges of unknown magnitudes q_1 and q_2 are placed as shown in figure 6. The electric field intensity is zero at point P, which is closer to q_1 than to q_2. Which of the following is true of the signs of the charges and their magnitudes?

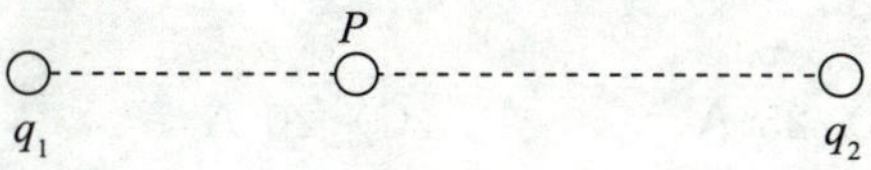

图 6

	signs	magnitudes
(A)	opposite	equal
(B)	opposite	$q_1 > q_2$
(C)	same	$q_1 > q_2$
(D)	same	$q_1 < q_2$

17. A 60 W, 120 V lightbulb has a resistance of

(A) 0.5 Ω (B) 2.0 Ω (C) 120 Ω (D) 240 Ω

Questions 18—19

Charge $+Q$ is located at the center of the hollow metal sphere shown in cross section in figure 7. The net charge on the sphere is zero.

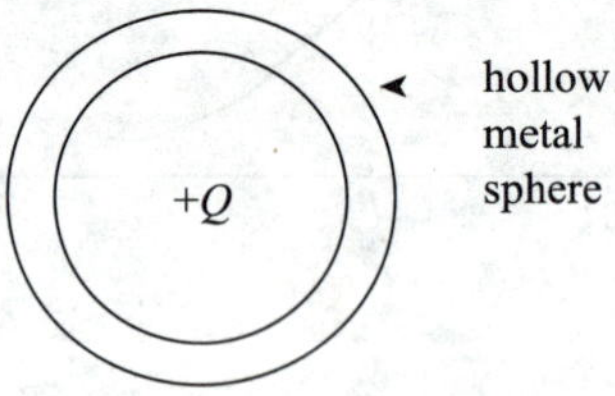

图 7

18. What are the induced charges, if any, on the inner and outer surfaces of the sphere?

inner surface	outer surface
(A) zero	zero
(B) $-Q$	zero
(C) $-Q$	$+Q$
(D) $+Q$	$-Q$

19. If the potential of the inner surface is V, what is the potential of the outer surface?

(A) less than zero (B) zero

(C) between zero and V (D) V

20. In the circuit shown in figure 8, what is the current in the 5 Ω resistor?

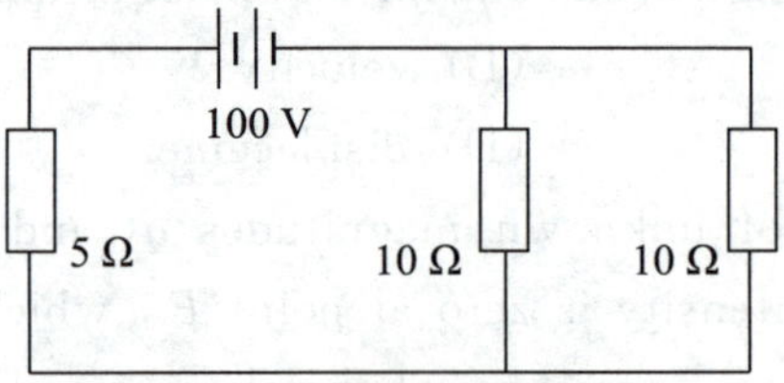

图 8

(A) 40 A (B) 25 A (C) 20 A (D) 10 A

21. Ball 1 and 2 are each thrown horizontally from the same height above the level ground, but ball 2 has greater initial velocity after leaving thrower's hand. If air resistance is negligible, how do the accelerations of the balls and the times it takes them to hit the ground compare?

acceleration a	the time to hit ground
(A) greater for ball 2	greater for ball 2
(B) greater for ball 2	equal
(C) equal	equal
(D) equal	less for ball 2

22. A block of mass 2 kg is suspended from a rope, as shown in figure 9. If the tension

in the rope is 10 N, the acceleration of the block is most nearly

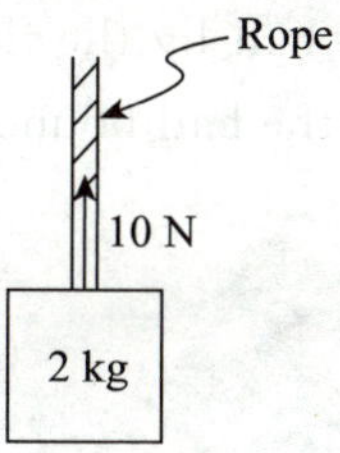

图 9

(A) zero (B) 2 m/s^2 up
(C) 2 m/s^2 down (D) 5 m/s^2 down

Question 23—24

A ball of mass m is attached to a vertical rod by two massless strings. The rod is rotated about its axis so that both strings are taut, with tensions T_1 and T_2, respectively. The strings and rod form the right triangle shown in figure 10. The ball rotates in a horizontal circle of radius r with speed v.

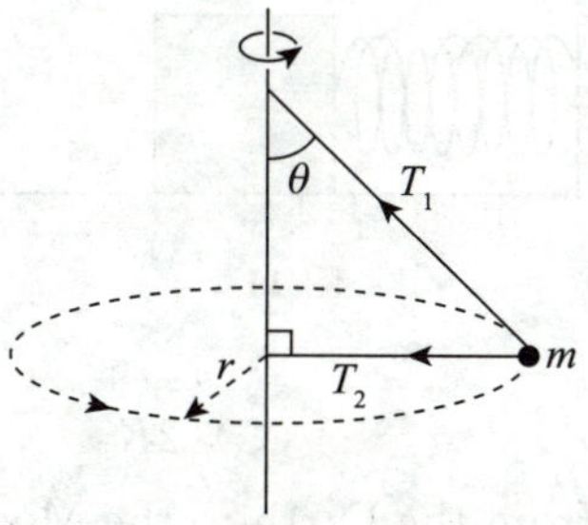

图 10

23. What is the magnitude of the net force on the ball?
(A) mv^2/r (B) mg (C) T_2 (D) T_1+mg

24. What is the tension T_1 in the upper string?
(A) $mg\cos\theta$ (B) $mg\sin\theta$ (C) $mg\tan\theta$ (D) $mg/\cos\theta$

25. A toy cannon of mass 1.0 kg is initially at rest on a horizontal surface when it launches a 0.05 kg projectile with a velocity of 10 m/s at an angle of 60° above the horizontal. What is the speed of the 1.0 kg cannon immediately after the projectile is released? Assuming that friction is negligible.
(A) 0.25 m/s (B) 0.35 m/s (C) 0.43 m/s (D) 0.50 m/s

Question 26—27

A ball is dropped and bounces off the floor. Its speed is the same immediately before and immediately after the collision.

26. Which of the following is true about the collision between the ball and the floor?
(A) The ball's momentum is conserved
(B) The ball's momentum changes direction but not magnitude

(C) The ball's momentum changes magnitude but not direction

(D) The impulse delivered to the ball by the floor is zero

27. How does the height to which the ball bounces compare to the height from which it was dropped?

(A) It is less

(B) It is the same

(C) It is greater

(D) It cannot be determined without knowing the value of the height from which the ball was dropped

28. A block on a horizontal surface of negligible friction is placed in contact with an ideal spring, as shown in figure 11. The block is moved to the left so that the spring is compressed a distance x from equilibrium and then released from rest. The block has kinetic energy K_1 when it separates from the spring. When the spring is compressed a distance $2x$ and the block is released from rest, the kinetic energy of the block when it separates from the spring is

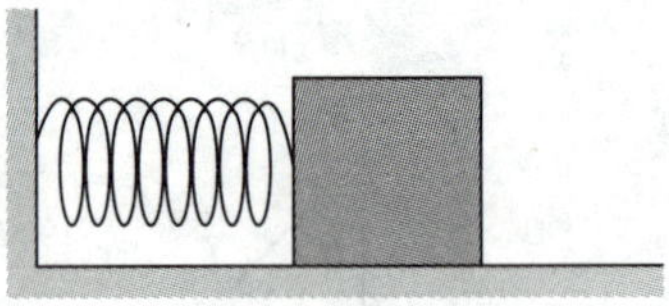

图 11

(A) $K_1/2$ (B) K_1 (C) $2K_1$ (D) $4K_1$

29. A spaceship is traveling from the Earth to the Moon. Which of the following is true of the gravitational force on the ship due to the two objects when the ship is equidistant from the Earth and the Moon?

(A) There is no net force because the ship is beyond the influence of the Earth and the Moon

(B) There is no net force because the forces exerted by the Earth and the Moon balance

(C) There is a net force because the force exerted by the Earth is smaller than that exerted by the Moon

(D) There is a net force because the force exerted by the Earth is greater than that exerted by the Moon

30. A ball is tossed straight up and later returns to the point from which it was launched. If the ball is subject to air resistance as well as gravity, which of the following statements is correct?

(A) The speed at which the ball returns to the point of launch is less than its speed when it was initially launched

(B) The time for the ball to fall is the same as the time for the ball to rise

(C) The force of air resistance is directed downward both when the ball is rising and when it is falling

(D) The net work done by air resistance on the ball during its flight is zero

31. A railroad car of mass m is moving with speed v when it collides with and connects to a second railroad car of mass $3m$, initially at rest. How do the speed and kinetic energy of the connected cars compare to those of the single car of mass m before the collision?

	speed	kinetic energy
(A)	less	less
(B)	less	the same
(C)	the same	less
(D)	the same	the same

32. A ball of mass m and momentum p has kinetic energy equal to which of the following?

(A) $\frac{1}{2}\frac{p^2}{m}$ (B) $\frac{p^2}{m}$ (C) $2\frac{p^2}{m}$ (D) $\frac{1}{2}\frac{m}{p^2}$

33. A simple pendulum consisting of a small object of mass m attached to a string of length l has a period T. A pendulum with which of the following combinations of object mass and string length will also have period T?

	object mass	string length
(A)	$m/2$	l
(B)	m	$l/4$
(C)	$\sqrt{2}m$	$l/\sqrt{2}$
(D)	$2m$	$4l$

34. A planet of mass m orbits a star of mass M, where $m \ll M$. The orbit is circular, its radius is r, and its period is T. True statements about the planet's orbit include which of the following?

Ⅰ. The orbital speed equals $\frac{2\pi r}{T}$

Ⅱ. The gravitational force equals $\frac{GMm}{r^2}$

Ⅲ. If the orbit radius r were greater, T would also be greater

(A) Ⅰ only (B) Ⅱ only (C) Ⅰ and Ⅲ only (D) Ⅰ, Ⅱ and Ⅲ

Question 35—36

A block of mass 3 kg slides along a horizontal surface that has negligible friction except for one section, as shown in figure 12. The block arrives at the rough section with a speed of 5 m/s and leaves it 0.5 s later with a speed of 3 m/s.

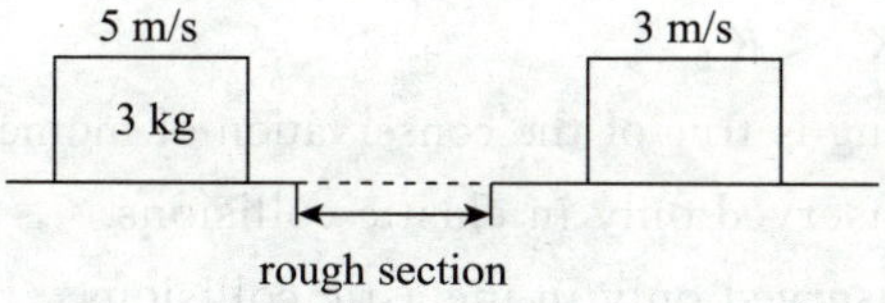

图 12

35. What is the magnitude of the work done by the frictional force exerted on the block by the rough section of the surface?

(A) 24 J (B) 16 J (C) 8 J (D) 6 J

36. What is the magnitude of the average frictional force exerted on the block by the rough section of the surface?

(A) 30 N (B) 12 N (C) 4.5 N (D) 4 N

37. An object attached to one end of a string moves in a circle at constant speed. Which of the following is correct?

(A) The object is accelerating as it moves

(B) The object's velocity is the same as its speed

(C) The object does not require a force to keep its state of circular motion

(D) If the string breaks, the object will keep its circular motion

38. Three point charges $+q$, $+Q$, and $-Q$ are fixed at the corners of an isosceles triangle in the plane of the page, as shown in figure 13. Which arrow in the figure represents the direction of the net electrostatic force on charge $+q$ due to the other two equal and opposite charges?

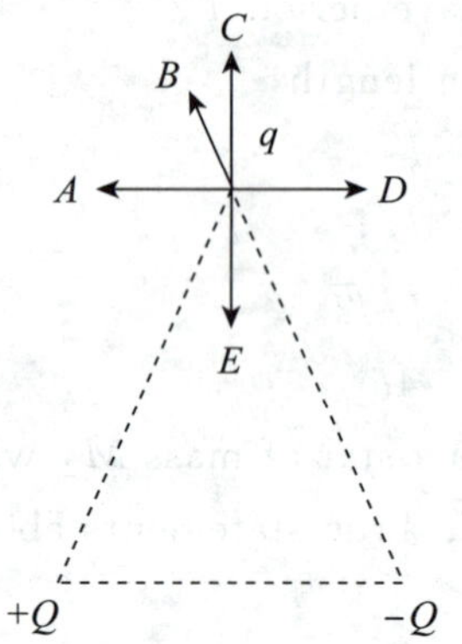

图 13

(A) A (B) B (C) C (D) D

39. Two skaters are initially at rest next each other on frictionless ice. Skater A pushes on skater B. If skater A has greater mass than skater B, which of the following correctly relates the magnitudes of their momentums P and their kinetic energies K after the push?

(A) $P_A = P_B$ and $K_A < K_B$

(B) $P_A = P_B$ and $K_A = K_B$

(C) $P_A = P_B$ and $K_A > K_B$

(D) $P_A < P_B$ and $K_A < K_B$

40. Which of the following is true of the conservation of momentum and kinetic energy?

(A) Momentum is conserved only in elastic collisions

(B) Momentum is conserved only in inelastic collisions

(C) Kinetic energy is conserved only in elastic collisions

(D) Both require the same conditions in order to be conserved

41. A force of constant magnitude F and fixed direction acts on an object of mass m that is initially at rest. If the force acts for a time interval Δt over a displacement Δx, what is the magnitude of the resultant change in the linear momentum of the object?

(A) $F\Delta t$ (B) $F\Delta x$ (C) $F\Delta t/m$ (D) $F\Delta x/m$

42. A ball is thrown with an initial speed of 20 m/s at an angle of 60° to the ground. If the air resistance is negligible, what is the ball's speed at the instant it reaches its maximum height from the ground?

(A) zero (B) 10 m/s (C) 14 m/s (D) 17 m/s

43. A box is a sudden push up a ramp. Friction between the box and the ramp is not negligible. Which of following diagrams best represents the direction of the actual forces acting on the box as it moves upward after the push?

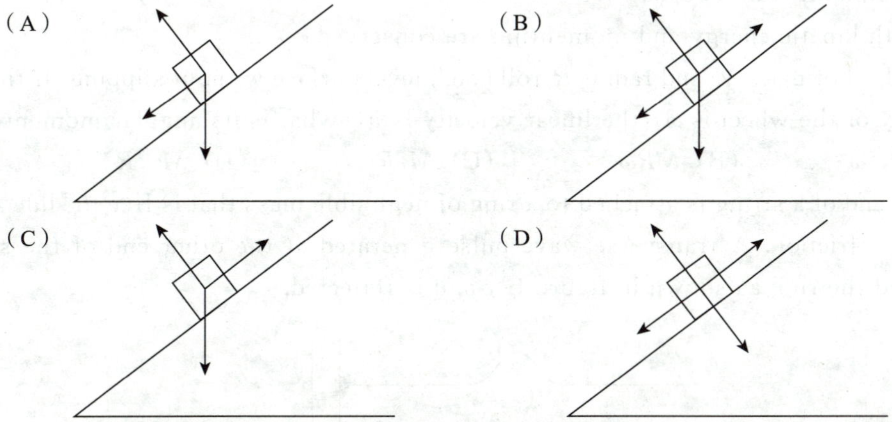

44. In a lab, a block weighing 80 N attached to a spring scale, and both are pulled to the right on a horizontal surface, as shown in figure 14. Friction between block and the surface is negligible. What is the acceleration of the block when the scale read 32 N ?

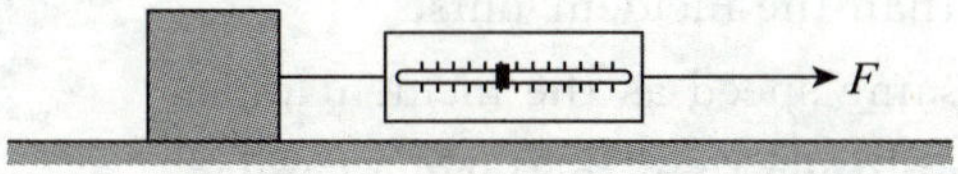

图 14

(A) 2.0 m/s² (B) 2.5 m/s² (C) 4.0 m/s² (D) 6.0 m/s²

45. The system shown in figure 15 is released from the rest. If friction is negligible, the acceleration of the 4.0 kg block sliding on the table is most nearly

(A) 0 (B) 1.7 m/s² (C) 3.3 m/s² (D) 5.0 m/s²

Directions: For each of the questions or incomplete statements below, two of the suggested answers will be correct. For each of these questions, you must select both correct choices to earn credit. No partial credit will be earned if only one correct choice is selected. Select the two that are best in each case and then enter both of the appropriate letters in the

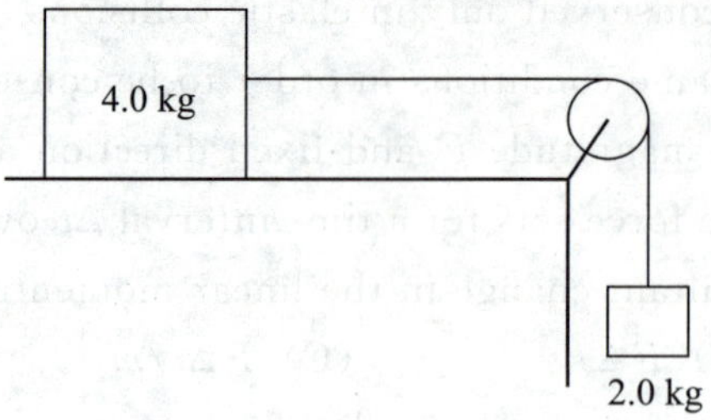

图 15

corresponding space on the answer sheet.

46. In an elastic collision, what physics quantities are conserved?

(A) Only potential energy is conserved

(B) Only kinetic energy is conserved

(C) The momentum is conserved

(D) Both kinetic energy and momentum are conserved

47. A wheel of mass M and radius R rolls on a level surface without slipping. If the angular velocity of the wheel is ω, the linear velocity is v, what is its angular momentum?

(A) $MR^2\omega$　　(B) $MR\omega$　　(C) MvR　　(D) Mv^2R

48. One end of a string is attached to a ring of negligible mass that is free to slide along a pole without friction. A transverse wave pulse generated at the other end of the string moves toward the ring as shown in figure 16, and is reflected.

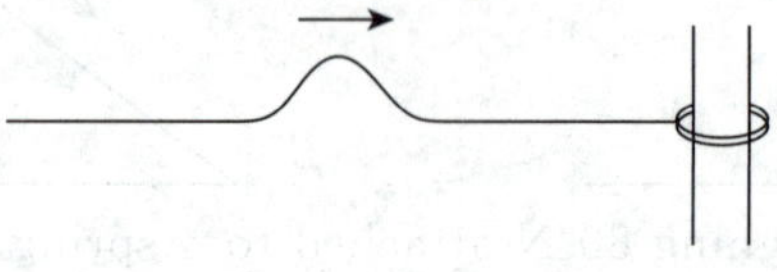

图 16

Properties of the reflected pulse include which of the following?

(A) It moves faster than the incident pulse

(B) It moves at the same speed as the incident pulse

(C) It has greater amplitude than the incident pulse

(D) It is on the same side of the string as the incident pulse

49. A spaceship is traveling from the Earth to the Moon. Which of the following is true of the gravitational force on the ship due to the two objects when the ship is equidistant from the Earth and the Moon?

(A) There is a net force because the mass of the Earth is large than the mass of the moon

(B) There is no net force because the forces exerted by the Earth and the Moon balance

(C) There is a net force because the force exerted by the Earth is smaller than that exerted by the Moon

(D) There is a net force because the force exerted by the Earth is greater than that exerted by the Moon

50. In the circuit (图 17), the resistors all have the same resistance. The battery, wires, and ammeter have negligible resistance. A closed switch also has negligible resistance. What statements about the circuits are true?

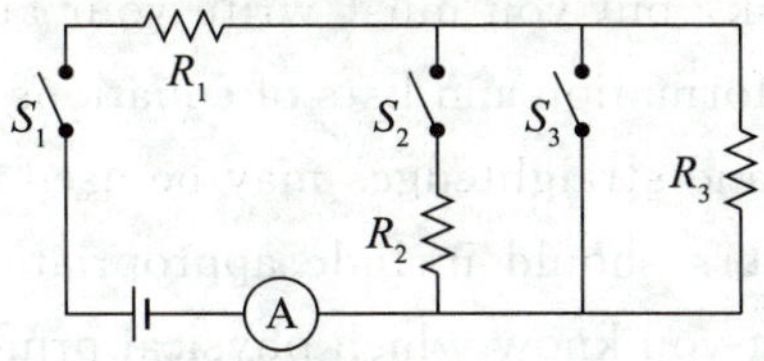

图 17

(A) Closing switches S_1 and S_2 will produce the greatest power dissipation in R_2

(B) Closing switches S_1, S_2 and S_3 will produce the greatest power dissipation in R_2

(C) Closing switches S_1 and S_3 will produce the greatest reading on the ammeter

(D) Closing switches S_1 and S_2 will produce the greatest reading on the ammeter

Answer for AP Physics 1 Practice (2)

1	B	17	D	33	A
2	B	18	C	34	D
3	C	19	D	35	A
4	D	20	D	36	B
5	C	21	C	37	A
6	D	22	D	38	D
7	B	23	A	39	A
8	D	24	D	40	C
9	C	25	A	41	A
10	D	26	B	42	B
11	B	27	B	43	A
12	B	28	D	44	C
13	C	29	D	45	C
14	D	30	A	46	C, D
15	A	31	A	47	A, C
16	D	32	A	48	B, D
				49	A, D
				50	A, C

Section Ⅱ: Free Response

DO NOT OPEN THIS BOOKLET UNTIL YOU ARE TOLD TO DO SO.

Instructions

The questions for Section Ⅱ are printed in this booklet. You may use any blank space in the booklet for scratch work, but you must write your answers in the spaces provided for each answer. A table of information and lists of equations that may be helpful are in the booklet. Calculators, rulers, and straightedges may be used in this section.

All final numerical answers should include appropriate units. Credit for your work depends on demonstrating that you know which physical principles would be appropriate to apply in a particular situation. Therefore, you should show your work for each part in the space provided after that part. If you need more space, be sure to clearly indicate where you continue your work. Credit will be awarded only for work that is clearly designated as the solution to a specific part of a question. Credit also depends on the quality of your solutions and explanations, so you should show your work.

Write clearly and legibly. Cross out any errors you make; erased or crossed-out work will not be scored. You may lose credit for incorrect work that is not crossed out.

Manage your time carefully. You may proceed freely from one question to the next.

You may review your responses if you finish before the end of the exam is announced.

Total Time

90 minutes

Number of Questions

5

Percent of Total Score

50%

Writing Instrument

Either pencil or pen with black or dark blue ink

Electronic Device

Calculator allowed

Suggested Time

Approximately 25 minutes each for questions 2 and 3 and 13 minutes each for questions 1, 4, and 5

Weight

Approximate weights:

Questions 2 and 3:

26% each

Questions 1, 4, and 5:

16% each

AP Physics 1—Practice (2)
Section Ⅱ
5 Questions
Time—90 minutes

Directions: Questions 2 and 3 are long free-response questions that require about 25 minutes each to answer and are worth 12 points each. Questions 1, 4, and 5 are short free-response questions that require about 13 minutes each to answer and are worth 7 points each. Show your work for each part in the space provided after that part.

1. (7 points, 13 minutes)

A 70 kg woman and her 35 kg son are standing at rest on an ice rink, as shown in figure 18. They push against each other for a time of 0.60 s, causing them to glide apart. The speed of the woman immediately after they separate is 0.55 m/s. Assume that during the push, friction is negligible compared with the forces the people exert on each other.

(a) Calculate the initial speed of the son after the push.

(b) Calculate the magnitude of the average force exerted on the son by the mother during the push.

(c) After the initial push, the friction that the ice exerts cannot be considered negligible, and the mother comes to rest after moving a distance of 7.0 m across the ice. If their coefficients of friction are the same, how far does the son move after the push?

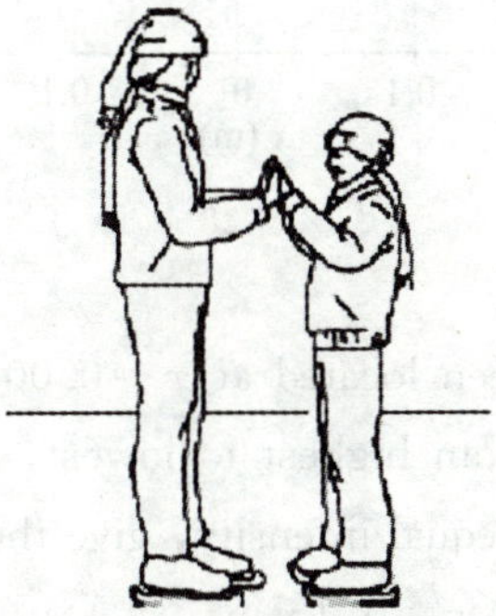

图 18

2. (12 points, 25 minutes)

A microwave source is placed behind two identical slits, as represented in diagram 19. The slit centers are separated by a distance of 0.20 m, and the slit widths are small compared to the slit separation but not negligible.

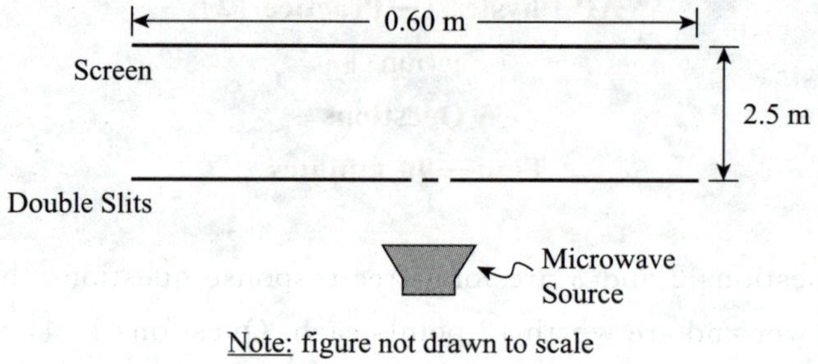

图 19

The microwave wavelength is 2.4×10^{-2} m. The resulting interference pattern is centered on a screen 0.60 m wide, located 2.5 m from the slits.

(a) Calculate the frequency of the microwave radiation.

(b) On graph 20, where the x-axis represents the distance along the screen and $x=0$ represents the center of the pattern, sketch the intensity of the interference pattern expected for that arrangement.

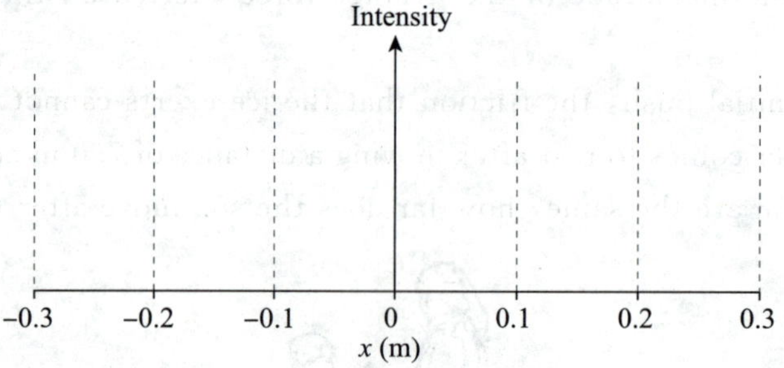

图 20

(c) Consider points on the screen located at $x=0.00$ m, $x=0.15$ m, and $x=0.30$ m. Rank the intensity at those points from highest to lowest, with number 1 corresponding to the highest intensity. If two points have equal intensity, give them the same ranking.

______ $x=0.00$ m　　______ $x=0.15$ m　　______ $x=0.30$ m

Justify your ranking.

(d) Suppose the microwave wavelength is decreased by a factor of three, to 0.80×10^{-2} m. Sketch the resulting interference pattern (图 21).

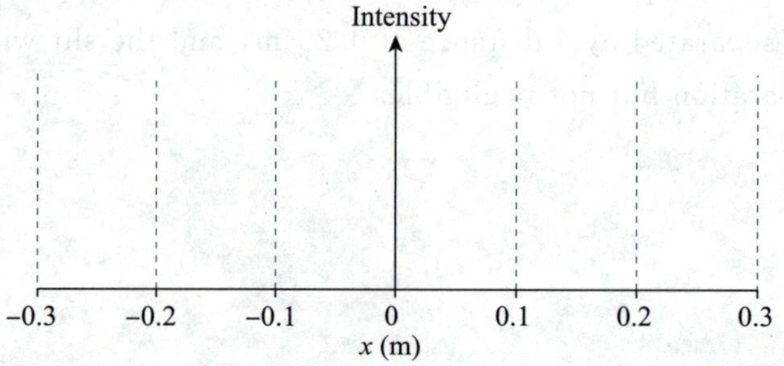

图 21

3. (12 points, 25 minutes.)

A large pan is filled to the top with oil of density ρ_0. A plastic cup of mass m_c, containing a sample of known mass m_s, is placed in the oil so that the cup and sample float, as shown in figure 22. The oil that overflows from the pan is collected, and its volume is measured. The procedure is repeated with a variety of samples of different mass, and the pan is refilled each time.

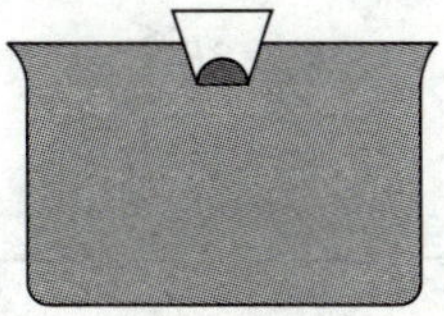

图 22

(a) On the dot (图 22) that represents the cup-sample system, draw and label the forces (not components) that act on the system when it is floating on the surface of the oil.

(b) Derive an expression for the overflow volume V_o (the volume of oil that overflows due to the floating system) in terms of ρ_0, m_s, m_c, and fundamental constants. If you need to draw anything other than what you have shown in part (a) to assist in your solution, use the space below. DO NOT add anything to the figure in part (a).

Assume that the following data (表 1) are obtained for the overflow volume V_o for several sample masses m_s:

表 1

Sample mass m_s (kg)	0.020	0.030	0.040	0.050	0.060	0.070
Overflow volume (m^3)	29×10^{-6}	38×10^{-6}	54×10^{-6}	62×10^{-6}	76×10^{-6}	84×10^{-6}

(c) Graph the data on the axes (图 23), plotting the overflow volume as a function of sample mass. Place numbers and units on both axes. Draw a straight line that best represents the data.

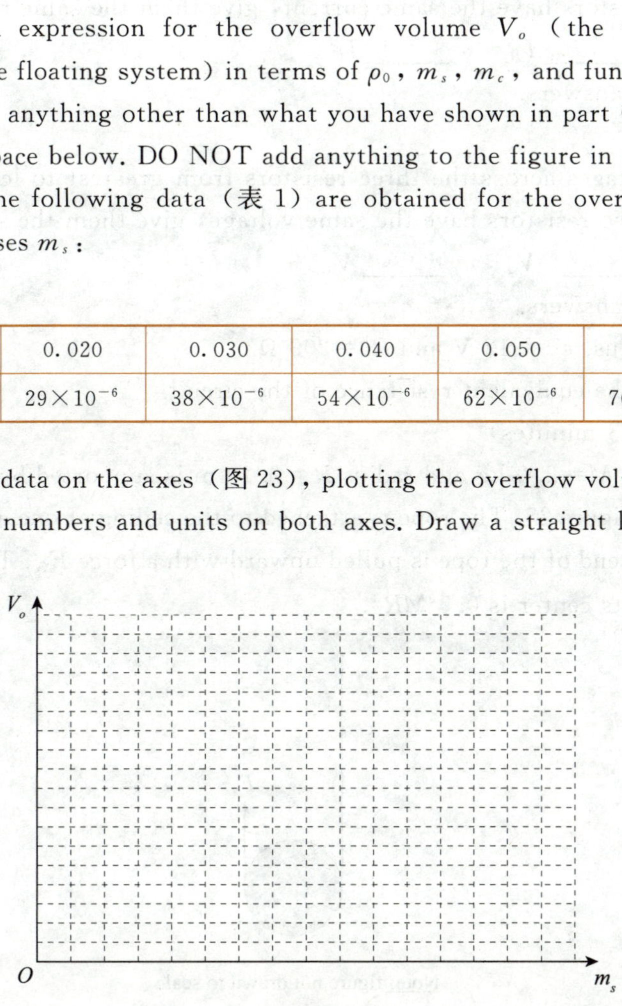

图 23

(d) Use the slope of the best-fit line to calculate the density of the oil.

4. (7 points, 13 minutes)

The circuit in figure 24 contains a battery with negligible internal resistance, a closed switch S, and three resistors, each with a resistance of R or $2R$.

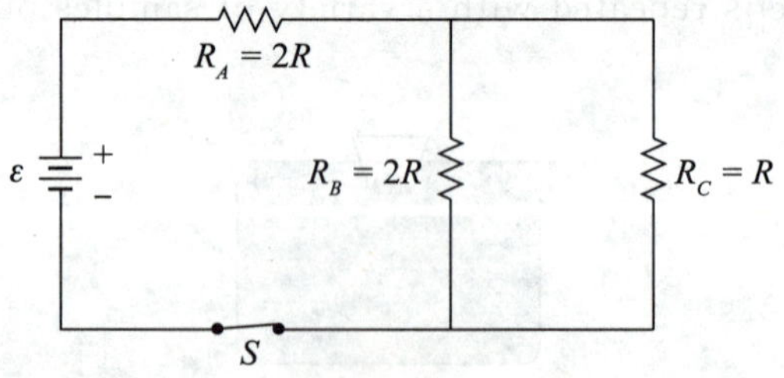

图 24

(a)

i. Rank the currents in the three resistors from greatest to least, with number 1 being greatest. If two resistors have the same current, give them the same ranking.

_____ I_A _____ I_B _____ I_C

ii. Justify your answers.

(b)

i. Rank the voltages across the three resistors from greatest to least, with number 1 being greatest. If two resistors have the same voltage, give them the same ranking.

_____ V_A _____ V_B _____ V_C

ii. Justify your answers.

For parts (c), use $\varepsilon = 12$ V and $R = 200$ Ω.

(c) Calculate the equivalent resistance of the circuit.

5. (7 points, 13 minutes)

A disk of mass $M=2.0$ kg and radius $R=0.10$ m is supported by a rope of negligible mass, as shown in figure 25. The rope is attached to the ceiling at one end and passes under the disk. The other end of the rope is pulled upward with a force F_A. The rotational inertia of the disk around its center is $0.5\ MR^2$.

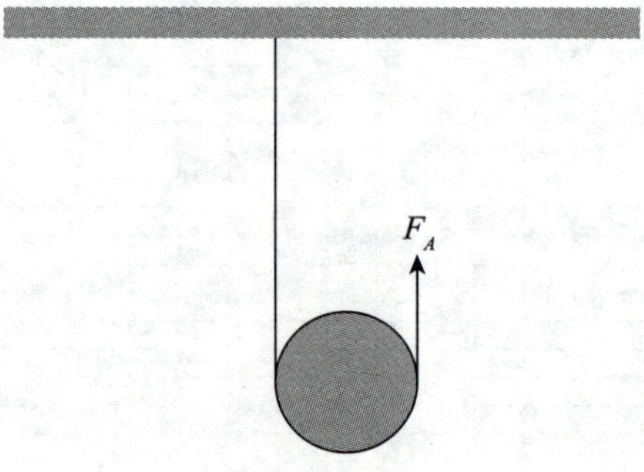

Note: figure not drawn to scale

图 25

(a) Calculate the magnitude of the force F_A necessary to hold the disk at rest.

At time $t=0$, the force F_A is increased to 12 N, causing the disk to accelerate upward. The rope does not slip on the disk as the disk rotates.

(b) Calculate the linear acceleration of the disk.

(c) Calculate the angular speed of the disk at $t=3.0$ s.

Scorning guidance for AP Physics 1 Practice (2)

Question 1

7 points total **Distribution of points**

(a) 2 points

For a correct expression of conservation of momentum 1 point

$$m_w v_w = m_s v_s$$

$$v_s = \frac{m_w m_w}{m_s} = \frac{70\ \text{kg} \times 0.55\ \text{m/s}}{35\ \text{kg}} = 1.1\ \text{m/s}$$ 1 point

(b) 2 points

For a correct use of the impulse-momentum relationship

$$F_{avg}\Delta t = m_s \Delta v_s$$ 1 point

$$F_{avg} = \frac{m_s \Delta v_s}{\Delta t} = \frac{35\ \text{kg} \times (1.1\ \text{m/s} - 0\ \text{m/s})}{0.60\ \text{s}} = 64\ \text{N}$$ 1 point

(c) 3 points

For indicating that the average force exerted on the mother by the son is equal in magnitude to that exerted on the son by the mother 1 point

For indicating that the average force exerted on the mother by the son is opposite in direction to that exerted on the son by the mother 1 point

For a correct justification invoking Newton's third law 1 point

Question 2

12 points total **Distribution of points**

(a) 2 points

For a correct relationship relating f to v and λ 1 point

$$v = f\lambda$$

$$f = v/\lambda, \text{ where } v = c$$

$$f = (3 \times 10^8\ \text{m/s})/(2.4 \times 10^{-2}\ \text{m})$$

For the correct answer with units 1 point

$$f = 1.25 \times 10^{10}\ \text{Hz}$$

(b) 4 points

The position of the maxima on the screen for a double-slit interference pattern is found from $x_m \approx \frac{m\lambda L}{d}$, where L is the distance to the screen, d is the slit separation, λ is the wavelength, and m is an integer.

To calculate the distance from the central maximum to the first secondary maximum, let $m=1$.

$$x_{m=1}=\frac{1\times 2.4\times 10^{-2}\ \text{m}\times 2.5\ \text{m}}{0.20\ \text{m}}=0.30\ \text{m}$$

For the graph symmetric with multiple peaks — 1 point

For the central maximum at $x=0$ — 1 point

For the first secondary maximum at either side at $x=\pm 0.3$ m — 1 point

For a reasonable curved shape with minima about halfway between the central and the first secondary maxima（图 26） — 1 point

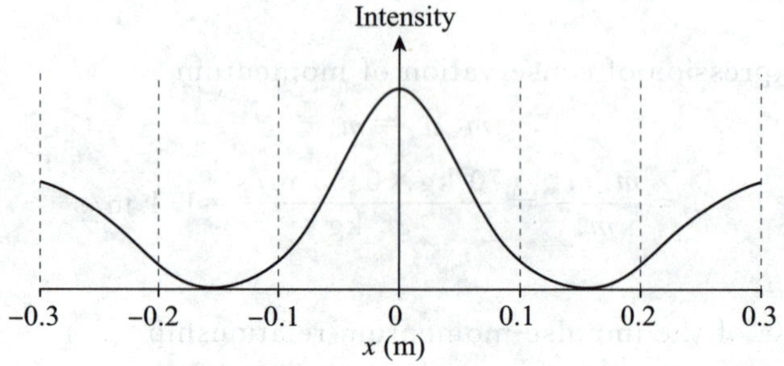

图 26

Relative heights of the peaks are not considered in the scoring of this question because that was considered in part (c).

(c) 3 points

__1__ $x=0.00$ m　　__3__ $x=0.15$ m　　__2__ $x=0.30$ m

For recognition that the intensity is greatest at $x=0.00$ m — 1 point

For recognition that the intensity is least at $x=0.15$ m — 1 point

For a correct justification of the correct answer — 1 point

Example: At $x=0.00$ m, constructive interference between the light from the two slits arriving in phase results in a central maximum of greatest intensity. At $x=0.15$ m, destructive interference between the light from the two slits arriving 180° out of phase (1/2 wavelength apart) results in a dark fringe of minimum intensity. At $x=0.30$ m, constructive interference also occurs, but the double-slit pattern is modulated by the single-slit pattern so that the first bright fringe is not as bright (has less intensity) than the central maximum.

(d) 3 points

In the equation $x_m \approx \frac{m\lambda L}{d}$ for the positions of the maxima, $x_m \propto \lambda$, so when the wavelength is reduced to $\lambda/3$, the distance of each maxima from the center is 1/3 the previous value.

For $m=1$, $x_{m=1}=0.30\ \text{m}/3=0.10$ m

For drawing more maxima than drawn in part (b) — 1 point

For showing the maxima at distances from the center that are 1/3 the distances shown in part (b) (with 0, ± 0.1m, ± 0.2m, and ± 0.3 m being the actual correct values) — 1 point

For the maxima decreasing in height with（图 27）increasing distance from the center 1 point

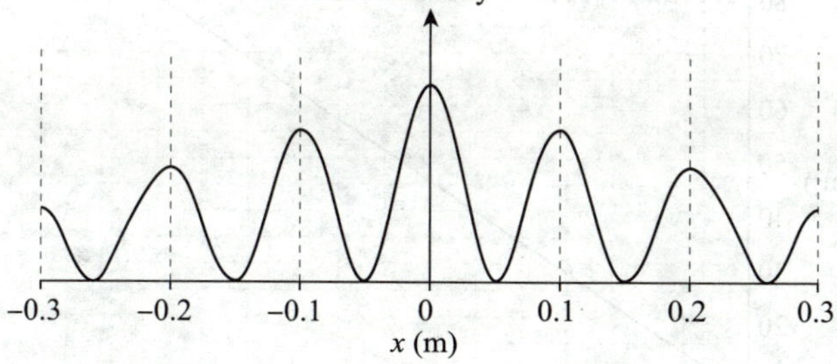

图 27

Question 3

12 points total **Distribution of points**

(a) 2 points

For a single upward force, appropriately labeled, representing the buoyant force 1 point

For downward gravitational force (or forces), appropriately labeled, representing the cup and the sample（图 28） 1 point

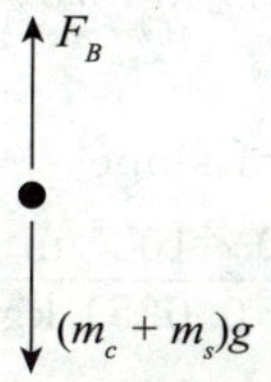

图 28

One earned point was deducted if any extraneous forces were present.

(b) 3 points

For any statement of equilibrium 1 point

$$F_B = F_g$$

For a correct substitution including both masses of m_c and m_s 1 point

$$\rho_o V_o g = (m_c + m_s)g$$

For a correct statement of the overflow volume, V_o 1 point

$$V_o = \frac{m_c + m_s}{\rho_o}$$

(c) 4 points

For data plotted correctly 1 point

For correct units on both axes 1 point

For numerical scales that are linear and allow the plotted data to extend over about half the grid area（图 29） 1 point

For a reasonable single straight best-fit line that does not go through (0, 0)　1 point

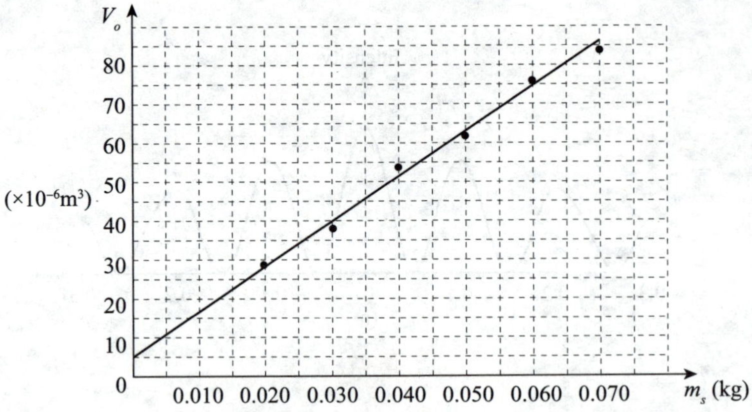

图 29

(d) 3 points

From part (b), $V_o = \frac{m_c}{\rho_o} + \frac{1}{\rho_o} m_s$

For properly calculating a slope using points on the straight line drawn, including data points only if they are on that line　1 point

Example: Using the two points (0.060 kg, 75×10^{-6} m^3) and (0.025 kg, 35×10^{-6} m^3) that are on the line in the graph above

For calculating a reasonable value of slope　1 point

$$\text{slope} = \frac{1}{\rho_o} = \frac{(75-35)\times10^{-6}\ \text{m}^3}{(0.060-0.025)\ \text{kg}} = 1.14\times10^{-3}\ \text{m}^3/\text{kg}$$

For an explicit or implicit indication of inverting the slope

For calculating a reasonable value for the oil's density (including units and four significant figures or less)　1 point

$$\rho_o = 8.8\times10^2\ \text{kg/m}^3$$

Question 4

7 points total　**Distribution of points**

(a) 2 points

(i) For ranking I_A as the greatest, I_C as the second greatest and I_B as the third　1 point

(ii) For a correct statement justifying that I_A is greatest (For example: The total current flows through R_A and gets divided between the other two resistors)

For correct statement justifying I_C is the second and I_B is the third　1 point

(b) 2 points

(i) For correct ranking V_A, V_B, V_C = 1, 2, 2　1 point

(ii) For correct justification that V_A is the greatest, $V_B = V_C$　1 point

(c) 3 points

Let R_{BC} as the total resistance R_B and R_C

$$\frac{1}{R_{BC}}=\frac{1}{R_B}+\frac{1}{R_C}=\frac{1}{2R}+\frac{1}{R}=\frac{1}{400\ \Omega}+\frac{1}{200\ \Omega}=\frac{3}{400\ \Omega}$$ 1 point

$$R_{BC}=\frac{400\ \Omega}{3}=133\ \Omega$$ 1 point

For correct form for determining the total resistance

$$R_{tot}=R_A+R_{BC}=2R+\frac{2R}{3}=400\ \Omega+133\ \Omega=533\ \Omega$$ 1 point

Question 5

7 points total **Distribution of points**

(a) 2 point

For a correct expression indicating that $F_{net}=0$ and correct answer 1 point

$$F_{net}=0$$

$$2F_A-Mg=0$$

$$F_A=\frac{Mg}{2}=\frac{2.0\ \text{kg}\times 9.8\ \text{m/s}^2}{2}=9.8\ \text{N}$$ 1 point

(b) 3 points

For correct expression of Newton's second law of translation motion

$$F_{net}=Ma$$

$$F_A+T-Mg=Ma \qquad ①$$ 1 point

For correct expression of Newton's second law of rotational motion

$$\tau=I\alpha$$

$$F_AR-TR=\left(\frac{MR^2}{2}\right)\alpha \quad \text{and} \quad \alpha=\frac{a}{R}$$

$$F_A-T=\frac{Ma}{2} \qquad ②$$ 1 point

From equation ① and ②, we have

$$2F_A-Mg=\frac{3}{2}Ma$$

$$a=\frac{2}{3}\left(\frac{2F_A}{M}-g\right)=\frac{2}{3}\left(\frac{2\times 12\ \text{N}}{2.0\ \text{kg}}-9.8\ \text{m/s}^2\right)=1.47\ \text{m/s}^2$$ 1 point

(c) 2 points

For using the relationship between linear and angular velocity

$$\omega=\omega_0+at \quad \text{and } \omega_0=0$$ 1 point

$$\omega=\frac{at}{R}=\frac{(1.47\ \text{m/s}^2)(3.0\ \text{s})}{0.10\ \text{m}}=44\ \text{rad/s}$$ 1 point

AP Physics 2 Practice Exam (1)

Section Ⅰ: Multiple Choice

DO NOT OPEN THIS BOOKLET UNTIL YOU ARE TOLD TO DO SO.

Instructions

Section Ⅰ of this exam contains 50 multiple-choice questions. Pages containing equations and other information are also printed in this booklet. Calculators, rulers, and straightedges may be used in this section.

Indicate all of your answers to the multiple-choice questions on the answer sheet.

No credit will be given for anything written in this exam booklet, but you may use the booklet for notes or scratch work.

For questions 1 through 45, select the single best answer choice for each question.

After you have decided which of the choices is best, fill in the appropriate letter in the corresponding space on the answer sheet.

For questions 46 through 50, select the two best answer choices for each question.

After you have decided which two of the choices are best, enter both letters in the corresponding space on the answer sheet.

Use your time effectively, working as quickly as you can without losing accuracy. Do not spend too much time on any one question. Go on to other questions and come back to the ones you have not answered if you have time. It is not expected that everyone will know the answers to all of the multiple-choice questions.

Your total score on Section Ⅰ is based only on the number of questions answered correctly.

Points are not deducted for incorrect answers or unanswered questions.

Total Time

90 minutes

Number of Questions

50

Percent of Total Score

50%

Writing Instrument

Pencil required

Electronic Device

Calculator allowed

AP Physics 2 Practice (1)

Section Ⅰ

50 Questions

Time—90 minutes

Note: To simplify calculations, you may use $g = 10 \text{ m/s}^2$ in all problems.

Directions: Each of the questions or incomplete statements below is followed by 4 suggested answers or completions. Select the one that is the best in each case and then enter the appropriate letter in the corresponding space on the answer sheet.

1. A proton is traveling to the left when it enters the space between two oppositely charged parallel plates, as shown in figure 1. Which of the four labeled paths will the proton take?

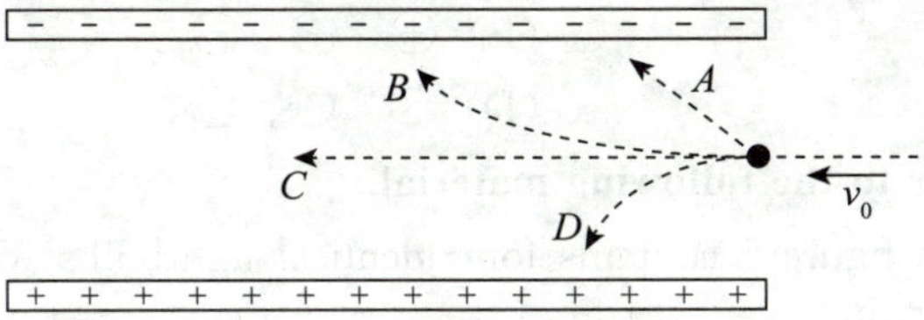

图 1

(A) A　　(B) B　　(C) C　　(D) D

2. Figure 2 represent two fields. The figure on the left represents the uniform gravitational field very near the Earth's surface. The figure on the right represents the uniform electric field E near the center of the region between very large parallel plates. Which of the following describes the shape of the isolines of potential for the gravitational field and the electric field in these regions?

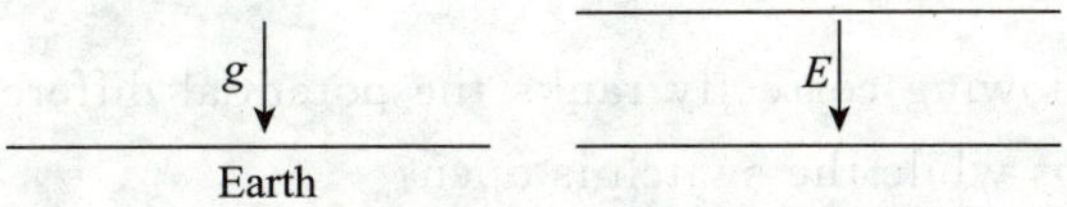

图 2

(A) Isolines are straight, vertical lines for both the gravitational and electric fields
(B) Isolines are straight, horizontal lines for both the gravitational and electric fields
(C) Gravitational isolines are straight, while electric isolines are curved
(D) Gravitational isolines are curved, while electric isolines are straight

3. Figure 3 shows a model of an electromagnetic wave, where E is the electric field and B is the magnetic field. In what direction is the energy of the wave transmitted?

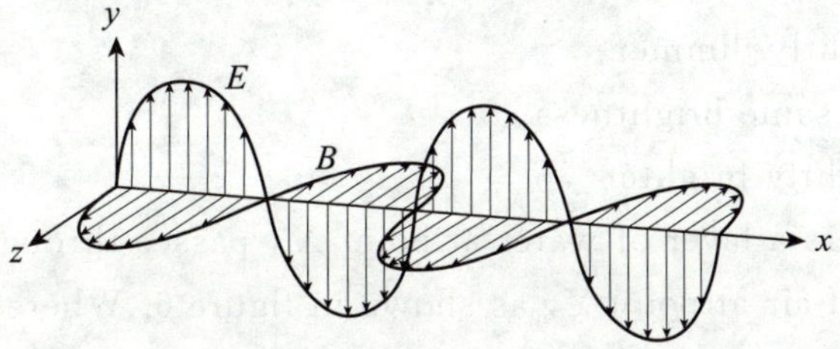

图 3

(A) Along the x-axis only

(B) Along the y-axis only

(C) Along the z-axis only

(D) In a direction that is at a nonzero angle to each of the axes

4. Three identical conducting spheres, $S1$, $S2$, and $S3$, are supported by insulating thread, as shown in figure 4. Initially, sphere $S1$ has a net positive charge and the other two spheres are uncharged.

Spheres $S1$ and $S2$ are brought into contact and then separated. Next, spheres $S2$ and $S3$ are brought into contact and then separated. Which of the following shows the signs of the final net charges on the spheres?

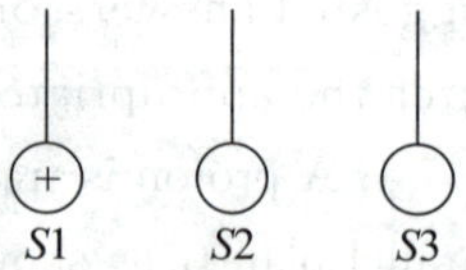

图 4

(A) ⊕ ⊕ ⊕　　(B) ⊕ ⊖ ⊖

(C) ○ ○ ⊕　　(D) ○ ⊕ ⊕

Questions 5—7 refer to the following material.

The circuit shown in figure 5 contains four identical lightbulbs with constant resistance, a capacitor C, which is initially uncharged, and a switch S. The switch is initially open.

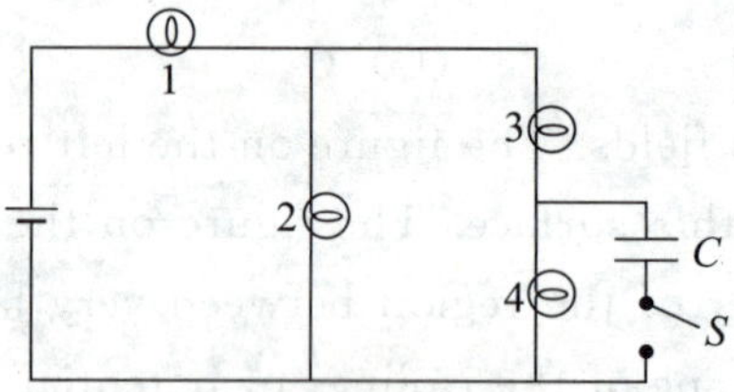

图 5

5. Which of the following correctly ranks the potential differences ΔV_1, ΔV_2, ΔV_3, and ΔV_4 across the bulbs while the switch is open?

(A) $\Delta V_1=\Delta V_2=\Delta V_3=\Delta V_4$　　(B) $\Delta V_1>\Delta V_2=\Delta V_3=\Delta V_4$

(C) $\Delta V_1>\Delta V_2>\Delta V_3=\Delta V_4$　　(D) $\Delta V_1>\Delta V_2>\Delta V_3>\Delta V_4$

6. Immediately after the switch is closed, the current in bulb 1 is I_a. What is the current in bulb 2 at that time?

(A) I_a　　(B) $2I_a/3$　　(C) $I_a/2$　　(D) $I_a/3$

7. After the switch has been closed a long time, how does the brightness of bulb 4 compare with its brightness before the switch was closed?

(A) Bulb 4 is much dimmer and is only barely lit

(B) Bulb 4 is slightly dimmer

(C) Bulb 4 is the same brightness

(D) Bulb 4 is slightly brighter

8. A light ray enters a layer of water at point X, passes through a layer of glass, and exits through a layer of air at point Y, as shown in figure 6. Where would the ray exit the layer of air if the glass was replaced with a material of higher index of refraction?

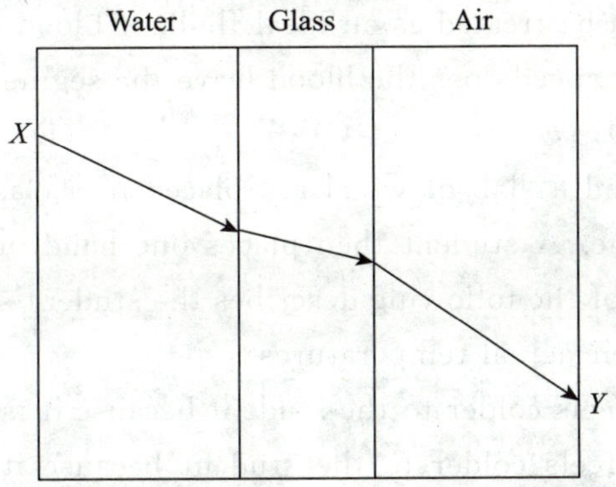

图 6

(A) At a point above point Y

(B) At a point below point Y

(C) At point Y

(D) The location cannot be determined without knowing how much higher the index of refraction of the new material is

9. Which of the following indicates the object distance s_o for which a spherical concave mirror of focal length f produces an upright image?

(A) $s_o < f$ (B) $s_o = f$ (C) $2f > s_o > f$ (D) $s_o > 2f$

10. Water is flowing with a speed of 9.0 m/s through a pipe of diameter 10 cm. The pipe widens to 30 cm as it goes up an 8.0 m step, as shown in figure 7. If the pressure at point A is 2×10^5 Pa, what is the pressure at point B?

(The density of water is 1.0×10^3 kg/m^3.)

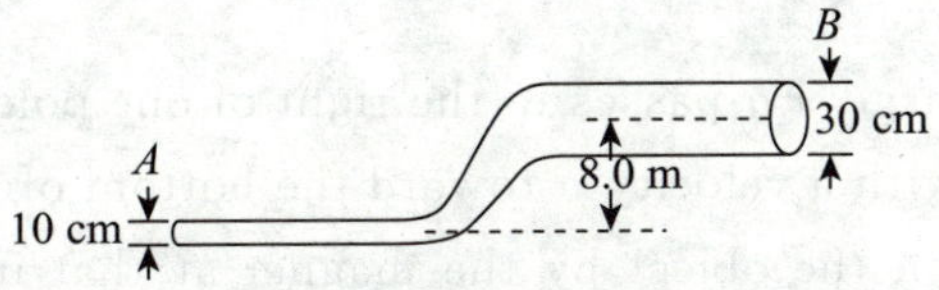

Note: figure not drawn to scale

图 7

(A) 1.2×10^5 Pa (B) 1.6×10^5 Pa

(C) 2.4×10^5 Pa (D) 3.2×10^5 Pa

11. Figure 8 represents a portion of a blood vessel with circular cross sections. The segment on the left has diameter 1.6 mm, and each of the segments on the right has diameter 0.4 mm.

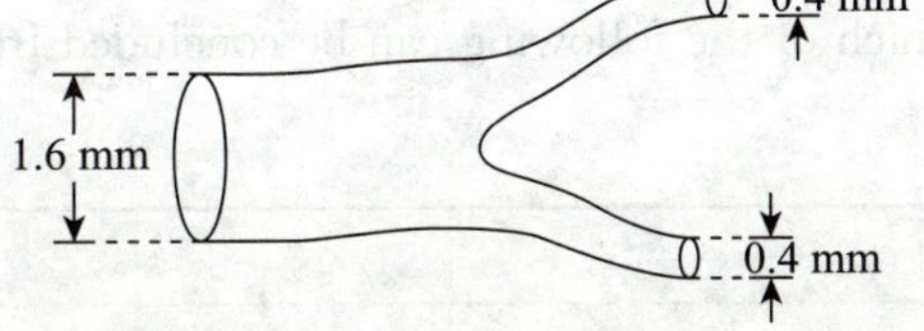

图 8

Assume the blood can be treated as an ideal fluid. If blood enters the segment on the left with speed v, at what speed does the blood leave the segments on the right?

(A) $2v$ (B) $4v$ (C) $8v$ (D) $16v$

12. A slab of metal and a slab of wood are placed in a classroom and allowed to sit undisturbed for a long time. A student then places one hand on the metal and the other hand on the wood. Which of the following describes the student's perception of the temperatures of the slabs and their actual temperatures?

(A) The metal slab feels colder to the student because it is at a lower temperature

(B) The metal slab feels colder to the student because it conducts thermal energy away from the student's hand faster, but the slabs have the same temperature

(C) The metal slab feels warmer to the student because it conducts thermal energy to the student's hand faster, but the slabs have the same temperature

(D) Both slabs feel the same to the student because they are at the same temperature

13. When one end of a cold metal spoon is placed upright in a cup of hot cocoa, the other end eventually gets warmer. On the scale of the molecules of the spoon, which of the following is the primary explanation of this phenomenon?

(A) Higher-energy molecules will, on average, rise to the top

(B) The hot molecules produce thermal radiation that is then absorbed by the colder molecules

(C) Higher-energy molecules hit lower-energy molecules and, on average, tend to reduce the difference in temperature between the ends of the spoon

(D) Thermal energy is a fluid that flows from high concentration (hot areas) to low concentration (cold areas)

14. An object with charge $+q$ passes to the right of one pole of a magnet and at a particular instant is moving with a velocity v toward the bottom of the page, as shown in figure 9. The force exerted on the object by the magnet at that instant is directed into the page. What is the direction of the force exerted on the magnet by the object?

(A) out of the page

(B) toward the right

(C) toward the top of the page

(D) no direction; the force is zero

图 9

15. A student was given five wires of the same length and diameter. The student connected the wires to the same battery one by one and measured the current through each wire. Table 1 shows the data collected. Which of the following can be concluded from the data?

表 1

wire	1	2	3	4	5
current (A)	0.5	1.0	2.0	2.5	3.0

(A) The wires are ohmic

(B) The wires are made of different materials

(C) The resistance of the wires depends on their size and shape

(D) The battery has internal resistance

16. In figure 10, the initially uncharged insulating rod on the left is free to rotate on an insulating stand. End X is then rubbed with a piece of fur. End Z of a second insulating rod is also rubbed with fur and then brought near the first rod. As the second rod is moved around, it is found that end Y of the first rod is attracted to end Z, and end X is repelled. Which of the following correctly describes the signs of the charges on ends X and Y, if any?

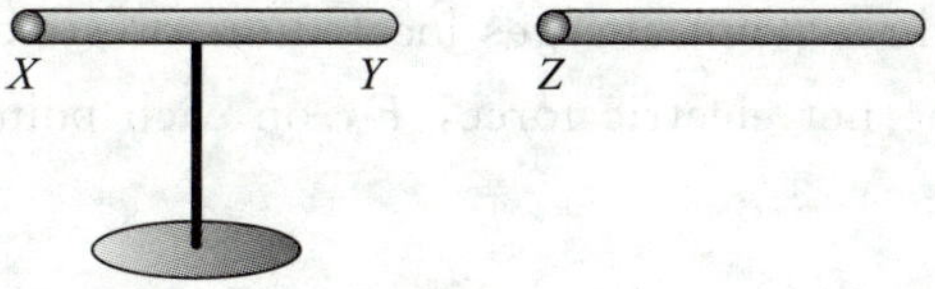

图 10

	End X	End Y
(A)	Opposite Z's charge	Same as Z's charge
(B)	Opposite Z's charge	No net charge
(C)	Same as Z's charge	Opposite Z's charge
(D)	Same as Z's charge	No net charge

17. A student has a positively charged insulating rod and two initially uncharged conducting spheres, A and B, with insulated handles. Which of the following ordered sequences of actions could the student use to produce a net positive charge on sphere A ?

(A) Bring the rod near the right side of A. Touch B to the left side of A. Remove B and then remove the rod

(B) Bring the rod near the right side of A. Touch B to the left side of A. Remove the rod and then remove B

(C) Bring the rod near the right side of B. Touch A to the left side of B. Remove B and then remove the rod

(D) Bring the rod near the right side of B. Touch A to the left side of B. Remove the rod and then remove B

18. Figure 11 shows a long conducting wire that lies in the plane of the page and carries an electric current I toward the right. At the instant shown, a positive point charge $+q$ is in the plane of the page and moving toward the bottom of the page. What is the direction of the magnetic force on the point charge at that instant?

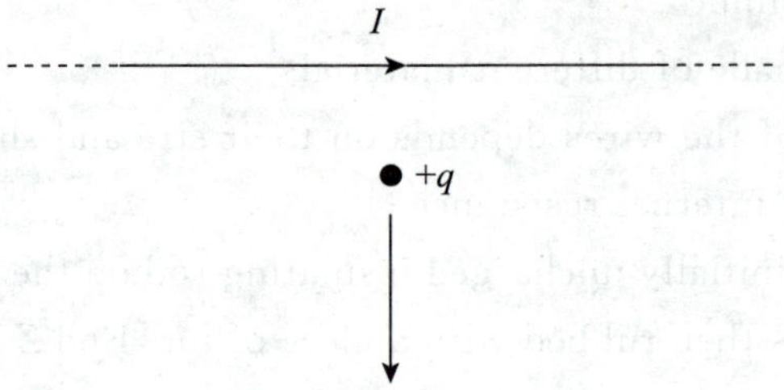

图 11

(A) into the page (B) out of the page

(C) toward the right (D) toward the left

19. Figure 12 shows three point charges located on an x-axis. Which of the following ranks the magnitude of the net electric force, F, on each point charge due to the other charges?

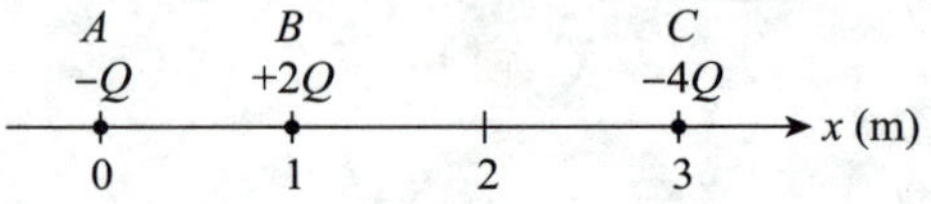

图 12

(A) $F_A = F_C > F_B$ (B) $F_B > F_C = F_A$

(C) $F_A = F_B > F_C$ (D) $F_A > F_C > F_B$

20. A block is submerged in a container of liquid and held under the surface of the liquid by a string connected to the bottom of the container, as shown in figure 13. The tension in the string is not zero. How does the buoyant force F_B exerted by the liquid on the block compare to the block's weight w ?

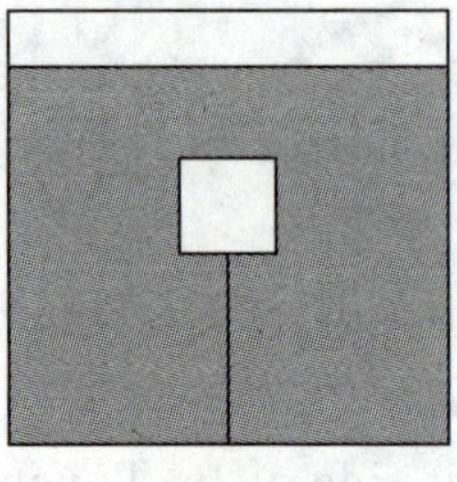

图 13

(A) $F_B < w$ (B) $F_B = w$ (C) $F_B > w$

(D) Either $F_B < w$ or $F_B > w$, depending on the density of the liquid

21. A variable resistor is connected to a battery of unknown potential difference. The power dissipated by the resistor and the current through the resistor are measured. Figure 14 shows a graph of the data and a best-fit line for the data. The potential difference provided by the battery is most nearly

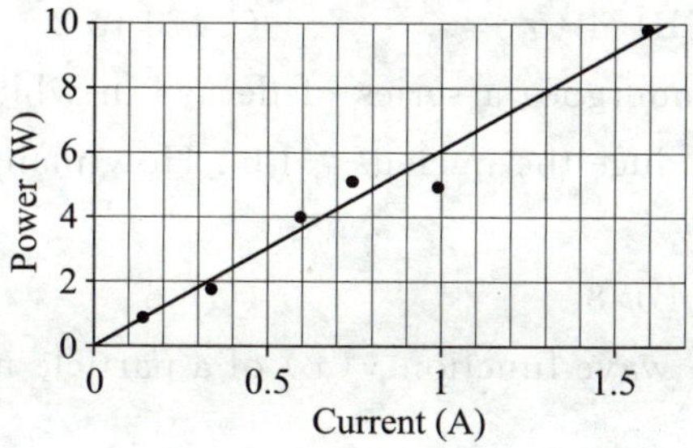

图 14

(A) 0.3 V (B) 3.0 V

(C) 5.0 V (D) 6.0 V

22. A scientist is studying the radioactive decay of an element and wants to determine the sign of some of the particles emitted. The scientist lets a beam of the particles, all with the same speed, pass through an opening in a shield and enter the field created by a wire carrying a constant current (图 15).

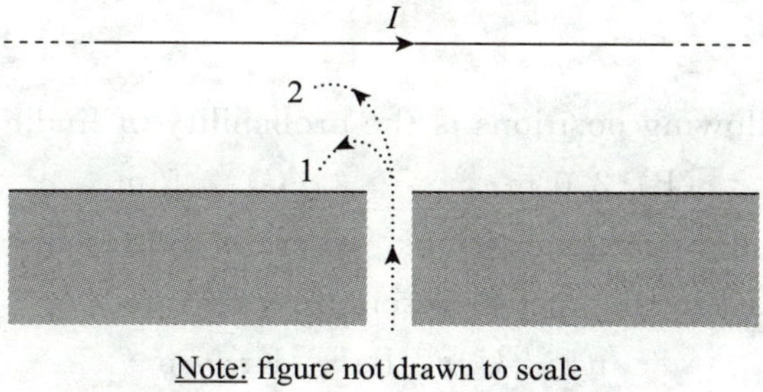

图 15

Paths 1 and 2 represent the observed paths of the particles. Based on the paths, which of the following could be the types of particles?

(A) 1 is electron, 2 is α particle (B) 1 is α particle, 2 is electron

(C) 1 is positron, 2 is proton (D) 1 is proton, 2 is positron

23. The energy states for an electron in a hypothetical atom are represented in figure 16. A cold sample of gas of such atoms is illuminated by a brief flash of light with a continuous spectrum. The atoms are then observed to emit light with discrete wavelengths. Which of the following transitions produces the longest wavelength light?

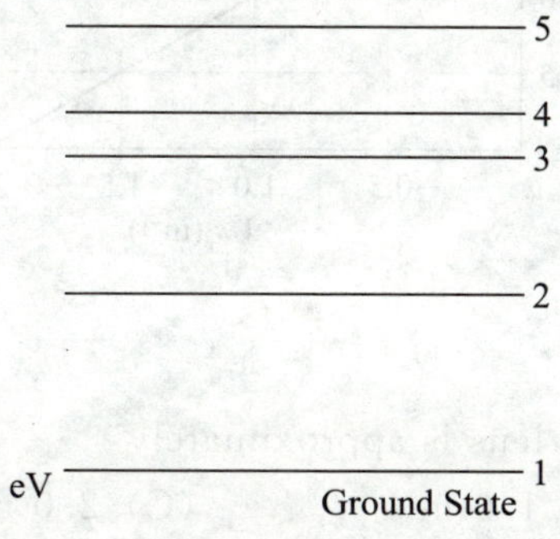

图 16

(A) 5 to 4 (B) 4 to 3 (C) 3 to 2 (D) 2 to 1

24. A nucleus of $^{238}_{92}U$ undergoes a series of decays in which it emits eight $^{4}_{2}He$ nuclei and some β^- particles to produce the nucleus $^{206}_{82}Pb$. How many β^- particles are emitted in the process?

(A) 10 (B) 8 (C) 6 (D) 5

25. Graph 17 shows the wave function $y(x)$ of a particle moving in the region $0<x<4$ m.

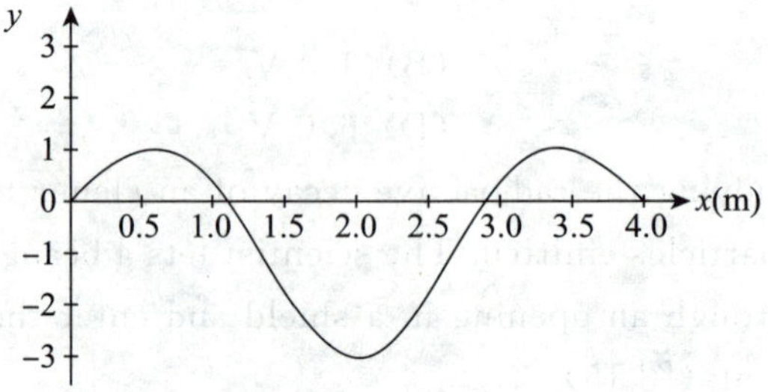

图 17

At which of the following positions is the probability of finding the particle greatest?

(A) 0.7 m (B) 2.0 m (C) 2.5 m (D) 4.0 m

26. Which of the following expressions, where m represents the mass of a particle, equals the energy released in the nuclear reaction represented?

$$n+^{235}_{92}U\rightarrow ^{141}_{56}Ba+^{92}_{36}Kr+3n$$

(A) $(m_{Ba}+m_{Kr}+m_U-3m_n)c^2$ (B) $(m_{Ba}+m_{Kr}-m_U-2m_n)c^2$

(C) $(m_U-m_{Ba}-m_{Kr}-3m_n)c^2$ (D) $(m_U-m_{Ba}-m_{Kr}-2m_n)c^2$

Questions 27—28 refer to the following material.

A group of students collected data using a lens. They varied the distance s_o of an object from the lens and measured the image distance s_i. Figure 18 is their graph of the inverse of the image distance as a function of the inverse of the object distance.

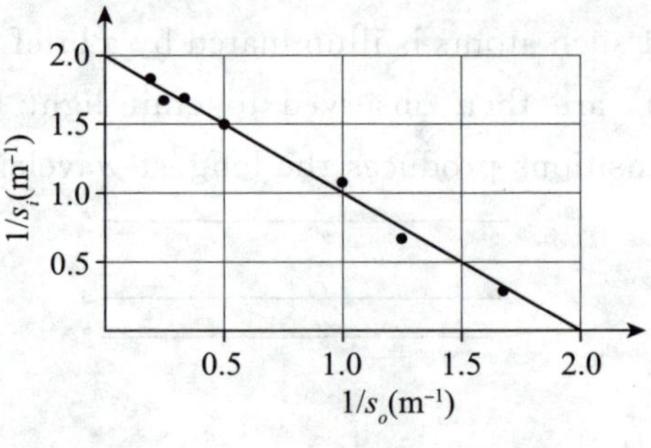

图 18

27. The focal length of the lens is approximately

(A) 0.5 m (B) 1.0 m (C) 2.0 m (D) 4.0 m

28. What is the magnitude of the image's magnification when the object is placed 2 m from the lens?

(A) 1/3 (B) 1 (C) 3

(D) The magnification is undefined because the image is an infinite distance from the lens

29. A partially evacuated vertical cylindrical container is covered by a circular lid that makes an airtight seal. The pressure in the room surrounding the container is 1.01×10^5 N/m^2, and the pressure inside the container is 0.41×10^5 N/m^2. The lid has radius 0.20 m and weight 200 N. The minimum upward applied force required to lift the lid is most nearly

(A) 7.5×10^3 N (B) 7.7×10^3 N

(C) 1.3×10^4 N (D) 1.8×10^4 N

30. A sealed container of air has been sitting on a table in a dark room for a very long time. The room is always kept at a constant temperature.

Which of the following best describes what will happen to the speeds and the average kinetic energy of the molecules of the air in the container as the container continues to sit on the table?

(A) Each molecule will continue to move at constant speed, keeping the average kinetic energy of the molecules constant

(B) Some molecules will speed up and others will slow down, keeping the average kinetic energy of the molecules constant

(C) The molecules will gradually slow down, decreasing their average kinetic energy

(D) The molecules will gradually speed up, increasing their average kinetic energy

31. Figure 19 shows a long, straight wire that has a steady current I in the $+y$-direction. A small object with charge $+q$ hangs from a thread near the wire. A student wants to investigate the magnetic force on the object due to the current but is not able to observe or measure changes in the tension in the string. Of the following actions that the student can take, which will allow the student to observe a reaction of the object due to the magnetic force on it?

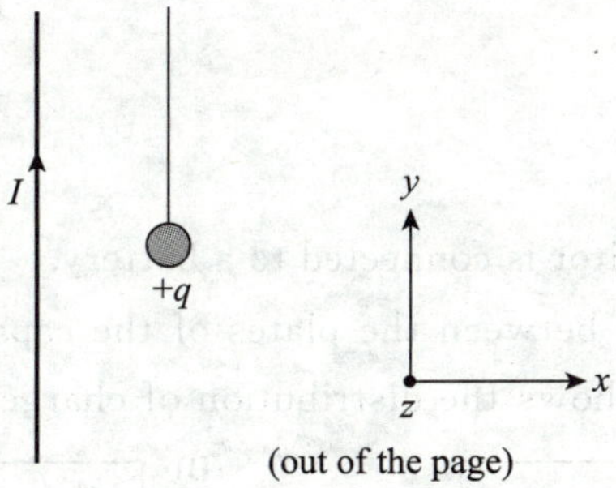

图 19

(A) Holding the object motionless

(B) Moving the object in a circle that is centered on the wire and in the x-z plane

(C) Moving the object in the $-x$-direction

(D) Moving the object in the $+y$-direction

Questions 32—33 refer to the following material.

Figure 20 shows the electric field in a region surrounding four charged particles, labeled 1, 2, 3, and 4, that are held in place. (Vectors with very large magnitude are not shown.)

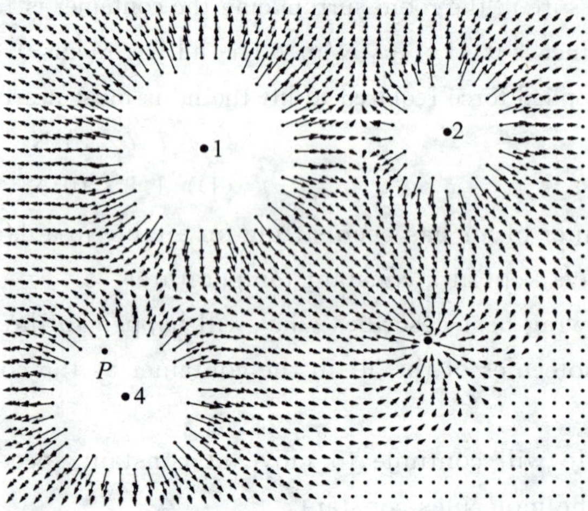

图 20

32. A small positive test charge is held at point P. Which of the following is closest to the direction of the force exerted on the test charge?

(A)  (B)  (C) 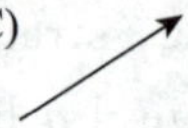 (D) 

33. Which of the following indicates the particles with charges of the greatest and the least magnitude?

	greatest	least
(A)	4	2
(B)	3	1
(C)	2	4
(D)	1	3

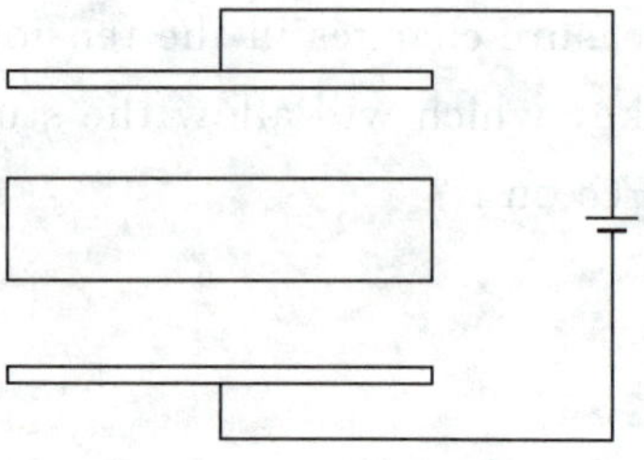

图 21

34. A parallel-plate capacitor is connected to a battery. A thick metal plate is located between the plates of the capacitor, as shown in figure 21. Which of the following best shows the distribution of charge on the thick plate?

(A) 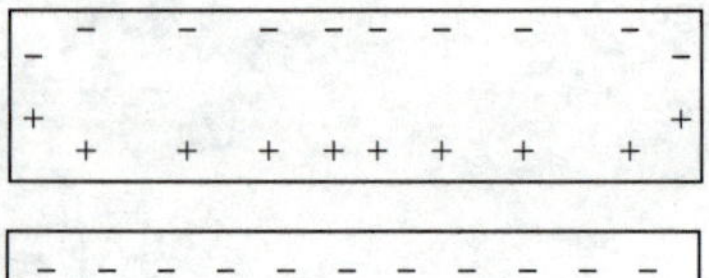

(B) 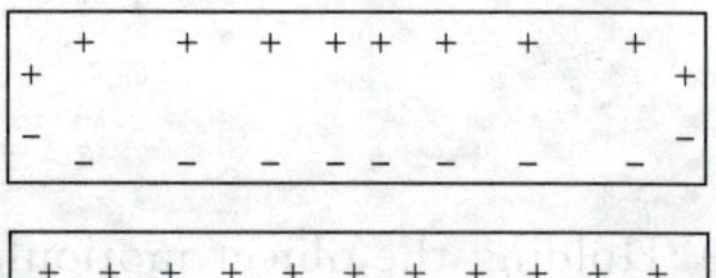

(C) 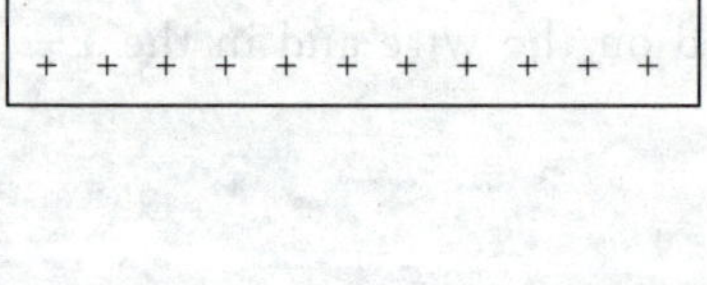

(D) 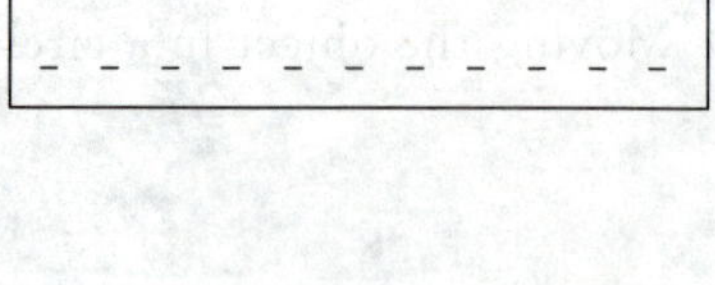

35. At the instant shown in figure 22, an electron is moving to the right along a straight line midway between two identical bar magnets. Which of the following describes the path of the electron as it continues to move?

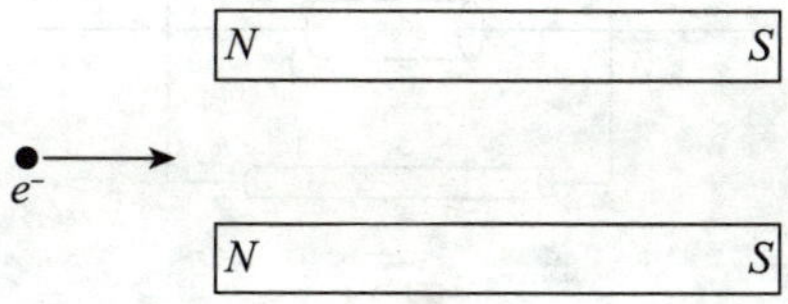

图 22

(A) It remains on the straight line and passes completely between the magnets

(B) It curves upward and strikes the top magnet

(C) It curves downward and reverses direction before it enters the space between the magnets

(D) It stops and reverses direction, then moves to the left along the same straight line

36. Which of the following procedures can a student follow to give a metallic sphere a positive charge by induction?

(A) Touching the sphere with a positively charged rod

(B) Placing a negatively charged rod near the sphere

(C) Placing a positively charged rod near the sphere, grounding the sphere, and then removing the connection to ground

(D) Placing a negatively charged rod near the sphere, grounding the sphere, and then removing the connection to ground

Questions 37—38 refer to the following material.

In the circuit shown in figure 23, the current through the ammeter is 20 mA and the voltmeter indicates 1. 0 V.

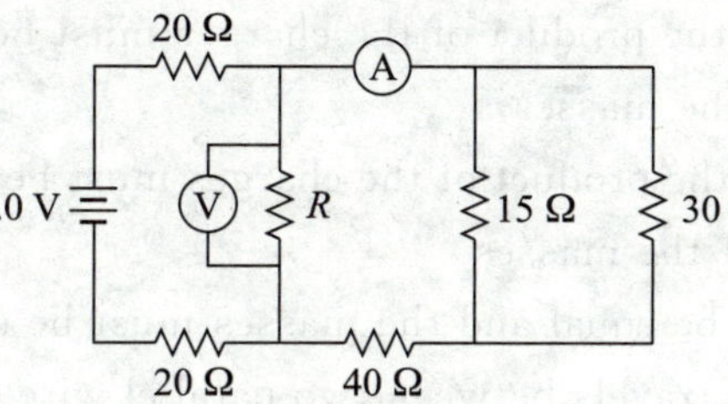

图 23

37. What is the current through the 40 Ω resistor?

(A) 7. 5 mA (B) 10 mA (C) 20 mA (D) 40 mA

38. The resistance of resistor R is most nearly

(A) 50 Ω (B) 33 Ω (C) 20 Ω (D) 14 Ω

39. Figure 24 represents a section of a circuit containing three resistors, X, Y, and Z, of different sizes but made of the same material. Which of the following correctly ranks the

current in the resistors?

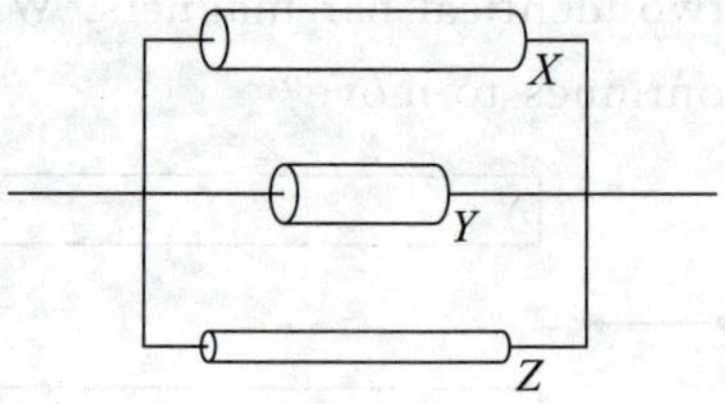

图 24

(A) $I_Z > I_X > I_Y$　　(B) $I_Z = I_X > I_Y$

(C) $I_Y = I_X = I_Z$　　(D) $I_Y > I_X > I_Z$

40. Three identical, small cork spheres are released from rest in a uniform electric field directed downward toward the floor. Sphere X is uncharged, but spheres Y and Z are charged. Sphere Z remains suspended in the field. Spheres X and Y fall downward, but sphere X takes a longer time to reach the floor. What are the signs of the charges on spheres Y and Z?

	sphere Y	sphere Z
(A)	negative	negative
(B)	negative	positive
(C)	positive	positive
(D)	positive	negative

41. Two charged objects are held at a distance d apart. Their mutual gravitational force is equal to their mutual electrostatic force. Which of the following must be true of the charges and masses of the objects, if charge is expressed in C and mass is expressed in kg?

(A) The magnitude of the product of the charges must equal to the magnitude of the product of the masses

(B) The magnitude of the product of the charges must be much less than the magnitude of the product of the masses

(C) The magnitude of the product of the charges must be much greater than the magnitude of the product of the masses

(D) The charges must be equal and the masses must be equal

42. A beam of electrons travels between two parallel wire coils, as shown in figure 25. When the coils carry no current, the electron beam is undeflected and hits the center of the screen, as indicated by the dashed line. When a constant current I in the indicated direction is created in the coils, the electron beam is deflected toward which edge of the screen?

(A) The top　　(B) The bottom

(C) The left　　(D) None; it is not deflected

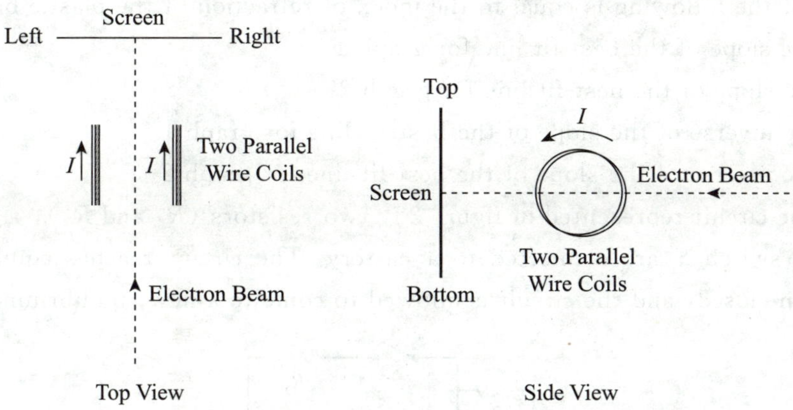

图 25

43. An absorption spectrum, observed for light traveling through interstellar gases, appears as a broad spectrum of light with isolated dark lines at frequencies characteristic of the molecules in the gases. Which of the following is primarily responsible for the location of the dark absorption lines in the spectrum?

(A) The characteristic velocities of the molecules in the gas

(B) The transition of electrons between discrete energy states within the gas molecules

(C) The percent of the gas molecules that were ionized by absorption of energy

(D) The fluctuations in density of molecules present in the interstellar gas cloud

44. A student performs an experiment to determine the index of refraction of a plastic block. The student shines a laser beam from air ($n_{air}=1$) into a piece of plastic ($n_{plastic}$) at an angle of incidence θ_1 and measures the angle of refraction θ_2, as shown in figure 26. Measurements are taken as θ_1 is increased. From the data the student produces the two graphs (图 27).

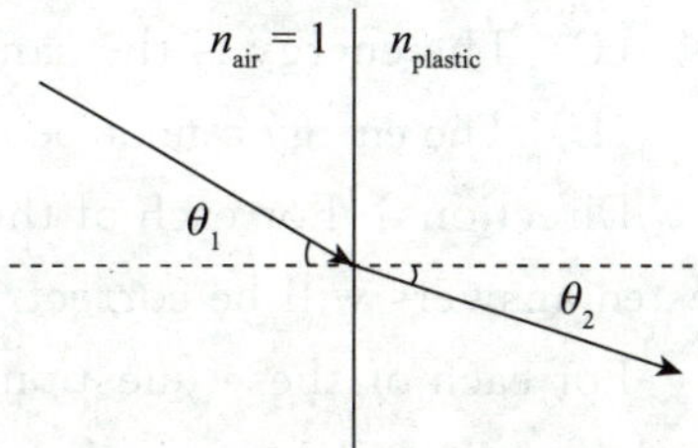

图 26

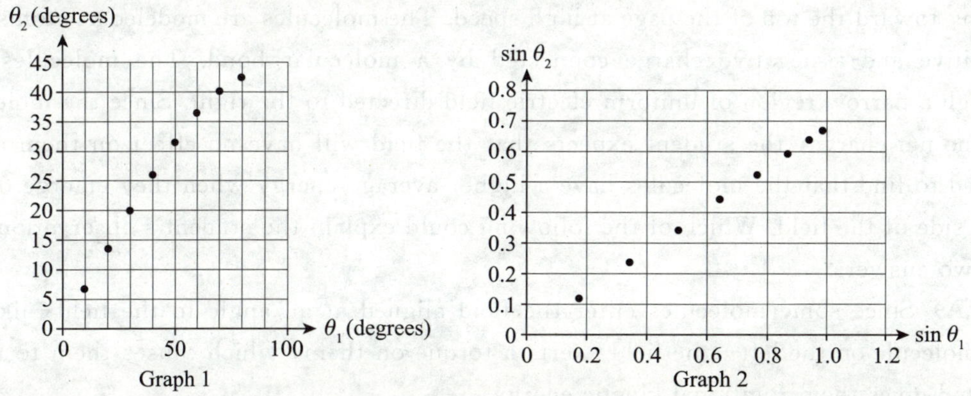

图 27

Which of the following is equal to the index of refraction of the plastic block?

(A) The slope of the best-fit line for graph 1

(B) The slope of the best-fit line for graph 2

(C) The inverse of the slope of the best-fit line for graph 1

(D) The inverse of the slope of the best-fit line for graph 2

45. In the circuit represented in figure 28, two resistors (R_1 and R_2), a capacitor C, and an open switch S are connected to a battery. The circuit reaches equilibrium. The switch is then closed, and the circuit is allowed to come to a new equilibrium.

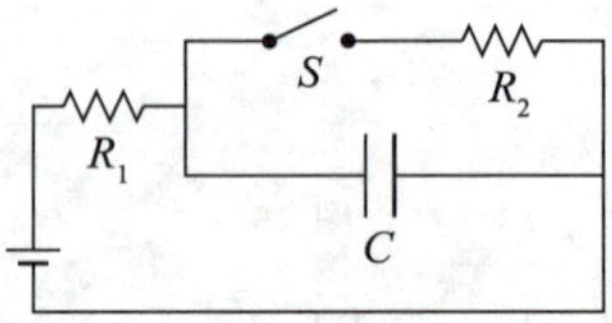

图 28

Which of the following is a true statement about the energy stored in the capacitor after the switch is closed compared with the energy stored in the capacitor before the switch is closed?

(A) The energy is greater

(B) The energy is less

(C) The energy is the same

(D) The energy cannot be determined without knowing the resistances of the resistors

Directions: For each of the questions or incomplete statements below, two of the suggested answers will be correct.

For each of these questions, you must select both correct choices to earn credit. No partial credit will be earned if only one correct choice is selected. Select the two that are best in each case and then enter both of the appropriate letters in the corresponding space on the answer sheet.

46. Figure 29 represents part of a student's experiment in which polar molecules are moving toward the top of the page at high speed. The molecules are modeled as atoms with a positive and a negative charge connected by a molecular bond. The molecules pass through a narrow region of uniform electric field directed to the right. Since the molecules have no net charge, the student expects that the field will have no effect on them and is puzzled to find that the molecules have a higher average energy when they emerge on the other side of the field. Which of the following could explain the student's observation? Select two answers.

(A) Since some molecules enter the field aligned at an angle to the field—like the molecule on the left—the field exerts a torque on them, which causes them to rotate and gives them rotational kinetic energy

(B) Since the molecules are polar, they behave as if they have a net charge. The field

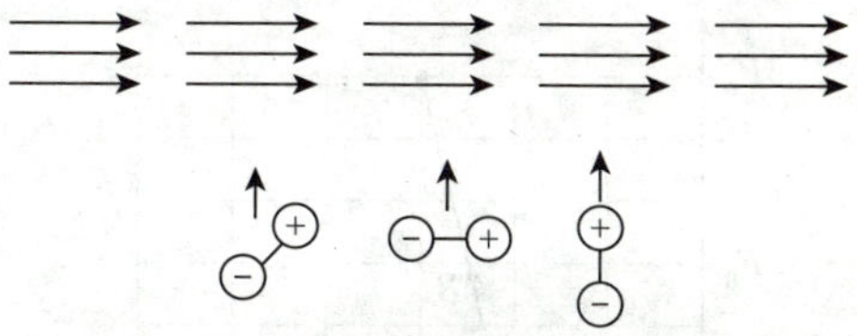

图 29

exerts a net force on them that does work and gives them linear kinetic energy

(C) Since some molecules enter the field aligned parallel to its direction—like the molecule in the middle—the field exerts forces in opposite directions on the two atoms, which stretches the molecular bonds and gives the molecules internal energy

(D) Since some molecules enter the field aligned perpendicular to its direction—like the molecule on the right—the field exerts forces in opposite directions on the two atoms. The force on the positive charge is greater, so the net force does work and gives them linear kinetic energy

47. A sample of an ideal gas can be taken from state i to state f via two processes, as shown in graph 30 of pressure P versus volume V. In one process the gas goes through state X, and in the other process the gas goes through state Y.

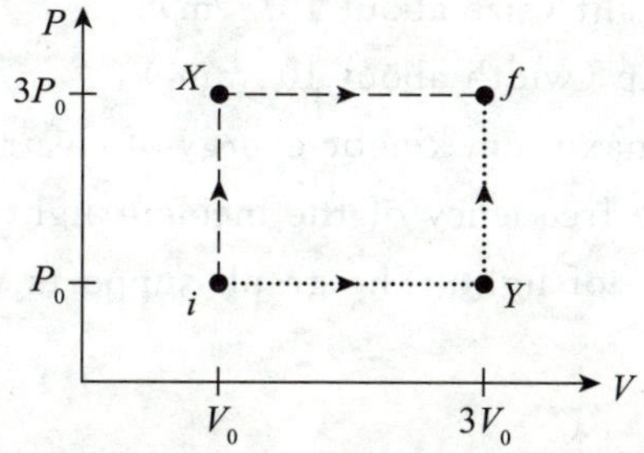

图 30

Which of the following will be the same for both processes? Select two answers.

(A) The change in the internal energy of the gas

(B) The temperature of the gas at the end of the process

(C) The thermal energy transferred to the gas by heating

(D) The work done on the gas

48. Identical samples of gas initially have pressure P_0, volume V_0, and temperature T_0. In some experiments, students take samples through each of the processes shown in graph 31. The final temperature is equal to the initial temperature for which of the processes? Select two answers.

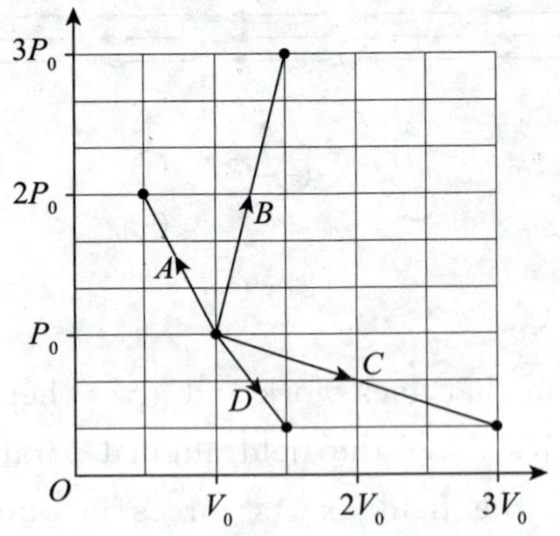

图 31

(A) A (B) B (C) C (D) D

49. A beam of electrons is moving at a speed of 6×10^6 m/s. If the beam is incident on each of the following objects, for which of the objects will diffraction be observed? Select two answers.

(A) A human hair (diameter about 10^{-5} m)

(B) A metal crystal (interatomic spacing about 10^{-10} m)

(C) A nanometer-width slit (size about 10^{-9} m)

(D) A narrow paper strip (width about 10^{-3} m)

50. Graph 32 shows the maximum kinetic energy of electrons released in the photoelectric effect as a function of the frequency of the incident light. This graph is often used as evidence of the particle nature of light. The graph supports which of the following statements? Select two answers.

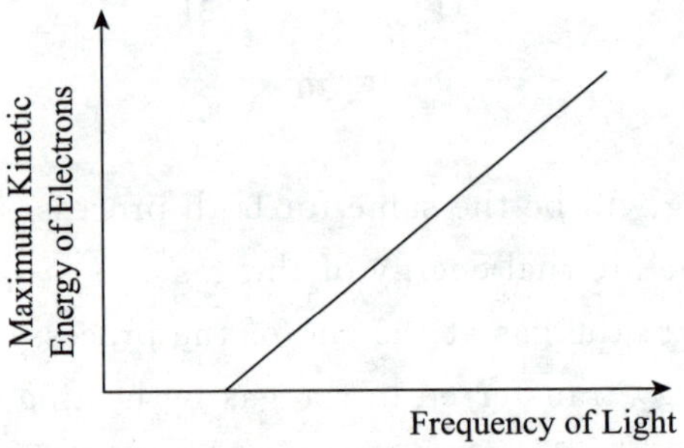

图 32

(A) The number of photoelectrons increases as the frequency increases

(B) There is a threshold frequency below which no photoemission occurs

(C) The energy of photoelectrons depends on the energy of the incoming light

(D) The intensity of the incoming light determines whether photoemission occurs

Section Ⅱ: Free Response

DO NOT OPEN THIS BOOKLET UNTIL YOU ARE TOLD TO DO SO.

Instructions

The questions for Section Ⅱ are printed in this booklet. You may use any blank space in the booklet for scratch work, but you must write your answers in the spaces provided for each answer. A table of information and lists of equations that may be helpful are in the booklet. Calculators, rulers, and straightedges may be used in this section.

All final numerical answers should include appropriate units. Credit for your work depends on demonstrating that you know which physical principles would be appropriate to apply in a particular situation. Therefore, you should show your work for each part in the space provided after that part. If you need more space, be sure to clearly indicate where you continue your work. Credit will be awarded only for work that is clearly designated as the solution to a specific part of a question. Credit also depends on the quality of your solutions and explanations, so you should show your work.

Write clearly and legibly. Cross out any errors you make; erased or crossed-out work will not be scored. You may lose credit for incorrect work that is not crossed out.

Manage your time carefully. You may proceed freely from one question to the next.

You may review your responses if you finish before the end of the exam is announced.

Total Time

90 minutes

Number of Questions

4

Percent of Total Score

50%

Writing Instrument

Either pencil or pen with black or dark blue ink

Electronic Device

Calculator allowed

Suggested Time

Approximately 25 minutes each for questions 2 and 3 and 20 minutes each for questions 1 and 4

Weight

Approximate weights:

Questions 2 and 3:

27% each

Questions 1 and 4:

23% each

AP Physics 2 Practice (1)
Section Ⅱ
4 Questions
Time—90 minutes

Directions: Questions 2 and 3 are long free-response questions that require about 25 minutes each to answer and are worth 12 points each. Questions 1 and 4 are short free-response questions that require about 20 minutes each to answer and are worth 10 points each. Show your work for each part in the space provided after that part.

1. (10 points, suggested time 20 minutes)

Three samples of a gas, X, Y, and Z, are prepared. Each sample contains the same number of molecules, but the samples are at different temperatures. The temperature of sample X is T_X, the temperature of sample Y is lower than that of sample X, and the temperature of sample Z is lower than that of sample Y ($T_X > T_Y > T_Z$).

(a) Graph 33 shows the distribution of the speeds of the molecules in sample Y. On the graph, sketch and label possible distributions for sample X and sample Z.

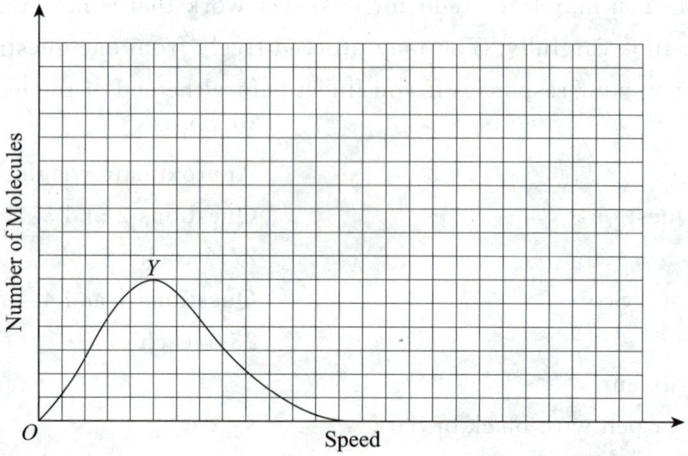

图 33

The three samples with initial temperatures $T_X > T_Y > T_Z$ are placed in thermal contact, with sample Z in the middle, as shown in figure 34, and the samples are insulated from their surroundings. The samples can exchange thermal energy but not gas molecules. The samples eventually reach equilibrium, with a final temperature greater than T_Y.

X	Z	Y

图 34

(b) In a few sentences, describe the change over time in the average kinetic energy of the molecules of each sample, from initial contact until they reach equilibrium. Explain how these changes relate to the energy flow between the pairs of samples that are in contact.

Sample X

Sample Y

Sample Z

(c) Indicate whether the net entropy of sample X increases, decreases, or remains the same as a result of the process of reaching equilibrium.

______ Increases ______ Decreases ______ Remains the same

Justify your answer at the microscopic level.

(d) For the three-sample system, indicate whether the entropy of the system increases, decreases, or remains the same.

______ Increases ______ Decreases ______ Remains the same

Justify your answer.

2. (12 points, suggested time 25 minutes)

Students are given some resistors with various resistances, a battery with internal resistance, and an ammeter. They are asked to determine the emf ε and internal resistance r of the battery using just this equipment. Working with the circuit shown in figure 35, they insert each resistor into the circuit and measure the current I in the circuit each time they insert a resistor. From their data, the students generate a graph of $1/I$ as a function of the resistance R of each resistor, as shown in figure 36.

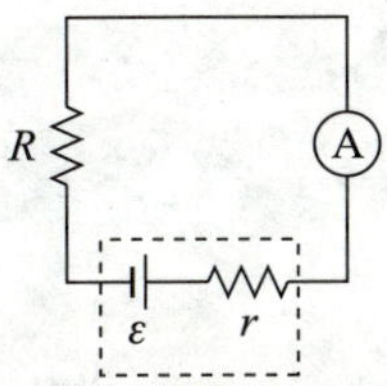

图 35

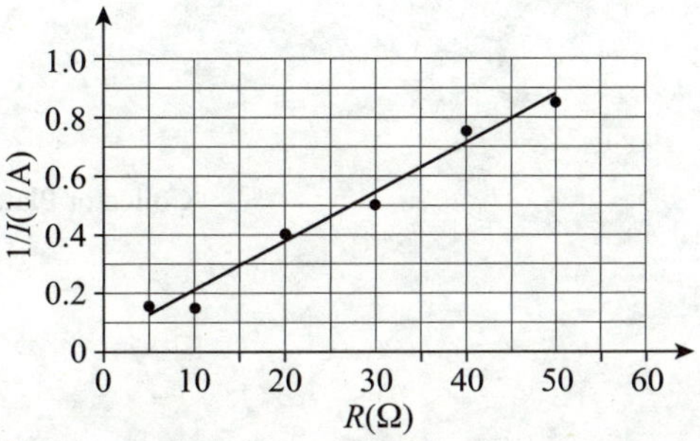

图 36

(a)

i. Write an algebraic equation describing the circuit that includes ε, R, r, and I.

ii. Use your equation and the graph to calculate the emf of the battery and the internal resistance of the battery.

The students are now given a voltmeter and a new resistor X to use with the resistors, battery, and ammeter they already have. They are asked to determine whether resistor X is ohmic.

(b)

i. Using standard symbols for circuit elements, as in the previously shown circuit, draw a diagram of a circuit that the students could use to determine whether resistor X is ohmic, including the appropriate placement of the meters. Clearly label your diagram.

ii. Describe the procedure you would use with your circuit to get enough data to determine whether resistor X is ohmic.

iii. What would you graph using your data, and what would you look for on your graph to determine whether resistor X is ohmic?

(c) Would your procedure or data analysis in part (b) need to be different if the internal resistance of the battery was nonohmic? Justify your answer.

3. (12 points, suggested time 25 minutes)

A particle with unknown mass and charge is projected into the apparatus shown in figure 37. The particle moves with constant speed v as it passes undeflected through a pair of parallel plates, as shown in the figure. The plates are separated by a distance d, and a constant potential difference V is maintained between them. A uniform magnetic field of magnitude B directed into the page exists both in the region between the plates and in the region to the right of the plates that is enclosed by the dashed lines. In the region to the right of the plates, the particle's path is circular with radius r. Assume the effects of gravity are negligible compared to other forces.

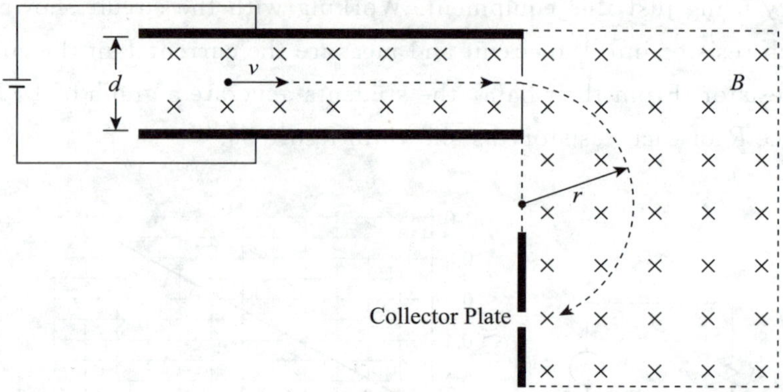

图 37

(a) Explain why the particle moves through the parallel plates undeflected in terms of the forces exerted on the particle.

(b) What is the sign of the charge on the particle? Justify your answer.

A magnetic field of 0.30 T is applied with the plate separation at 5.0×10^{-3} m. Singly ionized particles with various speeds enter the region between the plates, and only those with speed 2.0×10^{6} m/s are undeflected as they pass between the plates. These particles then reach the collector plate a distance of 0.42 m below the point at which they left the region between the parallel plates.

(c) Based on your explanation in part (a), derive an algebraic expression for the potential difference that must be applied to produce the motion of the undeflected particles. Use that expression to calculate the numerical value of the potential difference.

(d) By analyzing the circular part of the motion, derive an algebraic expression for the mass of the particles. Use that expression to calculate a numerical value for the mass.

(e) A scientist wants to use the apparatus to separate singly ionized atoms of ^{12}C and ^{14}C in order to use the ^{14}C in radio carbon dating. Describe how the motion of the two isotopes of carbon in both regions of the apparatus leads to their separation, appropriately relating your description to the algebraic equations you wrote in parts (c) and (d).

4. (10 points, suggested time 20 minutes)

Figure 38 labeled X, Y, and Z represent plane waves of the same wavelength incident on barriers that have openings of different sizes.

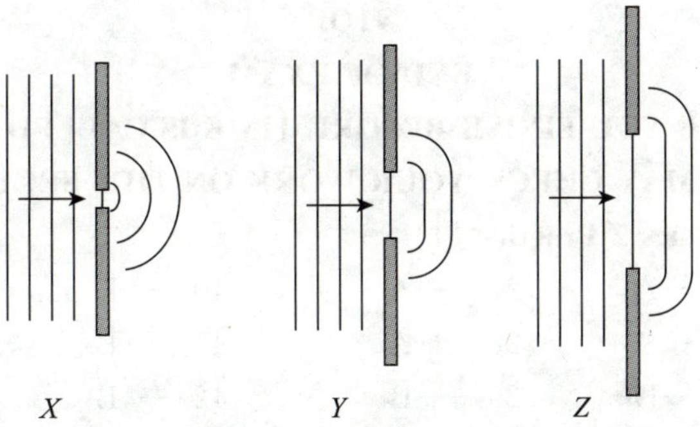

图 38

(a) One model of waves treats every point on a wave front as a point source. Give a clear, coherent, paragraph length description of how this model can be used to explain the shape of the wave fronts beyond the barriers.

Figure 39 represents another plane wave incident on a barrier with two identical openings, creating an interference pattern on the wall. Some positions in the pattern on the wall are labeled.

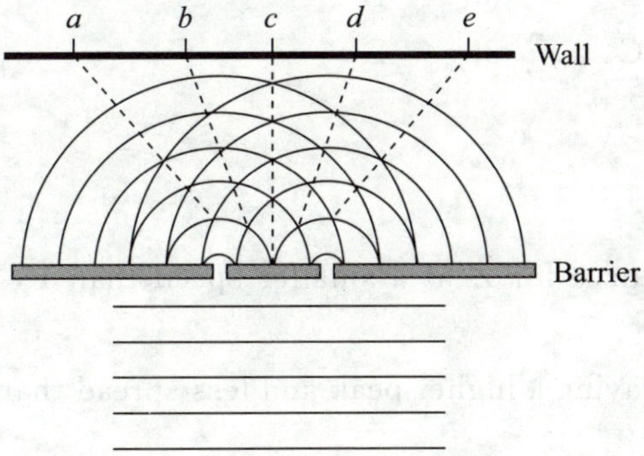

图 39

(b) In a few sentences, describe how the point-source model described in part (a) and the figure above can be used to explain the formation of the interference pattern on the wall.

(c) On the axes (图 40), sketch the intensity of the waves that are incident on the wall. The labels correspond to the positions noted in the figure above.

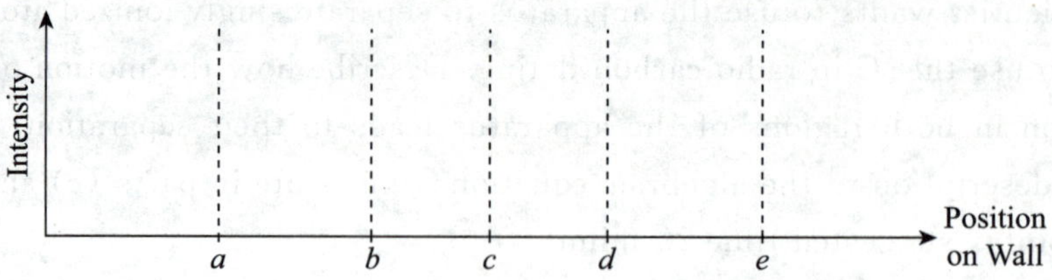

图 40

STOP

END OF EXAM

IF YOU FINISH BEFORE TIME IS CALLED,

YOU MAY CHECK YOUR WORK ON THIS SECTION.

Answer for AP Physics 2 Practice (1)

1——B	14——A	27——A	40——D
2——B	15——B	28——A	41——B
3——A	16——D	29——B	42——B
4——A	17——C	30——B	43——B
5——C	18——C	31——D	44——D
6——C	19——A	32——B	45——B
7——C	20——C	33——D	46——A, C
8——A	21——D	34——A	47——A, B
9——A	22——C	35——A	48——A, C
10——B	23——B	36——D	49——B, C
11——C	24——C	37——C	50——B, C
12——B	25——B	38——B	
13——C	26——C	39——D	

Question 1

10 points total **Distribution of points**

(a) 2 points

For the peak of the curve for Z at a smaller speed than Y, and X at a greater speed than Y 1 point

For the curve for Z having a higher peak and less spread than Y, and X having a lower peak and greater spread (图 41) 1 point

One earned point was deducted for correct curves that are not labeled

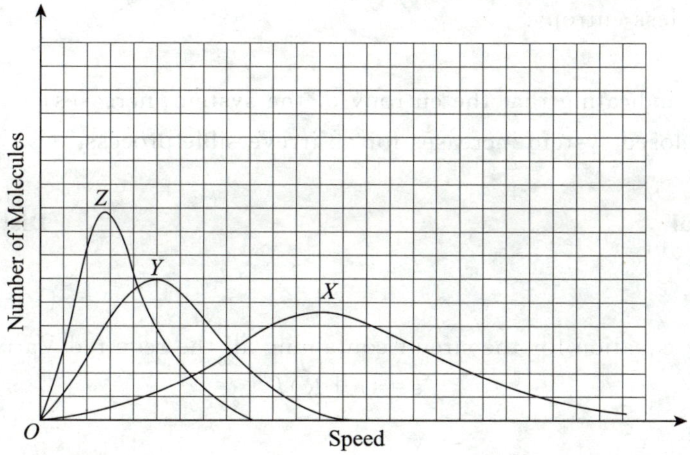

图 41

(b) 5 points

The kinetic energy of X decreases. It has the highest temperature and so to reach the same equilibrium temperature as the other samples it must lose energy, which flows into Z.

The kinetic energy of Y decreases and then increases. It has a higher temperature than Z, and so initially loses energy which flows into Z.

But it eventually must end up at a higher temperature than it initially had, so the net energy flow must be into Y. That can only happen if the direction of energy flow reverses.

The kinetic energy of Z could always increase, or it can increase and then decrease. Initially energy flows into it, since it has the lowest temperature. At some point energy begins to flow from Z to Y. Whether the temperature and thus the kinetic energy of Z continually increases or not depends on how much energy keeps flowing to it from X.

For exhibiting understanding that energy flows from systems at higher temperature to systems at lower temperature 1 point

For exhibiting understanding that higher temperature corresponds to higher kinetic energy 1 point

For exhibiting understanding that the energy flow stops 1 point

For exhibiting understanding that the energy flow for a sample can change Direction 1 point

For using the above understanding to indicate that the average kinetic energy of all three samples are the same when equilibrium is reached 1 point

(c) 2 points

For checking either the correct response or one consistent with the answer to part (b), with a reasonable attempt at justification 1 point

For a correct or consistent justification that relates the spread of the molecular distribution to entropy 1 point

For example: The entropy decreases. When the temperature goes down, the spread of the speeds and thus the kinetic energies of the individual molecules is less. This means less disorder and thus less entropy.

(d) 1 point

For correctly indicating that the entropy of the system increases, and explaining that the entropy of a closed system increases for an irreversible process. 1 point

Question 2

12 points total **Distribution of points**

(a) 4 points

(i) 1 point

For a correct equation for the circuit containing all the required Variables 1 point

$$\varepsilon=(r+R)I$$

(ii) 3 points

Slope-intercept method

The above equation can be re-arranged to express $1/I$ as a function of R

$$\frac{1}{I}=\frac{r}{\varepsilon}+\frac{R}{\varepsilon}$$

The emf is the inverse of the slope of the graph.

For using the slope of the line to determine the emf 1 point

For using data from the best-fit line 1 point

For example:

$$\text{Slope}=\frac{1}{\varepsilon}=\frac{(0.8-0.2)\left(\frac{1}{\text{A}}\right)}{(45-10)\ \Omega}=\frac{0.6}{35}\times\frac{1}{\text{V}}$$

$$\varepsilon=\frac{35}{0.6}\ \text{V}=58.3\ \text{V}$$

For using the intercept of the line to determine the battery resistance 1 point

Intercept$=r/\varepsilon$

For example, reading an intercept of 0.03(1/A)

$r=$ (Intercept) $\varepsilon=[0.03\ (1/\text{A})](58.3\ \text{V})=1.75\ \Omega$

(b) 6 points

(i) 2 points

For a design that allows the potential difference across resistor X to be varied 1 point

For an ammeter and a voltmeter connected correctly to measure current and potential difference for resistor X (图 42) 1 point

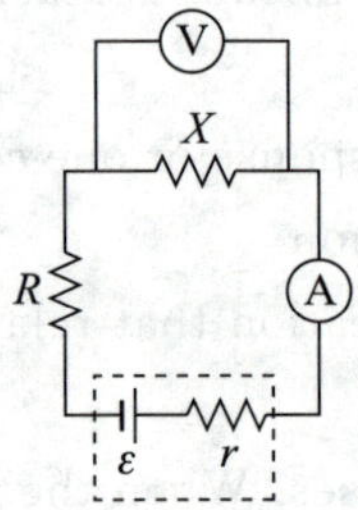

图 42

(ii) 2 points

For a correct method for varying the potential difference across resistor X 1 point

For the example circuit shown above, use each of the resistors in turn as resistor R.

For indicating appropriate measurements for each trial 1 point

For example, for each value of resistor R, measure the potential difference and current of resistor X.

(iii) 2 points

For indicating appropriate quantities to graph 1 point

For the example described above, graph current as a function of potential difference.

For indicating an appropriate property of the graph 1 point

For the example described above, the graph would be linear if resistor X was ohmic.

(c) 2 points

For correctly indicating "yes" or "no" consistent with circuit drawn, and some reasonable attempt at a correct justification 1 point

For the example described above, the correct response is "no".

For a correct justification 1 point

For the example described above, the measured values are those for only resistor X. Since the point is to change the resistance external to X, it does not matter how or why it changes. As long as the measurements are the quantities for X by itself the data is legitimate.

Question 3

12 points total **Distribution of points**

(a) 2 points

For any indication that both an electric and a magnetic force are exerted on the particle 1 point

For any indication that these forces are equal in magnitude and opposite in direction 1 point

For example: The plates create a vertical electric field that exerts a vertical force on the particle. The cross-product of the velocity and the magnetic field also results in a vertical force. No matter what the charge of the particle, these forces are in opposite directions. Since the charge moves in a straight line they must be equal in magnitude.

(b) 1 point

For indicating that the charge is negative and a correct justification 1 point

Examples:

The cross-product of the velocity and the magnetic field is directed upward. The particle curves downward as it leaves the plates, so it must be negative.

With the battery connection shown, the top plate is positive and the bottom one is negative. So the electric field is down. The particle would move down without the electric field, as it initially does when it leaves the plates, so the force from the field must be upward. That means the particle is negative.

(c) 3 points

$$F_B = qvB$$

For equating the magnitudes of the electric and magnetic forces 1 point

$$qE = qvB \text{ so } E = vB$$

For using the relationship between the field and potential for parallel plates, $E = V/d$ 1 point

$$V/d = vB$$
$$V = vBd$$

For correct substitutions 1 point

$$V = (2.0 \times 10^6 \text{ m/ s})(0.30 \text{ T})(5.0 \times 10^{-3} \text{ m})$$
$$V = 3\,000 \text{ V}$$

(d) 3 points

For equating the magnitudes of the centripetal and magnetic forces 1 point

$$mv^2/r = qvB$$
$$m = qBr/v$$

For realizing that the given distance to the collector plate is the diameter and using $r = 2D$ 1 point

For correct substitutions 1 point

$$m = (1.6 \times 10^{-19} \text{ C})(0.30 \text{ T})(0.21 \text{ m}) / (2.0 \times 10^6 \text{ m/s})$$
$$m = 5.04 \times 10^{-27} \text{ kg}$$

(e) 3 points

For any indication that the straight-line motion means that all the particles leaving the plates have the same speed regardless of charge and mass, referencing part (c) 1 point

For any indication that the radius of the circular motion depends on the charge and mass, as indicated by the equation $m = qBr/v$ 1 point from part (d)

For any indication that the radius depends only on the mass 1 point

Question 4

10 points total **Distribution of points**

(a) 5 points

For a correct description of X as essentially a point source 1 point

For demonstrating understanding that wider openings can be treated as multiple point sources 1 point

For demonstrating understanding that interference of wave fronts creates the pattern on the far side of the opening 1 point

For demonstrating understanding of the lack of interference at the edges 1 point

For coherently connecting the above ideas in a logical explanation 1 point

For example: In X, the size of the slit is comparable to the wavelength of the plane waves, so it acts most like a single point source and produces essentially spherical wave fronts. As the slit gets wider in Y and Z, it acts like an increasing number of point sources. Away from the edges, the spherical wave fronts interfere and the points of constructive interference result in the planar sections that reproduce the incident plane waves. Near the edges, there are no waves on one side to interfere, so the spherical fronts propagate.

(b) 3 points

For demonstrating understanding that the figure shows a set of wave fronts for each opening 1 point

For demonstrating understanding that maxima in the resulting interference pattern occur where the fronts intersect 1 point

For demonstrating understanding that the dashed lines lead to points of maxima in the resulting interference pattern on the wall 1 point

For example: Each opening creates a set of spherical wave fronts as shown.

Assuming the wave fronts denote maximum amplitude, where they cross are points of maximum constructive interference. The dashed lines connect these points, and indicate where the maxima of the pattern on the wall are located.

(c) 2 points

For maxima at the labeled positions on the wall and minima in between 1 point

For showing a decrease in the height of the maxima as position moves away from the central point c (图 43) 1 point

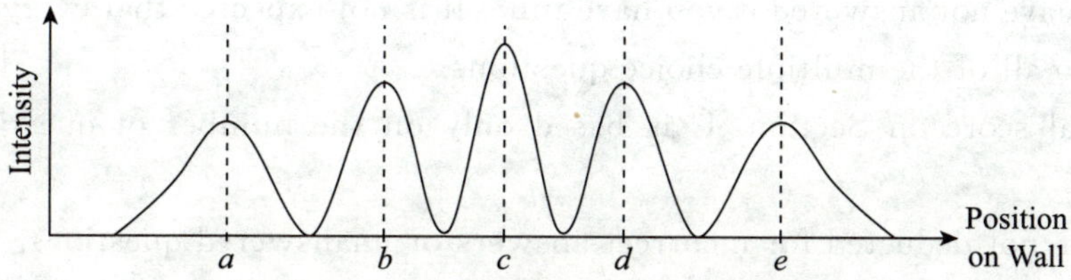

图 43

AP Physics 2 Practice Exam (2)

Section Ⅰ: Multiple Choice

DO NOT OPEN THIS BOOKLET UNTIL YOU ARE TOLD TO DO SO.

Instructions

Section Ⅰ of this exam contains 50 multiple-choice questions. Pages containing equations and other information are also printed in this booklet. Calculators, rulers, and straightedges may be used in this section.

Indicate all of your answers to the multiple-choice questions on the answer sheet.

No credit will be given for anything written in this exam booklet, but you may use the booklet for notes or scratch work.

For questions 1 through 45, select the single best answer choice for each question.

After you have decided which of the choices is best, fill in the appropriate letter in the corresponding space on the answer sheet.

For questions 46 through 50, select the two best answer choices for each question.

After you have decided which two of the choices are best, enter both letters in the corresponding space on the answer sheet.

Use your time effectively, working as quickly as you can without losing accuracy. Do not spend too much time on any one question. Go on to other questions and come back to the ones you have not answered if you have time. It is not expected that everyone will know the answers to all of the multiple-choice questions.

Your total score on Section Ⅰ is based only on the number of questions answered correctly.

Points are not deducted for incorrect answers or unanswered questions.

Total Time

90 minutes

Number of Questions

50

Percent of Total Score

50%

Writing Instrument

Pencil required

Electronic Device

Calculator allowed

AP Physics 2 Practice (2)

Section Ⅰ

50 Questions

Time—90 minutes

Note: To simplify calculations, you may use $g = 10\ m/s^2$ in all problems.

Directions: Each of the questions or incomplete statements below is followed by 4 suggested answers or completions. Select the one that is the best in each case and then enter the appropriate letter in the corresponding space on the answer sheet.

1. A vertical force of 30 N is applied uniformly to a flat button with a radius of 1 cm that is lying on a table. Which of the following is the best order of magnitude estimate for the pressure applied to the button?

(A) 10 Pa (B) 10^2 Pa (C) 10^3 Pa (D) 10^5 Pa

2. A ball that can float on water has mass 5.00 kg and volume $2.50 \times 10^{-2}\ m^3$. What is the magnitude of the downward force that must be applied to the ball to hold it motionless and completely submerged in fresh water of density $1.00 \times 10^3\ kg/m^3$?

(A) 20.0 N (B) 25.0 N (C) 200.0 N (D) 30.0 N

3. Water flows through the pipe shown in figure 1. At the larger end, the pipe has diameter D and the speed of the water is v_1. What is the speed of the water at the smaller end, where the pipe has diameter d ?

图 1

(A) V_1 (B) $\left(\frac{d}{D}\right)v_1$ (C) $\left(\frac{D}{d}\right)v_1$ (D) $\left(\frac{D^2}{d^2}\right)v_1$

4. The floor of a building is made from a square, solid piece of concrete. When the temperature of the floor increases from 20℃ to 28℃, each side of the square expands by 0.4 cm. If the temperature of the floor were to decrease from 20℃ to 8℃, by how much would each side of the square contract?

(A) 0.2 cm (B) 0.4 cm (C) 0.6 cm (D) 1.0 cm

5. The temperature of an ideal gas is directly proportional to which of the following?

(A) average translational kinetic energy of the molecules

(B) average velocity of the molecules

(C) average potential energy of the molecules

(D) average momentum of the molecules

Question 6—8

In the circuit shown in figure 2, the resistors all have the same resistance. The battery, wires, and ammeter have negligible resistance. A closed switch also has negligible resistance.

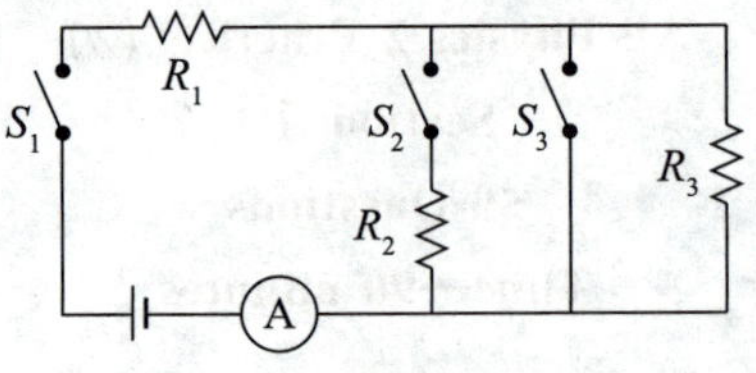

图 2

6. Closing which of the switches will produce the greatest power dissipation in R_2?

(A) S_1 only　　(B) S_2 only

(C) S_1 and S_2 only　　(D) S_1 and S_3 only

7. Closing which of the switches will produce the greatest reading on the ammeter?

(A) S_1 only　　(B) S_2 only

(C) S_3 only　　(D) S_1 and S_3

8. Closing which of the switches will produce the greatest voltage across R_3?

(A) S_1 only

(B) S_2 only

(C) S_1 and S_2 only

(D) S_1 and S_3 only

9. A ray of light in glass that is incident on an interface with ice, as shown in figure 3, is partially reflected and partially refracted. The index of refraction n for each of the two media is given in the figure. How do the angle of reflection and the angle of refraction compare with the angle of incidence θ?

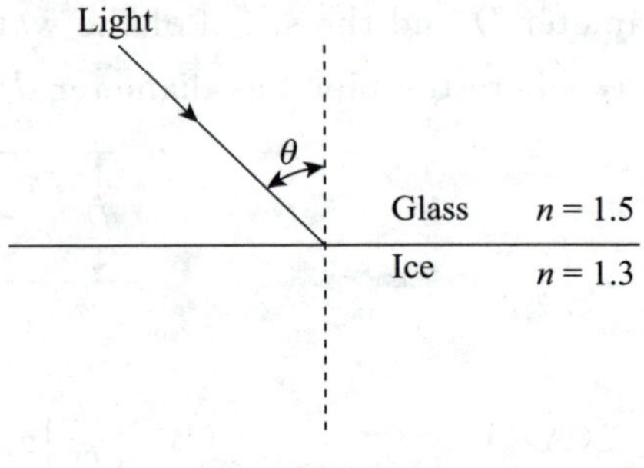

图 3

	angle of reflection	angle of refraction
(A)	same	larger
(B)	same	smaller
(C)	smaller	same
(D)	smaller	smaller

Question 10—11

A sample of an ideal gas is taken through the circle $abca$, as shown in diagram 4.

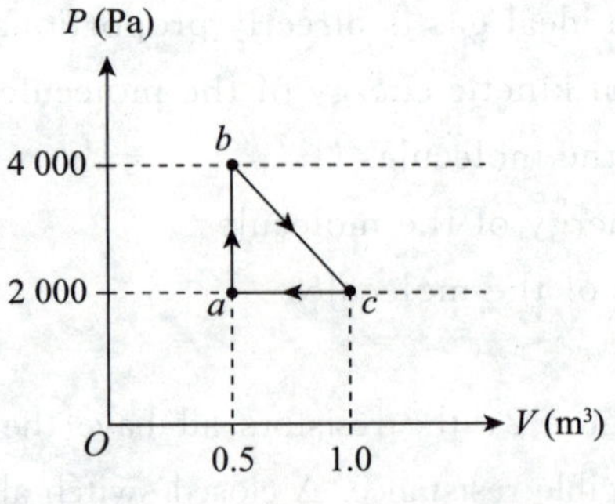

图 4

10. What is the change in internal energy of the gas for the process bc?

(A) $-1\,500$ J (B) -500 J (C) 0 J (D) $+500$ J

11. If 1 000J of heat is added to the gas during process ab, what is the change in internal energy of the gas for the process ab?

(A) $-2\,000$ J (B) $-1\,000$ J (C) 0 J (D) $+1\,000$ J

12. At room temperature and one atmosphere of pressure, which of following materials is the best conductor of heat?

(A) air (B) helium (C) silver (D) water

13. The equation of continuity for an incompressible fluid, $A_1V_1=A_2V_2$, is essentially an expression of the conservation of which of the following quantities?

(A) energy (B) time

(C) linear momentum (D) mass

14. A long, straight wire in the plane of the page carries a current I toward the right, as shown in figure 5. What is the direction of the magnetic field at point P which is in the plane of the page and above the wire?

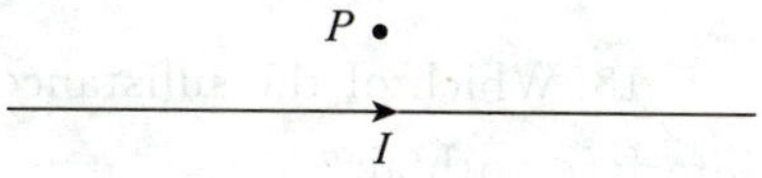

图 5

(A) out of the page (B) into of the page

(C) toward the left (D) toward the right

15. Two initially uncharged metal spheres are mounted on insulating stands and placed in contact with each other, as shown in figure 6. A student brings a positively charged insulating rod near sphere Ⅰ. Sphere Ⅱ is then removed, and finally the rod is moved away. What is the net charge on sphere Ⅰ?

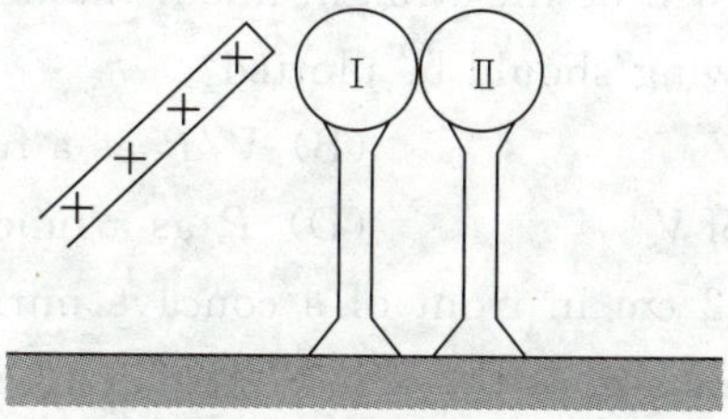

图 6

(A) no charge

(B) a positive charge

(C) a negative charge

(D) either a positive or a negative charge, depending on the sizes of the spheres

16. The magnetic flux for each turn of a 30-turn coil increases from zero to 0.60 Tm^2 in 3 s. The average emf induced in the coil is

(A) 6 V (B) 18 V (C) 54 V (D) 90 V

17. It takes 5.00×10^{-14} J of work to bring a particle with charge $+Q$ from infinity to a distance of 10.0 cm from positive point charge Y. How much work would be required to

bring the particle from infinity to a distance 5.0 cm from Y if the charge on the particle were $Q/2$ instead?

(A) 20.0×10^{-14} J (B) 10.0×10^{-14} J

(C) 5.00×10^{-14} J (D) 2.50×10^{-14} J

Questions 18—19 refer to the following table (表 1).

表 1 **Index of Refraction of Various Substances**

Acetone	1.36
Carbon tetrachloride	1.46
Diamond	2.42
Glass	1.52
Water	1.33

18. Which of the substances in the table would have the smallest critical angle for an interface with air?

(A) acetone (B) carbon tetrachloride

(C) diamond (D) glass

19. If light has a speed v in water, what would be its speed in diamond?

(A) $\frac{v}{2.42}$ (B) $1.33v$ (C) $2.42v$ (D) $\frac{1.33}{2.42}v$

20. A series of measurements are made of the pressure P and the volume V of the sample of nitrogen gas kept at a constant temperature. It is desired to represent the data graphically so that the graph will be the straight line if the behavior of the gas is ideal. Accordingly, which of the following should be plotted?

(A) P as a function of V (B) V/P as a function of V

(C) P/V as a function of V (D) P as a function of $1/V$

21. An object is placed 12 cm in front of a concave mirror. The mirror forms a real inverted image that is the same size as the object, also at 12 cm in front of the mirror. The focal length of the mirror is

(A) 3.0 cm (B) 4.0 cm (C) 6.0 cm (D) 12 cm

22. At the point of observation, the time between successive crests of a passing wave is 0.2 s. Which of the following must be true of the wave?

(A) Its wavelength is 0.2 m (B) Its wavelength is 5 m

(C) Its period is 5 s (D) Its frequency is 5 Hz

23. The two wave pulses, X and Y, show in figure 7 are moving toward each other in the same medium. Which the following interference patterns (图 8) could result at some instant as the pulse pass through each other?

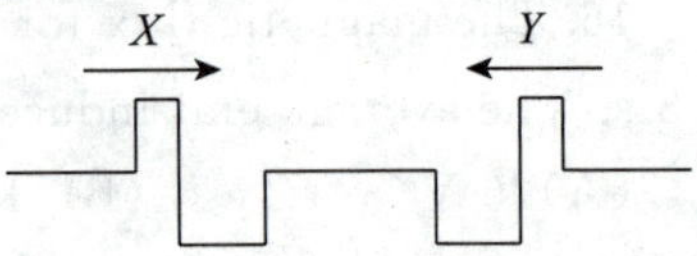

图 7

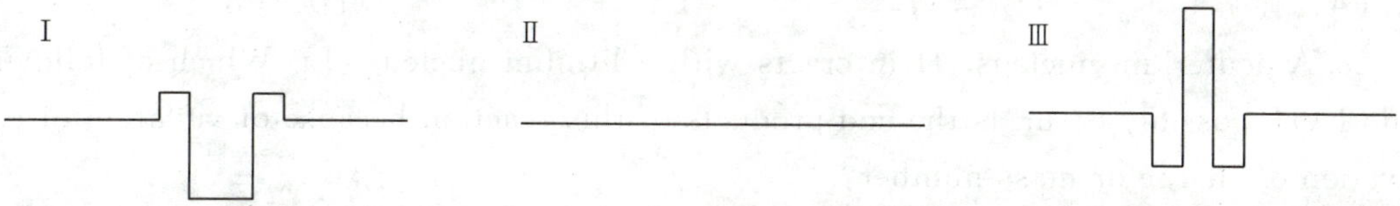

图 8

(A) Ⅰ only (B) Ⅱ only
(C) Ⅰ and Ⅲ only (D) Ⅱ and Ⅲ only

24. Which of the following experimental observations provides the best support for the statement "light behaves like a wave"?

(A) Light can be reflect by a mirror
(B) Light is scatted when passing through smoke
(C) Monochromatic light forms bright and dark bands after passing through two narrow slits
(D) White light can be broken into component colors by a prism

25. When a light wave passes from air into glass, quantities that remain constant include which of the following?

Ⅰ. frequency
Ⅱ. wavelength
Ⅲ. speed

(A) Ⅰ only (B) Ⅱ only
(C) Ⅰ and Ⅲ only (D) Ⅰ, Ⅱ and Ⅲ

26. Figure 9 shows a portion of the energy level diagram for hydrogen. Possible transitions among the energy levels are represented by arrows Ⅰ, Ⅱ and Ⅲ. Which of following lists the transitions in order of their associated wavelengths of light, from shortest wavelength to longest wavelength?

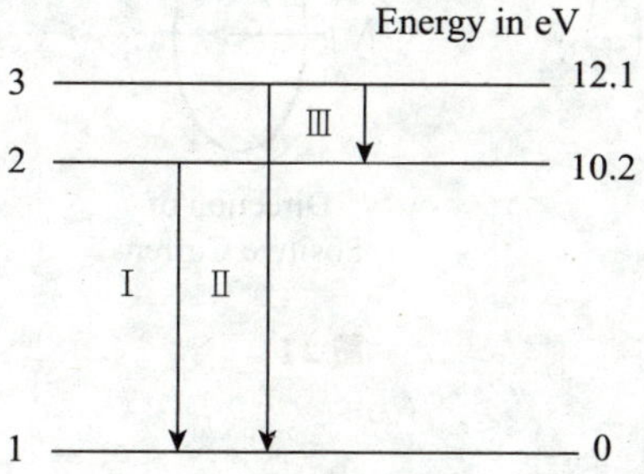

图 9

(A) Ⅰ, Ⅱ, Ⅲ (B) Ⅰ, Ⅲ, Ⅱ (C) Ⅱ, Ⅰ, Ⅲ (D) Ⅱ, Ⅲ, Ⅰ

27. $^{238}_{92}U$ undergoes a series of reaction in which it emits eight $^{4}_{2}He$ nuclei and six electrons. What is the isotope that results from this series of reactions?

(A) $^{222}_{66}\mathrm{Dy}$　　(B) $^{206}_{81}\mathrm{Ti}$　　(C) $^{206}_{82}\mathrm{Pb}$　　(D) $^{207}_{82}\mathrm{Pb}$

28. A deuterium nucleus $^{2}_{1}\mathrm{H}$ interacts with a lithium nucleus $^{7}_{3}\mathrm{Li}$. Which of following could NOT possibly occur as the end products of this reaction because of violation of conservation of charge or mass number?

(A) $^{4}_{2}\mathrm{He}+^{3}_{1}\mathrm{He}+^{2}_{1}\mathrm{H}$　　(B) $2^{3}_{1}\mathrm{H}+^{1}_{1}\mathrm{H}+2^{1}_{0}\mathrm{n}$

(C) $2^{4}_{2}\mathrm{He}+^{1}_{0}\mathrm{n}$　　(D) $^{7}_{3}\mathrm{Li}+^{2}_{1}\mathrm{H}$

29. The pressure exerted on the bottom of a dam by the water in the reservoir created by the dam depends on the

(A) shape of the dam　　(B) area of the dam

(C) depth of the water at the dam　　(D) surface area of the reservoir

30. A 3.0 kg block hanging from scale is submerged in a beaker of water until the spring scale reads 20 N. What is the buoyant force on the block?

(A) 10 N　　(B) 17 N

(C) 37 N　　(D) 50 N

31. Three particles having charges of equal magnitude Q are fixed at the corners of an equilateral triangle as shown in figure 10. Two of the charges are negative; the other is positive. Which of the following vectors best represents the direction of the resultant electric field at point P, the center of the triangle?

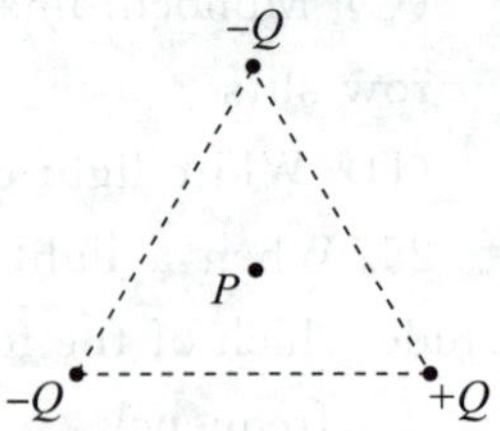

图 10

(A) 　　(B) 　　(C) 　　(D) 

32. The magnet shown in figure 11 is initially far away from the conducting loop and is moved at a constant speed toward and completely through the loop. If T is the time at which the magnet is halfway through the loop, which of the following graphs best represents the induced current I in the loop as a function of time t?

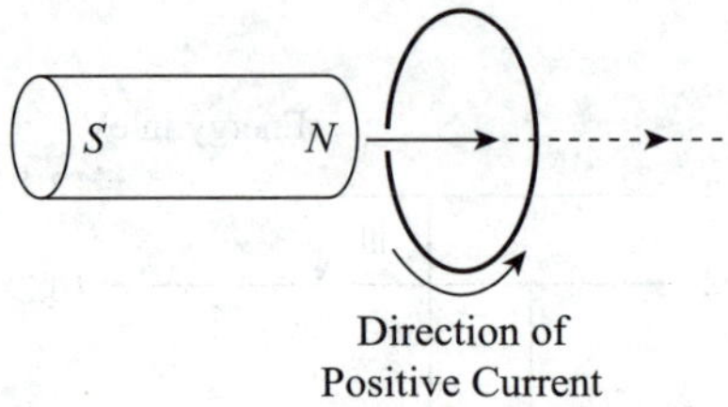

图 11

(A)

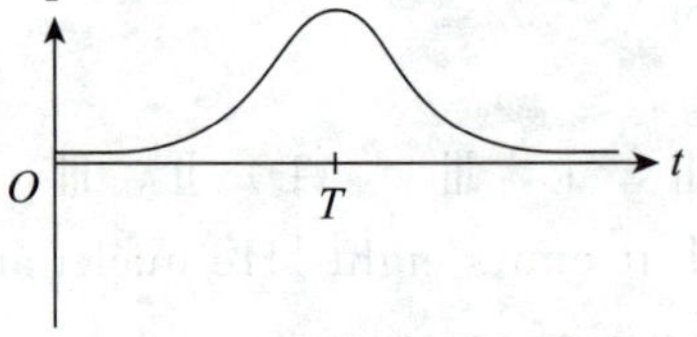

(B)

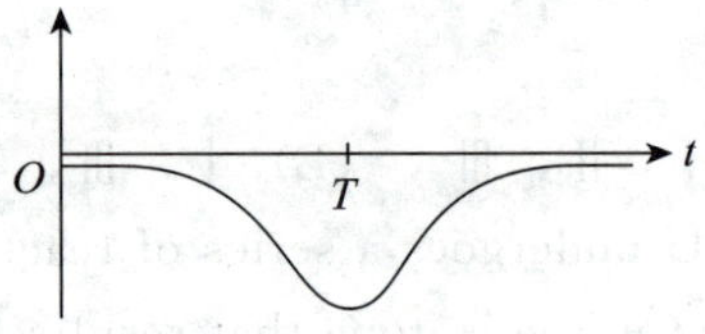

(C)

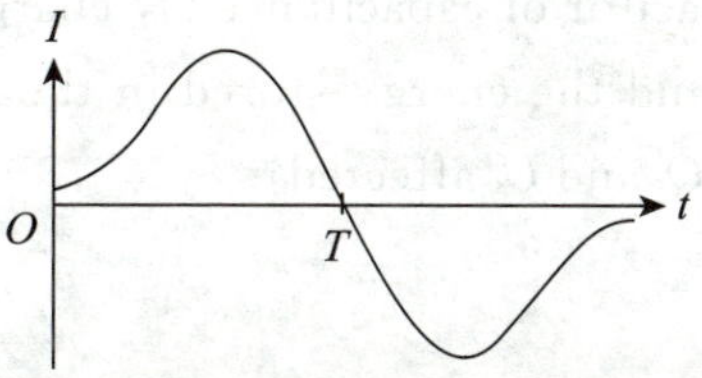

(D)

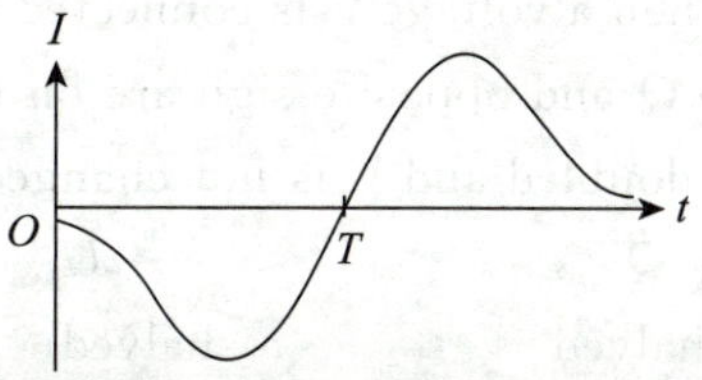

33. The index of refraction of air increases linearly with its density. A ray of light passes through an interface from vacuum into air. Statements that are true as the air is heated at constant pressure include which of the followings?

Ⅰ. The frequency of refracted light in the air decreases

Ⅱ. The wavelength of refracted light in the air decreases

Ⅲ. The speed of light in the air increases

(A) Ⅱ only (B) Ⅲ only (C) Ⅰ and Ⅱ (D) Ⅱ and Ⅲ

34. As shown in figure 12, two lenses are arranged so that a narrow laser beam of parallel rays spreads into a wider beam of a parallel rays. If the lenses are interchanged, the light in region 3 will be a beam of

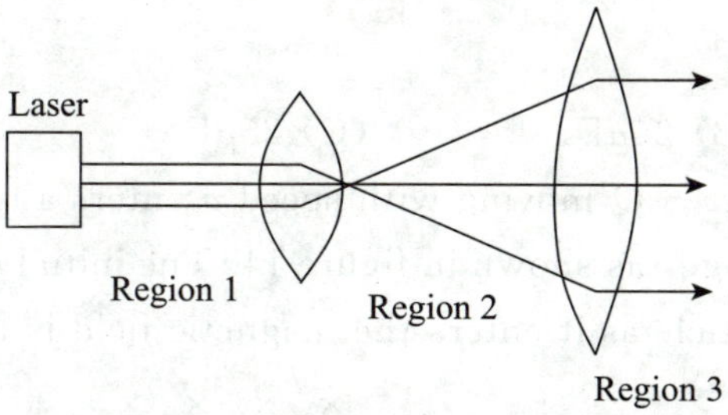

图 12

(A) parallel rays narrower than the beam in the region 1

(B) parallel rays wider than the beam in region 1

(C) parallel rays about the same width as in region 1

(D) diverging rays

35. Traveling waves are generated on a string fixed at both ends. The string has a length L, a linear mass density μ, and a tension T. Which of the following will cause the wave speed to increase?

Ⅰ. Using the same string but increase the tension

Ⅱ. Using a longer string with the same μ and T

Ⅲ. Using a string with the same L and T but a smaller μ

(A) Ⅰ only (B) Ⅱ only (C) Ⅲ only (D) Ⅰ or Ⅲ

36. The wavelength of yellow sodium light in vacuum is approximately 6.0×10^{-7} m. The frequency of the light is most nearly

(A) 2.0×10^{-15} Hz (B) 1.8 Hz

(C) 1.7×10^{6} Hz (D) 5.0×10^{14} Hz

37. When a voltage V is connected across a capacitor of capacitance C, charges of equal magnitude Q and opposite sign are on the plates, and the energy stored in the capacitor is U. If C is doubled and V is not changed, how are Q and U affected?

	$\underline{Q}$	$\underline{U}$
(A)	halved	halved
(B)	halved	quartered
(C)	not changed	not changed
(D)	doubled	doubled

38. In diagram 13, what is the equivalent capacitance between P and Q?

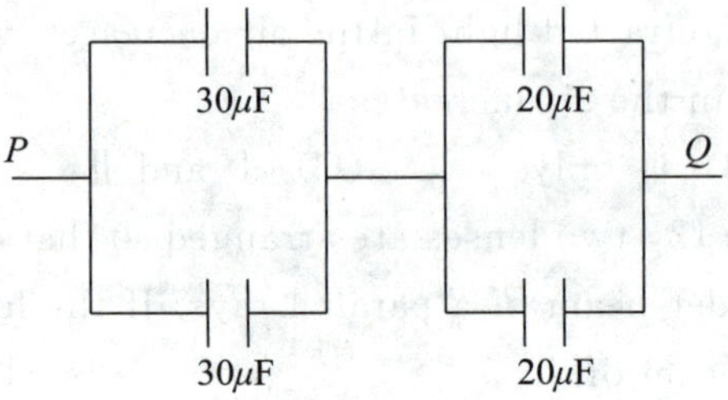

图 13

(A) 6μF　　(B) 24μF　　(C) 25μF　　(D) 50μF

39. A particle of charge $+Q$ moving with speed v enters a region of constant magnetic field B directed into the page, as shown in figure 14. The initial direction and magnitude of the acceleration of the particle as it enters the magnetic field is toward the

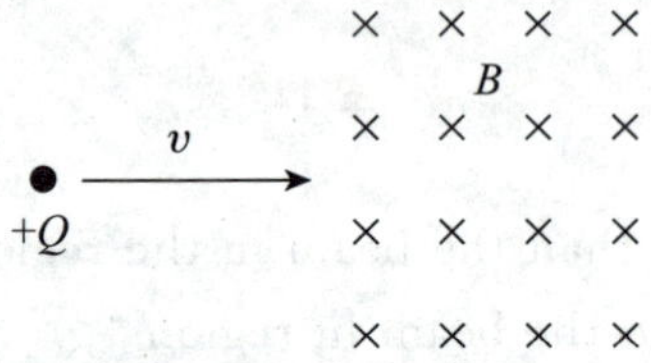

图 14

(A) bottom of the page and proportional to B
(B) bottom of the page and proportional to v
(C) top of the page and inversely proportional to B
(D) top of the page and proportional to both B and v

40. Current I flows into and out of the page in two parallels, long straight wires that are fixed at two vertices of an equilateral triangle, as shown in figure 15. The net magnetic field at the third vertex (point P) due to these currents is in which of the following directions?

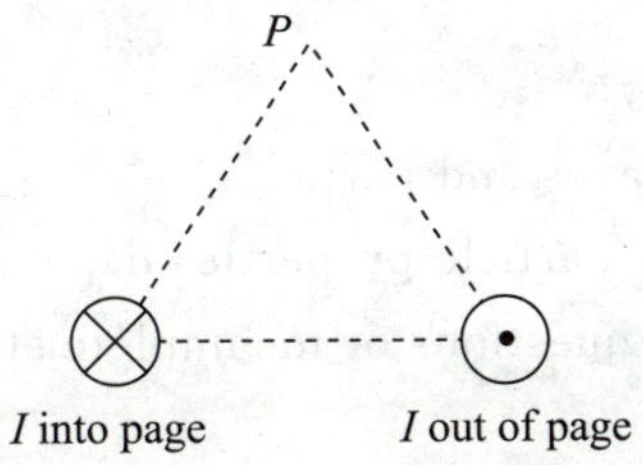

图 15

(A) downwards (B) upwards (C) to the right (D) to the left

41. One section of wire in the circuit shown in figure 16 lies in a region containing a magnetic field B directed into the page. The magnetic force on the section of wire is directed

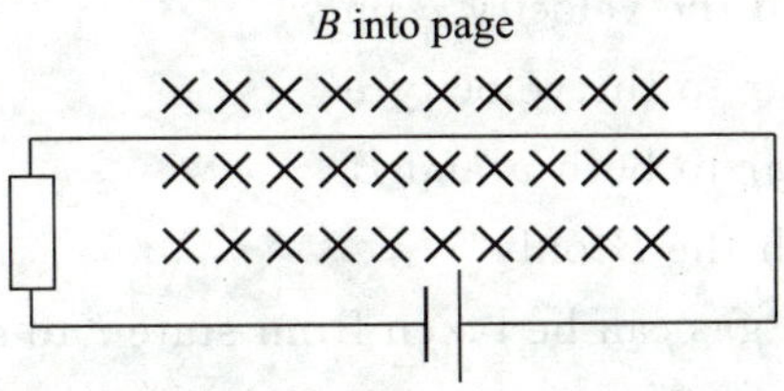

图 16

(A) into the page (B) toward the right
(C) toward the top of the page (D) toward the bottom of the page

42. A student wants to determine experimentally, without disconnecting any wires in the circuit, the DC current moving through a copper wire. Which of the following items of laboratory equipment would be sufficient to make the necessary measurements for this determination?

(A) magnetic field sensor only (B) magnetic field sensor and meterstick
(C) bar magnet and meterstick (D) stop watch and meterstick

43. The maximum energy of photoelectrons emitted from a metal surface depends on which of the following?

Ⅰ. the intensity of the incident light
Ⅱ. the frequency of the incident light
Ⅲ. the work function of the metal

(A) Ⅰ only (B) Ⅲ only (C) Ⅰ and Ⅱ only (D) Ⅱ and Ⅲ only

44. For the hydrogen atom, how does the potential energy U of the atom change as the energy level n increases?

(A) the potential energy always increases
(B) the potential energy always decreases
(C) the potential energy increases then decreases
(D) the potential energy decreases then increases

45. The present theory of the nature of light suggests that it

(A) is a wave only

(B) consists of particle only

(C) is a mixture of both waves and particles

(D) exhibits either wave or particle properties depending on the situation

Directions: For each of the questions or incomplete statements below, two of the suggested answers will be correct.

For each of these questions, you must select both correct choices to earn credit. No partial credit will be earned if only one correct choice is selected. Select the two that are best in each case and then enter both of the appropriate letters in the corresponding space on the answer sheet.

46. A charged particle q with velocity v is always experienced a magnetic force F, when it enters into a magnetic field B. The direction of this force is

(A) always parallel with the velocity v only

(B) always perpendicular to the velocity v

(C) always perpendicular to both v and B

(D) always parallel with the B only

47. A sample of an ideal gas can be taken from state i to state f via two processes, as shown in graph 17 of pressure P versus volume V. In one process the gas goes through state X, and in the other process the gas goes through state Y.

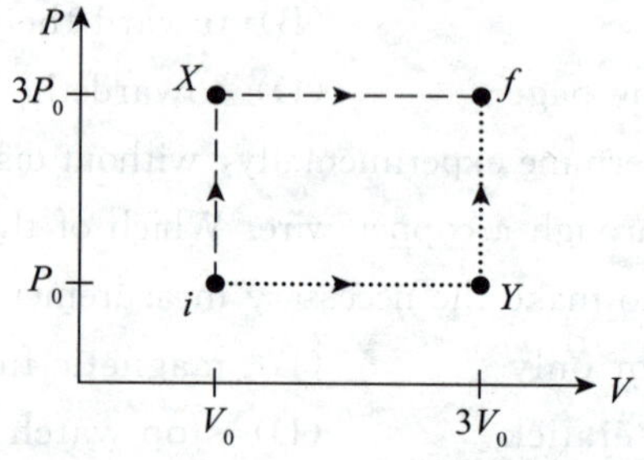

图 17

Which of the following will be the same for both processes? Select two answers.

(A) the change in the internal energy of the gas

(B) the temperature of the gas at the end of the process

(C) the thermal energy transferred to the gas by heating

(D) the work done on the gas

48. In Young's two slits interference experiment, if you hope to increase the distance between two bright (or dark) lines in the interference patterns, which of the following are the possible ways to obtain this result?

(A) increase the wavelength

(B) increase the distance between the slits and the screen

(C) increase the distance between the two slits

(D) increase all of the above three

49. Figure 18 (a) shows the photocurrent versus intensity, figure 18 (b) shows the photocurrent versus frequency of incident light. Following are the conclusions based on the two figures. Which two are correct?

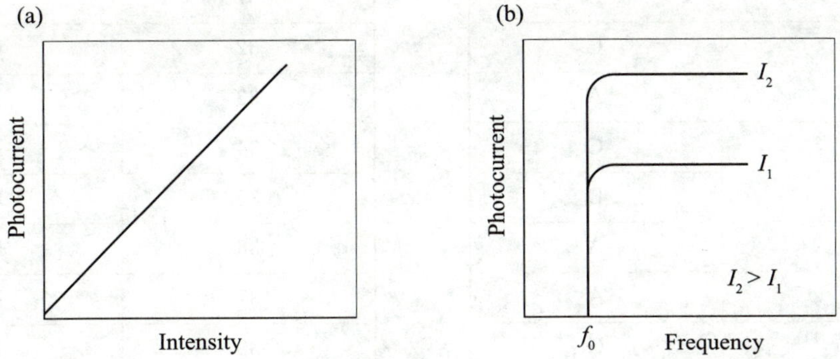

图 18

(A) The photocurrent increases with the intensity of the incident light, no matter what frequency it is

(B) The photocurrent increases with the intensity of the incident light only when the frequency of the incident light is above a certain frequency f_0 which called threshold frequency

(C) The intensity (brightness) of the light has no effect on the threshold frequency. No matter how intense the incident light, if it is below the threshold frequency, not a single photoelectron is emitted (figure (b))

(D) The higher the intensity of the incident light is, the higher the threshold frequency

50. Among the following four experiments, which two show the wave property of the light?

(A) reflect and refract experiments

(B) Yong's double slits interference experiment

(C) photoelectric effect experiment

(D) diffraction experiment

Answer for AP Physics 2 Practice（2）

1	D
2	C
3	D
4	C
5	A
6	C
7	D
8	A
9	A
10	C
11	D
12	C
13	D
14	A
15	C
16	A
17	C
18	C
19	D
20	D
21	C
22	D
23	C
24	C
25	A
26	C
27	C
28	B

29	C
30	A
31	A
32	C
33	B
34	A
35	D
36	D
37	D
38	B
39	D
40	A
41	D
42	B
43	D
44	A
45	D
46	B，C
47	A，B
48	A，B
49	B，C
50	B，D

Section Ⅱ: Free Response

DO NOT OPEN THIS BOOKLET UNTIL YOU ARE TOLD TO DO SO.

Instructions

The questions for Section Ⅱ are printed in this booklet. You may use any blank space in the booklet for scratch work, but you must write your answers in the spaces provided for each answer. A table of information and lists of equations that may be helpful are in the booklet. Calculators, rulers, and straightedges may be used in this section.

All final numerical answers should include appropriate units. Credit for your work depends on demonstrating that you know which physical principles would be appropriate to apply in a particular situation. Therefore, you should show your work for each part in the space provided after that part. If you need more space, be sure to clearly indicate where you continue your work. Credit will be awarded only for work that is clearly designated as the solution to a specific part of a question. Credit also depends on the quality of your solutions and explanations, so you should show your work.

Write clearly and legibly. Cross out any errors you make; erased or crossed-out work will not be scored. You may lose credit for incorrect work that is not crossed out.

Manage your time carefully. You may proceed freely from one question to the next.

You may review your responses if you finish before the end of the exam is announced.

Total Time

90 minutes

Number of Questions

4

Percent of Total Score

50%

Writing Instrument

Either pencil or pen with black or dark blue ink

Electronic Device

Calculator allowed

Weight

Approximate weights:

Questions 2 and 3:

27% each

Questions 1 and 4:

23% each

Suggested Time

Approximately 25 minutes each for questions 2 and 3 and 20 minutes each for questions 1 and 4

AP Physics 2 Practice (2)
Section Ⅱ
4 Questions
Time—90 minutes

Directions: Questions 2 and 3 are long free-response questions that require about 25 minutes each to answer and are worth 12 points each. Questions 1 and 4 are short free-response questions that require about 20 minutes each to answer and are worth 10 points each. Show your work for each part in the space provided after that part.

1. (10 points, suggested time 20 minutes)

A cylindrical container is fitted with a frictionless piston that is initially locked in place. The cylinder contains a fixed amount of an ideal gas that is initially at room temperature and atmospheric pressure.

(a) The cylinder is placed in a hot-water bath. On the axes (图 19), sketch a graph of pressure versus temperature for the process the gas undergoes as a result, and indicate the direction of the process on the graph.

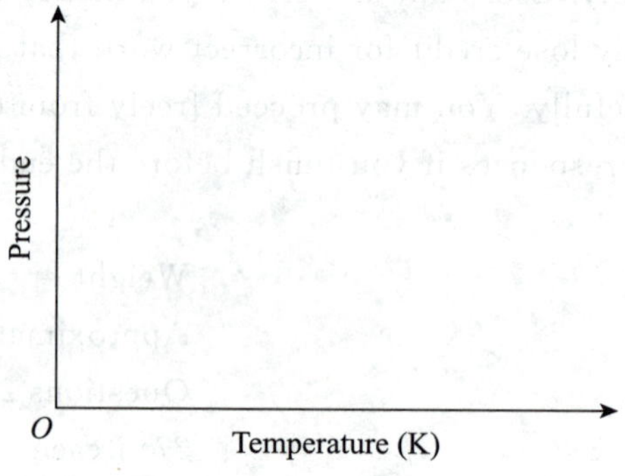

图 19

(b) The cylinder is removed from the hot-water bath. After equilibrium is reached, the lock is removed so the piston is free to move. Indicate whether the piston moves up, moves down, or remains stationary.

______ Moves up ______ Moves down ______ Remains stationary

Justify your answer.

(c) When the system is again at equilibrium, the piston is pushed down very slowly.

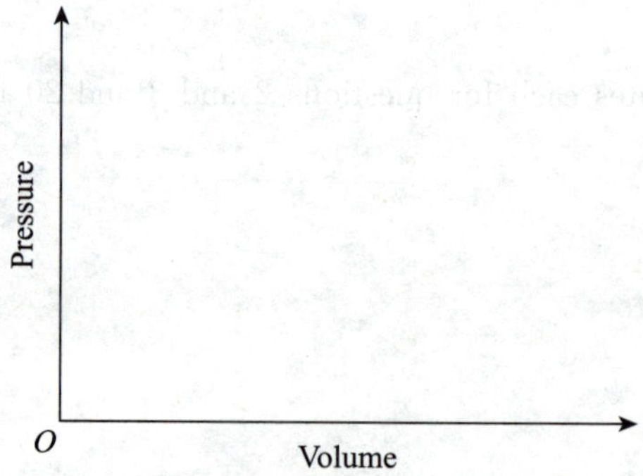

图 20

On the axes (图 20), sketch a graph of pressure versus volume for the process the gas undergoes as a result, and indicate the direction of the process on the graph. Label this process "C".

(d) Now the piston is pulled up quickly, so no heat is added to or removed from the gas during the process. On the axes above, sketch a graph of pressure versus volume for the process the gas undergoes as a result, and indicate the direction of the process on the graph. Label this process "D".

2. (12 points, suggested time 25 minutes)

The circuit shown in figure 21 contains a battery with negligible internal resistance, a closed switch S, and three resistors, each with a resistance of R or $2R$.

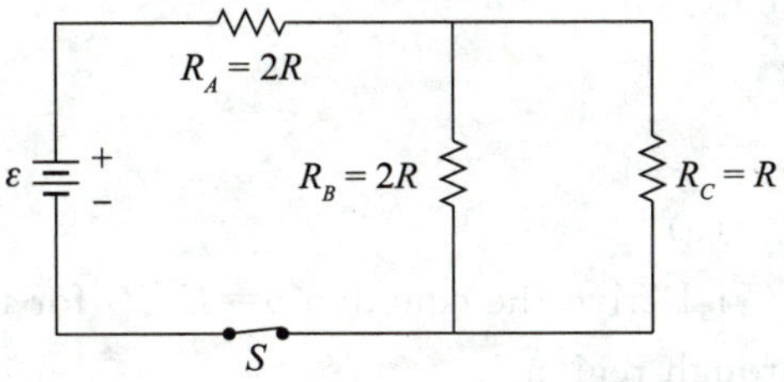

图 21

(a)

i. Rank the currents in the three resistors from greatest to least, with number 1 being greatest. If two resistors have the same current, give them the same ranking.

______ I_A ______ I_B ______ I_C

ii. Justify your answers.

(b)

i. Rank the voltages across the three resistors from greatest to least, with number 1 being greatest. If two resistors have the same voltage across them, give them the same ranking.

______ V_A ______ V_B ______ V_C

ii. Justify your answers.

For parts (c) and (d), use $\varepsilon = 12$ V and $R = 200\ \Omega$.

(c) Calculate the equivalent resistance of the circuit.

(d) Calculate the current in resistor R_C.

3. (12 points, suggested time 25 minutes)

Diagram 22 illustrates a velocity selector, labeled region 1. It consists of two parallel conducting plates, with charges on the plates as indicated creating an electric field of magnitude E directed toward the top of the page. A uniform magnetic field of magnitude B_1 directed out of the page exists between the plates. The magnitude of the magnetic field can be adjusted so that only particles of a particular speed pass through the selector in a straight line. A radioactive source ε to the left of the selector emits charged particles, each having the same charge $+q$ and moving to the right in the plane of the page. The effect of gravity can be neglected throughout the problem.

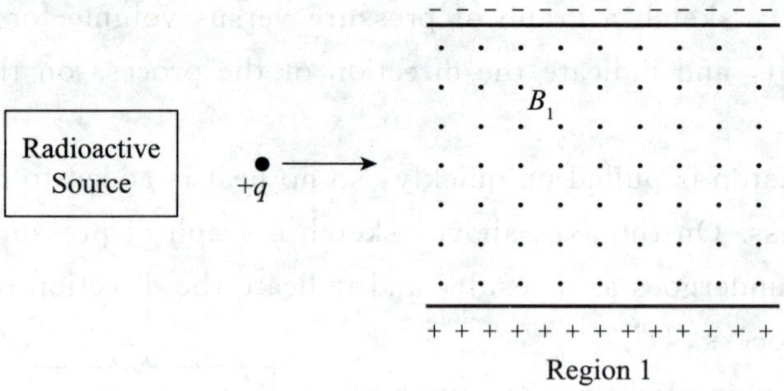

图 22

(a)

i. Derive the equation $v=E/B_1$ for the speed v of particles that move in a straight line through region 1.

ii. Some particles are emitted from the source with speeds greater than E/B_1. Which of the following describes the initial path of one of these particles immediately after entering region 1 ?

______ It curves toward the top of the page.

______ It curves into the page.

______ It curves toward the bottom of the page.

______ It curves out of the page.

______ It moves in a straight line.

Explain your reasoning.

A constant magnetic field of magnitude B_2 directed into the page is now added in region 2 to the right of region 1, as represented in figure 23. Suppose a particle leaves the radioactive source, travels through region 1 in a straight line, and enters region 2. For each of the following, express algebraic answers in terms of E, B_1, B_2, q, and fundamental constants, as appropriate.

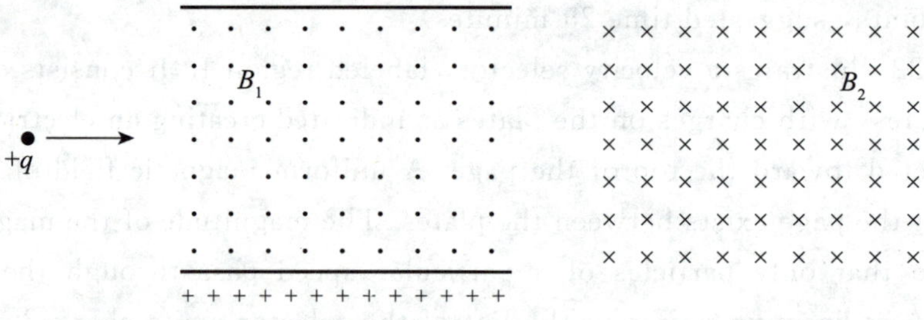

图 23

(b) Determine an expression for the initial magnetic force on the particle in region 2 and state its direction.

Describe the changes, if any, in the magnitude and direction of the magnetic force as the particle moves in region 2.

(c) Describe the path of the particle in region 2.

(d) Derive an expression for the charge-to-mass ratio q/m of the particle. Specifically note any quantities not previously defined that are included in your answer.

4. (10 points, suggested time 20 minutes)

In an experiment a beam of red light of wavelength 675 nm in air passes from glass into air, as shown in figure 24. The incident and refracted angles are θ_1 and θ_2, respectively. In the experiment, angle θ_2 is measured for various angles of incidence θ_1, and the sines of the angles are used to obtain the line shown in figure 25.

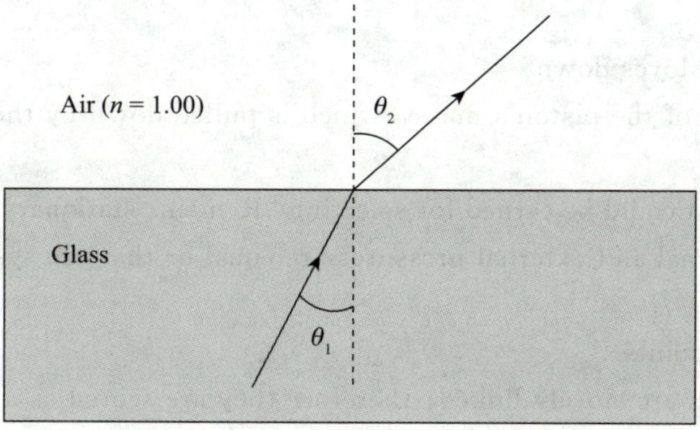

图 24

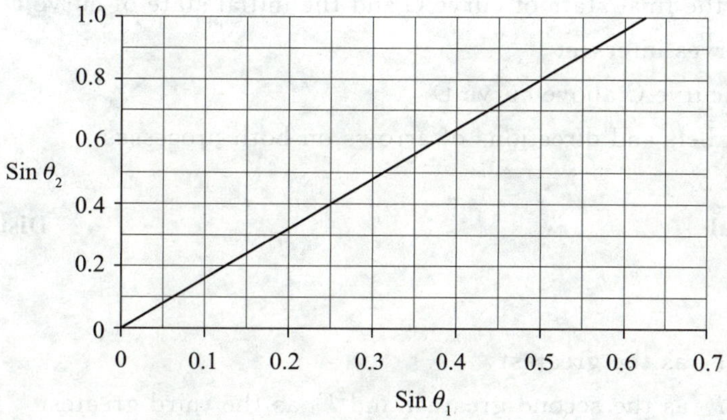

图 25

(a) Assuming an index of refraction of 1.00 for air, use the graph to determine a value for the index of refraction of the glass for the red light. Explain how you obtained this value.

(b) For this red light, determine the following.

i. The frequency in air.

ii. The speed in glass.

iii. The wavelength in glass.

Scorning guideline for AP Physics 2 Practice (2)

Question 1

10 points total **Distribution of points**

(a) 3 points

For a graph showing pressure proportional to temperature (i. e., a straight line segment that, if extended, would pass through the origin) 1 point

For showing that the initial pressure and initial temperature are not zero 1 point

For a final state that is at a higher pressure and temperature than the initial state (regardless of the shape of the path) 1point

(b) 2 points

For selecting "Moves down" 1 point

For recognition of the piston's mass, which is pulled down by the force of gravity 1 point

Note: One point could be earned for selecting "Remains stationary" with a clear explanation that the internal and external pressures are equal or that the system returned to the original pressure P_0

(c) and (d) 5 points

These two parts are closely linked; therefore they are scored as a unit.

For drawing curve C as concave up, with a negative slope 1 point

For drawing curve D as concave up, with a negative slope 1 point

For drawing the final state of curve C and the initial state of curve D as the only point where the two curves intersect 1 point

For drawing curve C above curve D 1 point

For correct labels and directions of arrows on both processes 1 point

Question 2

12 points total **Distribution of points**

(a) 4 points

(i) 2 points

For ranking I_A as the greatest 1 point

For ranking I_C as the second greatest and I_B as the third greatest 1 point

(ii) 2 points

For a correct statement justifying that I_A is the greatest (For example: The total current flows through R_A and gets divided between the other two resistors.) 1 point

For a correct statement justifying that I_C is the second and I_B is the third 1 point

(For example: R_B and R_C share the current in the parallel segment, and that current divides between R_B and R_C so that the smaller resistor R_C carries the most current.)

(b) 3 points

(i) 1 point

For the correct ranking V_A, V_B, V_C=1, 2, 2 1 point

(ii) 2 points

For a correct justification that V_A is the greatest (For example, because no resistor is greater than R_A, and R_A has the full current through it. Or, because R_A is greater than the equivalent parallel resistance of R_B and R_C.) 1 point

For a correct justification that $V_B = V_C$ (For example, since the voltage across the parallel resistors R_B and R_C is the same.) 1 point

(c) 3 points

Let R_{BC} be the resistance of the parallel combination of R_B and R_C.

For one correct form for determining R_{BC} 1 point

$$\frac{1}{R_{BC}}=\frac{1}{R_B}+\frac{1}{R_C}=\frac{1}{2R}+\frac{1}{R}=\frac{1}{400\ \Omega}+\frac{1}{200\ \Omega}=\frac{3}{400\ \Omega}$$

$$R_{BC}=\frac{2R}{3}=\frac{400\ \Omega}{3}=133\Omega$$

For one correct form for determining the total resistance 1 point

$$R_{tot}=R_A+R_{BC}=2R+\frac{2R}{3}=400\ \Omega+133\ \Omega$$

For the correct numerical value 1 point

$$R_{tot}=533\ \Omega$$

(d) 2 points

For one correct form for the current I_A, using R_{tot} from part (c) 1 point

$$I_A=\frac{\varepsilon}{R_{tot}}=\frac{12\ \text{V}}{533\ \Omega}=0.022\ 5\text{A}$$

For the correct value of V_C

$$V_C=\varepsilon-V_A=\varepsilon-I_AR_A=12\ \text{V}-(0.022\ 5\ \text{A})(400\ \Omega)=3.0\ \text{V}$$

$$I_C=\frac{V_C}{R_C}=\frac{3\ \text{V}}{200\ \Omega}$$

For the correct numerical value of I_C 1 point

$$I_C=0.015\ \text{A}$$

Alternate solution Alternate points

Using $R_B=2R_C$ and $V_B=V_C$, so that $I_BR_B=I_CR_C$

$$I_C=\frac{R_B}{R_C}I_B=\frac{2R_C}{R_C}I_B=2I_B$$

For the correct numerical value of I_{tot}

$$I_{tot}=\varepsilon/R_{tot}=0.022\ 5\ \text{A}$$

For correctly relating I_{tot} to I_C 1 point

$$I_{tot}=I_B+I_C=\frac{I_C}{2}+I_C=\frac{3I_C}{2}$$

For the correct numerical value of I_C 1 point

$$I_C=\frac{2I_{tot}}{3}=\frac{2}{3}\ (0.022\ 5\ \text{A})\ =0.015\ \text{A}$$

Question 3

12 points total **Distribution of points**

(a) 6 points

i. 3 points

For the correct identification of the magnetic force on the positively charged particles as they enter region 1 1 point

$F_B=qvB_1$, directed toward the bottom of the page

For the correct identification of the electric force on the positively charged particles as they enter region 1 1 point

$F_E=qE$, directed toward the top of the page

$$F_{net}=\sum F=qE-qvB=ma$$

For the particle to move at constant velocity, $a=0$

For the correct application of Newton's laws to obtain the equation needed to relate the speed of the particles to the electric and magnetic field strengths 1 point

$$qE-qvB_1=0$$

$$v=E/B_1$$

ii. 3 points

For selecting "It curves toward the bottom of the page." 1 point

For an appropriate explanation 2 point

Example: The electric force $F_E=qE$, directed upward, is the same for all speeds of the particle. However, the magnetic force $F_B=qvB_1$, directed downward, increases with increasing speed. Therefore if the speed of the particle is greater than that in part (a) i, then $F_B>F_E$. The resultant force on the particle as it enters region 1 is toward the bottom of the page, causing the particle to curve in that direction.

The justification point could be earned only if the correct answer was selected.

(b) 2 points

$$F_B=qvB_2$$

For correctly substituting $v=E/B_1$ from part (a) i 1 point

$$F_B=\frac{qEB_2}{B_1}$$

For correctly indicating that the direction is up or toward the top of the page 1 point

(c) 2 points

For including in the description that the force is constant 1 point

For including in the description that the direction of the force changes 1 point

Example: The magnitude $F_B=qvB_2$ does not change because the particle is in a constant magnetic field B_2 and the angle between the velocity and magnetic force remains constant at 90°.

But as the particle curves the direction of its velocity changes, and the direction of the magnetic force also changes because it remains perpendicular to the velocity vector.

(d) 2 points

For indicating that the net force provides the centripetal acceleration

$$F_{net}=\frac{mv^2}{R}$$

For indicating that the net force is due to the magnetic field

$$F_{net}=qvB_2$$

$$qvB_2=\frac{mv^2}{R}$$ 1 point

Substituting $v=\frac{E}{B_1}$, from part (a) i

$$qvB_2=\frac{m}{R}\frac{E^2}{B_1^2}$$

For the correct answer

$$\frac{q}{m}=\frac{E}{B_1B_2R}$$

For stating that R, a previously undefined quantity, is the radius of the circular arc in which the particle moves when in the magnetic field B_2. 1 point

Question 4

10 points total **Distribution of points**

(a) 3 points

For using Snell's law or indicating that the index of refraction is the slope of the graph 1 point

$$n_1\sin\theta_1=n_2\sin\theta_2$$

For substituting or otherwise indicating use of data from the given graph 1 point

$$n_1\times0.5=1\times0.8 \quad \text{or} \quad \text{Slope}=\frac{0.8}{0.5}$$

For the correct answer 1 point

$$n_1=1.60$$

Alternate Solution Alternate points

For using the expression for the critical angle 1 point

$$\sin\theta_c=\frac{n_2}{n_1}$$

For using data from the graph 1 point

The critical angle occurs when $\sin\theta_2=1$. From the graph, $\sin\theta_1=\sin\theta_c=0.625$

$$0.625=\frac{1}{n_1}$$

For the correct answer 1 point

$$n_1=1.60$$

Two points were awarded for inverting the values from the graph to obtain $n_2=0.625$

(b) 7 points

i. 2 points

For using the correct equation with the correct substitutions 1 point

$$f=v/\lambda$$

$$f=(3\times10^8\ \text{m/s})/(675\times10^{-9}\ \text{m})$$

For the correct answer 1 point

$$f=4.44\times10^{14}\ \text{Hz}$$

ii. 2 points

For using the correct equation with the correct substitutions 1 point

$$v=c/n$$

$$v=(3\times10^8\ \text{m/s})/1.60$$

For the correct answer

$$v=1.88\times10^8\ \text{m/s}$$ 1 point

If an incorrect answer was carried through to obtain a speed greater than 3×10^8 m/s, only one point was awarded for the solution. The second point could be earned if there was some indication that the student realized that the value must be incorrect, because it could not be greater than the speed of light.

iii. 3 points

For using the correct equation with correct substitutions (consistent with previous answers) 1 point

$$\lambda=v/f$$

$$\lambda=(1.88\times10^8\ \text{m/s})/(4.44\times10^{14}\ \text{Hz})$$

For the correct answer 1 point

$$\lambda=423\times10^{-9}\ \text{m}=423\ \text{nm}$$

For correct units on the three answers in part (b) 1 point

Appendixes
附　录

1 The constants on Physics 物理常数

Physics Constants and Conversions 物理常量及单位转换表

Name	Symbol	Numerical Value and Unit
Mass of the Earth		5.98×10^{24} kg
Mass of the Moon		7.35×10^{22} kg
Mean radius of the Earth		6.37×10^{6} m
Mean radius of the Moon		1.74×10^{6} m
Mean distance—the Earth to the Moon		3.84×10^{8} m
Mean distance—the Earth to the Sun		1.50×10^{11} m
Universal gravitational constant	G	6.67×10^{-11} $N\cdot m^2/kg^2$
Acceleration due to gravity near the Earth's surface	g	9.81 m/s^2
1 elementary charge	e	1.60×10^{-19} C
1 coulomb (C)		6.25×10^{18} elementary charges
1 electronvolt (eV)		1.60×10^{-19} J
Vacuum permittivity	ε_0	8.85×10^{-12} $C^2/N\cdot m^2$
Coulomb's law constant	$k=\frac{1}{4\pi\varepsilon_0}$	9.0×10^{9} $N\cdot m^2/C^2$
Vacuum permeability	μ_0	$4\pi\times10^{-7}$ $(T\cdot m)/A$
Magnetic constant	$k'=\frac{\mu_0}{4\pi}$	10^{-7} $(T\cdot m)/A$
Density of water		1.00×10^{3} kg/m^3

续前表

Name	Symbol	Numerical Value and Unit
1 atmosphere pressure	atm	1.0×10^5 N/m^2 or 1.0×10^5 Pa
Specific heat capacity of water		4.18×10^3 J/ (kg · K)
Speed of light in a vacuum	c	3.00×10^8 m/s
Speed of sound in air at STP		3.31×10^2 m/s
1 unified atomic mass unit (u)		1.66×10^{-27} kg or 9.31×10^2 MeV/c^2
Rest mass of the electron	m_e	9.11×10^{-31} kg
Rest mass of the proton	m_p	1.67×10^{-27} kg
Rest mass of the neutron	m_n	1.67×10^{-27} kg
Avogadro's number	N_0	6.02×10^{23} mol^{-1}
Universal gas constant	R	8.31 J / (mol · K)
Boltzmann's constant	k_B	1.38×10^{-23} J/K
Planck's constant	h	6.63×10^{-34} J · s or 4.14×10^{-15} eV · s

Table of Prefixes for Powers of 10 十进倍数和分数单位的词头表

Prefix	Symbol	Factor	Short name	Prefix	Symbol	Factor	Short name
yotta	Y	10^{24}	Septillion	yocto	y	10^{-24}	Septillionth
zetta	Z	10^{21}	Sextillion	zepto	z	10^{-21}	Sextillionth
exa	E	10^{18}	Quintillion	atto	a	10^{-18}	Quintillionth
peta	P	10^{15}	Quadrillion	femto	f	10^{-15}	Quadrillionth
tera	T	10^{12}	Trillion	pico	p	10^{-12}	Trillionth
giga	G	10^{9}	Billion	nano	n	10^{-9}	Billionth
mega	M	10^{6}	Million	micro	μ	10^{-6}	Millionth
kilo	k	10^{3}	Thousand	milli	m	10^{-3}	Thousandth
hecto	h	10^{2}	Hundred	centi	c	10^{-2}	Hundredth
deka	da	10^{1}	Ten	deci	d	10^{-1}	Tenth
(none)	(none)	10^{0}	One	(none)	(none)	10^{0}	One

Table of SI Base Units 国际基本单位

Quantity	Name	Symbol
Mass 质量	Kilogram 千克	kg
Distance/Length 长度	Metre 米	m
Time 时间	Second 秒	s
Electric current 电流	Ampere 安培	A
Amount of substance 物质的量	Mole 摩尔	mol
Temperature 热力学温度	Kelvin 开尔文	K
Light intensity/Luminous intensity 发光强度	Candela 坎德拉	cd

Table of Derived Units 国际导出单位

Derived quantity	Name	Symbol	Expression in other SI units
Frequency 频率	hertz 赫[兹]	Hz	/
Force 力	newton 牛[顿]	N	/
Pressure 压强	pascal 帕[斯卡]	Pa	N/m^2
Energy; work; quantity of heat 能量；功；热	joule 焦[耳]	J	N·m
Energy 能量	electron-volt 电子伏特	eV	C·V
Power 功率	watt 瓦[特]	W	J/s
Quantity of electricity 电荷量	coulomb 库[仑]	C	/
Electric potential difference; Electromotive force 电位；电压；电动势	volt 伏[特]	V	J/C
Capacitance 电容	farad 法[拉]	F	C/V
Electric resistance 电阻	ohm 欧[姆]	Ω	V/A
Electric conductance 电导	siemens 西[门子]	S	A/V
Magnetic flux 磁通量	weber 韦[伯]	Wb	V·s
Magnetic flux density 磁通量密度、磁感应强度	tesla 特[斯拉]	T	Wb/m^2
Inductance 电感	henry 亨[利]	H	Wb/A
Celsius temperature 摄氏温度	degree Celsius 摄氏度	℃	/

2 The words on Physics 物理专业英语词汇

Vector and scalar 矢量和标量

vector	['vektə]	*n.*	矢（向）量
scalar	['skelə]	*n.*	标量
vector algebra	['vektə 'ældʒəbrə]	*n. ph.*	矢量代数
two-dimensional	[tu di'menʃənl]	*n. ph.*	两维
physical quantity	['fizikəl 'kwɔntiti]	*n. ph.*	物理量
magnitude	['mægni.tjud]	*n.*	量、大小
direction	[di'rekʃən]	*n.*	方向
commutative law	[kə'mjutətiv lɔ:]	*n. ph.*	交换法则
vector addition	['vektə ə'diʃən]	*n. ph.*	矢量和
geometric	[ˌdʒiə'metrik]	*adj.*	几何学的
vector subtraction	['vektə səb'trækʃən]	*n. ph.*	矢量差
vector resolving	['vektə ri'zɔlviŋ]	*n. ph.*	矢量分解
standard basis vector	['stændəd 'beisis 'vektə]	*n. ph.*	标准基本向量
horizontal basis vector	[ˌhɔri'zɔntl 'beisis 'vektə]	*n. ph.*	水平基本向量
vertical basis vector	['vətikəl 'beisis 'vektə]	*n. ph.*	垂直基本向量
horizontal vector	[ˌhɔri'zɔntl 'vektə]	*n. ph.*	水平向量
vertical vector	['və:tikəl 'vektə]	*n. ph.*	垂直向量
unit vector	['junit 'vektə]	*n. ph.*	单位矢（向）量
vector component	['vektə kəm'pəunənt]	*n. ph.*	分矢量、矢量分量
dot (scalar) product	[dɔt ('skelə) 'prɔdəkt]	*n. ph.*	点乘
cross product	[krɔs 'prɔdəkt]	*n. ph.*	差乘
perpendicular	[ˌpəpən'dikjələ]	*adj.*	垂直的

Kinematics 运动学

kinematics	[ˌkinə'mætiks]	*n.*	运动学
motion	['məuʃən]	*n.*	运动
frame of reference	[freim ɔv 'refrəns]	*n. ph.*	参照系
position	[pə'ziʃən]	*n.*	位置
displacement	[dis'pleismənt]	*n.*	位移
distance	['distəns]	*n.*	距离
pythagorean Theorem	[paiˌθægə'ri:ən 'θiərəm]	*n. ph.*	勾股定理

track-and-field	[træk ænd fi:ld]	*n. ph.*	田径
oval	['əuvəl]	*n.*	椭圆
stationary	['steiʃənəri]	*adj.*	静止的
at rest	[æt rest]	*n. ph.*	静止
speed	[spi:d]	*n.*	速率
velocity	[vi'lɔsiti]	*n.*	速度
initial	[i'niʃəl]	*adj.*	最初的
final	['fainəl]	*adj.*	最终的
average speed	['ævəridʒ spi:d]	*n. ph.*	平均速率
average velocity	['ævəridʒ vi'lɔsiti]	*n. ph.*	平均速度
acceleration	[ækˌselə'reʃən]	*n.*	加速度
uniformly-accelerated motion		*n. ph.*	匀加速运动
relative motion	['relətiv 'məuʃən]	*n. ph.*	相对运动
relative velocity	['relətiv vi'lɔsiti]	*n. ph.*	相对速度
straight	[streit]	*adj.*	直的
straight-line motion	[streit lain 'məuʃən]	*n. ph.*	直线运动
rectilinear motion	[ˌrekti'liniə 'məuʃən]	*n. ph.*	直线运动
non-linear motion	['nɔn 'liniə 'məuʃən]	*n. ph.*	非直线运动
constant	['kɔnstənt]	*adj.*	恒定的
equation	[i'kweiʃən]	*n.*	(公式)等式
equal	['ikwəl]	*adj.*	相等的
axis/axes	['æksis]	*n.*	轴
position vs. time graph		*n. ph.*	位置—时间图
velocity vs. time graph		*n. ph.*	速度—时间图
acceleration vs. time graph		*n. ph.*	加速度—时间图
curve	[kə:v]	*n.*	曲线
instantaneous velocity	[ˌinstən'teiniəs vi'lɔsiti]	*n. ph.*	瞬（即）时速度
slope	[sləup]	*n.*	斜率
gradient	['greidiənt]	*n.*	斜度（率）
free fall	[fri: fɔ:l]	*n. ph.*	自由落体
air resistance	[εə ri'zistəns]	*n. ph.*	空气阻力
terminal velocity	['tə:minəl vi'lɔsiti]	*n. ph.*	终速/自由沉降速度
projectile motion	[prə'dʒektail 'məuʃən]	*n. ph.*	抛射运动
parabolic motion	[ˌpærə'bɔlik 'məuʃən]	*n. ph.*	抛物线运动
trajectory	[trə'dʒektəri]	*n.*	轨道
launch angle (angle of projection)	[lɔ:ntʃ 'æŋgəl]	*n. ph.*	发（抛）射角
range	[reindʒ]	*n.*	射程
particle	['pɑ:tikl]	*n.*	粒子

Free-body analysis 隔离体方法受力分析

force	[fɔ:s]	*n.*	力
gravitational force	[ˌgrævi'teiʃnəl fɔ:s]	*n. ph.*	万有引力
gravity	['græviti]	*n.*	地心引力
mechanics	[mi'kæniks]	*n.*	力学、机械学
action-reaction pair	['ækʃən ri'ækʃən pɛə]	*n. ph.*	力与反作用力
balanced forces	['bælənst fɔ:siz]	*n. ph.*	平衡力
exert a force	[ig'zət ə fɔ:s]	*n. ph.*	施力
weight	[weit]	*n.*	重力
tension	['tenʃən]	*n.*	拉力
net force	[net fɔ:s]	*n.*	净力
resultant force	[ri'zʌltənt fɔ:s]	*n. ph.*	合外力
free-body diagram/ force diagram	[fri: 'bɔdi 'daiəˌgræm]	*n. ph.*	受力分析图
coordinate system	[kəu'ɔ:dineit 'sistəm]	*n. ph.*	坐标系
coordinate axis	[kəu'ɔ:dineit 'æksis]	*n. ph.*	坐标轴
normal force	['nɔ:məl fɔ:s]	*n. ph.*	支持/垂直力
friction	['frikʃən]	*n.*	摩擦力
static friction	['stætik 'frikʃən]	*n. ph.*	静摩擦力
kinetic (sliding) friction	[kai'netik ('slaidiŋ) 'frikʃən]	*n. ph.*	（滑）动摩擦力
coefficient of friction	[ˌkəuə'fiʃənt ɔv 'frikʃən]	*n. ph.*	摩擦系数
pulley	['puli:]	*n.*	滑轮
ramp	[ræmp]	*n.*	斜坡
inclined plane	[in'klaind plein]	*n. ph.*	斜面
breaking strength	['breikiŋ streŋθ]	*n. ph.*	断裂强度
torque/moment	[tɔ:k 'məumənt]	*n. ph.*	扭（力）矩
line of action of forces	[lain ɔv 'ækʃən ɔv fɔ:siz]	*n. ph.*	力的作用线
couple	['kʌpəl]	*n.*	力偶
clockwise	['klɔkwaiz]	*n.*	顺时针
anticlockwise/counterclockwise	[ˌænti:'klɔkwaiz/ˌkauntə'klɔkˌwaiz]	*n. ph.*	逆时针
equilibrium of forces	[ˌi:kwə'libri:əm ɔv fɔ:siz]	*n. ph.*	力平衡
concurrent force	[kən'kə:rənt fɔ:s]	*n. ph.*	共点力
non-concurrent force	[nɔn kən'kə:rənt fɔ:s]	*n. ph.*	非共点力
rigid body	['ridʒid 'bɔdi]	*n. ph.*	刚性物体
translational equilibrium	[træns'leiʃənəl ˌi:kwə'libri:əm]	*n. ph.*	平动平衡

rotational equilibrium	[rəu'teiʃənəl ˌi:kwə'libri:əm]	*n. ph.*	转动平衡
static equilibrium	['stætik ˌi:kwə'libri:əm]	*n. ph.*	静平衡
compression	[kəm'preʃən]	*n.*	弹簧压力

Newton's laws of motion 牛顿运动定律

dynamics	[dai'næmiks]	*n.*	动力学
Newton's laws of motion		*n. ph.*	牛顿运动定律
inertia	[i'nə:ʃjə]	*n.*	惯性
momentum	[məu'mentəm]	*n.*	动量
mass	[mæs]	*n.*	质量
（uniform）circular motion		*n. ph.*	（匀速率）圆周运动率
centripetal acceleration	[sen'tripitl ækˌselə'reʃən]	*n. ph.*	向心加速度
breaking strength	['breikiŋ streŋθ]	*n. ph.*	断裂强度
radius/radius of curvature	['reidjəs ɔv 'kə:vəˌtʃə]	*n. ph.*	半径
perimeter/circumference	[pə'rimitə / sə'kʌmfərəns]	*n. ph.*	周长
arc	[ɑ:k]	*n.*	弧
loop	[lu:p]	*n.*	圈、回路
torque/moment	[tɔ:k 'məumənt]	*n.*	扭（力）矩
line of action of forces	[lain ɔv 'ækʃən ɔv fɔ:siz]	*n. ph.*	力的作用线
pivot	['pivət]	*n.*	枢轴
center of mass	['sentə ɔv mæs]	*n. ph.*	质心
center of weight	['sentə ɔv weit]	*n. ph.*	重心
couple	['kʌpəl]	*n.*	力偶
translation	[træns'leiʃən]	*n.*	平动/平移
rotation	[rəu'teiʃən]	*n.*	转动
principle of moment	['prinsəpəl ɔv 'məumənt]	*n. ph.*	力矩原理
lever arm (or moment arm)	['li:və ɑ:m]	*n. ph.*	力臂

Momentum and its conservation 动量和动量守恒

momentum	[məu'mentəm]	*n.*	动量
momentum—time graph	[məu'mentəm taim græf]	*n. ph.*	动量—时间图
linear momentum	['liniə məu'mentəm]	*n. ph.*	线动量
angular momentum	['æŋgjələ məu'mentəm]	*n. ph.*	角动量

directly proportional	[di'rektli prə'pɔ:ʃənəl]	*n. ph.*	正比
inversely proportional	['in'və:sli prə'pɔ:ʃənəl]	*n. ph.*	反比
impulse	['im͵pʌls]	*n.*	冲量
impulse—momentum theorem		*n. ph.*	冲量—动量原理
principle conservation of linear momentum		*n. ph.*	线动量守恒原理
isolated	['aisə͵leitid]	*n.*	隔离/封闭
collision	[kə'liʒən]	*n.*	碰撞
(perfectly or completely or totally) elastic collision	[i'læstik kə'liʒən]	*n. ph.*	弹性碰撞
inelastic collision	[ini'læstik kə'liʒən]	*n. ph.*	非弹性碰撞
perfectly（completely）inelastic collision	['pə:fiktli ini'læstik kə'liʒən]	*n. ph.*	完全非弹性碰撞
head on collision	[hed ɔn kə'liʒən]	*n. ph.*	对心碰撞

Work and energy, the conservation of energy 功和能，能量守恒

work	[wə:k]	*n.*	功
constant force	['kɔnstənt fɔ:s]	*n. ph.*	恒力
variable force	['vɛəriəbl fɔ:s]	*n. ph.*	变力
energy	['enədʒi]	*n.*	能量
power	['pauə]	*n.*	功率
potential energy	[pə'tenʃəl 'enədʒi]	*n. ph.*	势能
gravitational potential energy	[͵grævi'teiʃnəl pə'tenʃəl 'enədʒi]	*n. ph.*	重力势能
elastic potential energy	[i'læstik pə'tenʃəl 'enədʒi]	*n. ph.*	弹性势能
kinetic energy	[kai'netik 'enədʒi]	*n. ph.*	动能
thermal energy	['θə:məl 'enədʒi]	*n. ph.*	热能
electrical energy	[i'lektrikəl 'enədʒi]	*n. ph.*	电能
mass energy	[mæs 'enədʒi]	*n. ph.*	质量能
radiant energy	['reidiənt 'enədʒi]	*n. ph.*	辐射能
light energy	[lait 'enədʒi]	*n. ph.*	光能
sound energy	[saund 'enədʒi]	*n. ph.*	声能
hydroelectric energy	[͵haidrəui'lektrik 'enədʒi]	*n. ph.*	水电能
wind energy	[wind 'enədʒi]	*n. ph.*	风能
tidal energy	['taidl 'enədʒi]	*n. ph.*	潮汐能
geothermal energy	[͵dʒi(:)əu'θə:məl 'enədʒi]	*n. ph.*	地热能
mechanical energy	[mi'kænikəl 'enədʒi]	*n. ph.*	机械能

first law of thermodynamics	[fə:st lɔ: ɔv ˌθə:məudai'næmiks]	*n. ph.*	热力学第一定理
law of conservation of energy	[lɔ: ɔv ˌkɔnsə'veiʃən ɔv 'enədʒi]	*n. ph.*	能量守恒原理
the work-energy theorem	[ðə wə:k 'enədʒi 'θiərəm]	*n. ph.*	功能原理
conservation of mechanical energy	[ˌkɔnsə'veiʃən ɔv mi'kænikəl 'enədʒi]	*n. ph.*	机械能守恒
conservative forces	[kən'sə:vətiv fɔ:siz]	*n. ph.*	保守力
non-conservative forces	[nɔn kən'sə:vətiv fɔ:siz]	*n. ph.*	非保守力

Gravitation and circular motion 万有引力和圆周运动

Newton's law of gravitation		*n. ph.*	牛顿万有引力定律
Kepler's laws	['kepləs lɔ:z]	*n. ph.*	开普勒定律
elliptical orbit	[i'liptikəl 'ɔ:bit]	*n. ph.*	椭圆轨道
circular orbit	['sə:kjulə 'ɔ:bit]	*n. ph.*	圆形轨道
focus	['fəukəs]	*n.*	焦点
solar system	['səulə 'sistəm]	*n. ph.*	太阳系
semi-major axis	['semi 'meidʒə 'æksis]	*n. ph.*	半长轴
gravitational field	[ˌgrævi'teiʃnəl fi:ld]	*n. ph.*	重力（万有引力）场
gravitational force	[ˌgrævi'teiʃnəl fɔ:s]	*n. ph.*	万有引力
gravitational attraction	[ˌgrævi'teiʃnəl ə'trækʃən]	*n. ph.*	万有引力
eccentricity	[ˌeksen'trisiti]	*n.*	离心率
orbit speed	['ɔ:bit spi:d]	*n. ph.*	轨道速度 (率)
linear speed	['liniə spi:d]	*n. ph.*	线速度（率）
universal gravitational constant	[ˌju:ni'və:səl ˌgrævi'teiʃnəl 'kɔnstənt]	*n. ph.*	万有引力常数
revolution	[ˌrevə'lu:ʃən]	*n.*	转动
centripetal force	[sen'tripitl fɔ:s]	*n. ph.*	向心力
uniform sphere	['ju:nifɔ:m sfiə]	*n. ph.*	（质量）均匀球体
absolute gravitational acceleration	['æbsəˌlut ˌgrævi'teiʃnəl ækˌselə'reiʃən]	*n. ph.*	绝对重力加速度
geosynchronous orbit	[ˌdʒi:əu'siŋkrənəs 'ɔ:bit]	*n. ph.*	地球同步卫星轨道
altitude	['æltiˌtud]	*n.*	海拔
rotation period	[rəu'teiʃən 'piəriəd]	*n. ph.*	旋转（自转）周期
equator	[i'kweitə]	*n.*	赤道
universal speed	[ˌju:ni'və:səl spi:d]	*n. ph.*	宇宙速度
barycenter	['bæriˌsentə]	*n.*	重心

Rotational motion of a rigid body 刚体的定轴转动

rigid body	['ridʒid bɔdi]	*n.*	刚体
reference line	['refrəns lain]	*n. ph.*	参考（基准）线
angular displacement	['æŋgjələ dis'pleismənt]	*n. ph.*	角位移
angular velocity	['æŋgjələ vi'lɔsiti]	*n. ph.*	角速度
angular acceleration	['æŋgjələ ækˌselə'reiʃən]	*n. ph.*	角加速度
rotational inertia	[rəu'teiʃənəl i'nə:ʃjə]	*n. ph.*	转动惯量
parallel-axis theorem	['pærəlel 'æksis 'θiərəm]	*n. ph.*	平行轴原理
angular momentum	['æŋgjələ məu'mentəm]	*n. ph.*	角动量
the theorem of angular momentum		*n. ph.*	角动量定理

Simple harmonic motion 简谐运动

simple harmonic motion (SHM)	['simpəl hɑ:'mɔnik 'məuʃən]	*n. ph.*	简谐运动
spring	[spriŋ]	*n.*	弹簧
ideal (linear) spring	[ai'diəl spriŋ]	*n. ph.*	理想弹簧
compression	[kəm'preʃən]	*n.*	弹簧压力
spring-block oscillator	[spriŋ blɔk 'ɔsileitə]	*n. ph.*	弹簧振子
periodic motion	[ˌpiəri'ɔdik 'məuʃən]	*n. ph.*	周期运动
Hooke's law	[huks lɔ:]	*n. ph.*	胡克定律
spring (or force) constant	[spriŋ 'kɔnstənt]	*n. ph.*	弹簧常数
equilibrium position	[ˌi:kwə'libri:əm pə'ziʃən]	*n. ph.*	平衡位置
restoring force	[ri'stɔ:iŋ fɔ:s]	*n. ph.*	回复（恢复）力
oscillation region	[ˌɔsə'leiʃən 'ridʒən]	*n. ph.*	振动区域
amplitude of oscillation	['ampliˌtud ɔv ˌɔsə'leiʃən]	*n. ph.*	振幅
cycle of oscillation	['saikəl ɔv ˌɔsə'leiʃən]	*n. ph.*	循环
period of oscillation	['piəriəd ɔv ˌɔsə'leiʃən]	*n. ph.*	振动周期
frequency of oscillation	['fri:kwənsi ɔv ˌɔsə'leiʃən]	*n. ph.*	振动频率
Hertz	[hə:ts]	*n.*	赫兹
effective spring constant	[i'fektiv spriŋ 'kɔnstənt]	*n. ph.*	有效弹簧常数
(simple) pendulum	[('simpəl) 'pendjuləm]	*n. ph.*	（单）摆
initial phase	[i'niʃəl feiz]	*n. ph.*	初相位
phase angle	[feiz 'æŋgl]	*n. ph.*	相位角

Fluid mechanics 流体力学

fluid	['flu:id]	*n.*	流体

liquid	['likwid]	*n.*	液体
gas	[gæs]	*n.*	气体
density	['densiti]	*n.*	密度
specific gravity (sp. gr)	[spi'sifik 'græviti]	*n. ph.*	比重
pressure	['preʃə]	*n.*	压强
Pascal	[pæ'skæl]	*n.*	帕斯卡
atmospheric pressure(or atmosphere)	[ˌætmə'sferik 'preʃə]	*n. ph.*	大气压
hydrostatic pressure	[ˌhaidrə'stætik 'preʃə]	*n. ph.*	流体静压
container	[kən'teinə]	*n.*	容器
total (absolute) pressure	['təutəl 'preʃə]	*n. ph.*	总（绝对）压强
gauge pressure	[geidʒ 'preʃə]	*n. ph.*	测量压强
buoyancy (buoyant force)	['bɔiənsi]	*n. ph.*	浮力
Archimedes' principle	[ˌɑ:ki'mi:di:z 'prinsəpəl]	*n. ph.*	阿基米德原理
submerged volume	[səb'mə:dʒd 'vɔlju:m]	*n. ph.*	浸没体积
flow rate	[fləu reit]	*n. ph.*	流量率
flow speed	[fləu spi:d]	*n. ph.*	流速
cross-sectional area	[krɔs 'sekʃənəl 'ɛəriə]	*n. ph.*	横截面
continuity equation	[ˌkɔnti'njuiti i'kweiʃən]	*n. ph.*	连续方程
Bernoulli's equation	[bə:'nu:lis i'kweiʃən]	*n. ph.*	伯努利方程
viscosity	[vi'skɔsiti]	*n.*	黏性
streamline/laminar flow	['stri:mlain / 'læminə fləu]	*n. ph.*	流线流
turbulent flow	['tə:bjulənt fləu]	*n. ph.*	湍流
Torricelli's theorem	[ˌtɔri'tʃelis 'θiərəm]	*n. ph.*	托里切利原理
efflux speed	['eflʌks spi:d]	*n. ph.*	流出速度
Bernoulli effect	[bə:'nu:li i'fekt]	*n. ph.*	伯努利效应

Thermodynamics 热力学

heat/thermal energy	[hi:t / 'θə:məl 'enədʒi]	*n. ph.*	热能
temperature	['tempəritʃə]	*n.*	温度
thermometer	[θə'mɔmitə]	*n.*	温度计
temperature scales	['tempəritʃə skeils]	*n. ph.*	温标
degree Fahrenheit (°F)	[di'gri: 'færənˌhait]	*n. ph.*	华氏度
degree Celsius (℃)	[di'gri: 'selsi:əs]	*n. ph.*	摄氏度
Kelvin (K)	['kelvin]	*n.*	开尔文
conversion	[kən'və:ʃən]	*n.*	转换（换算）
absolute temperature scale	['æbsəˌlut 'tempəritʃə skeil]	*n. ph.*	绝对温标
triple point of water	['tripl pɔint ɔv 'wɔtə]	*n. ph.*	水的三相点

heat transfer	[hi:t træns'fə:]	*n. ph.*	热传递
substance	['sʌbstəns]	*n.*	物质
phase change	[feiz tʃeindʒ]	*n. ph.*	（状）态变
solid	['sɔlid]	*n.*	固体
vapor	['veipə]	*n.*	蒸汽
melt	[melt]	*n.*	熔化
liquefy	['likwəˌfai]	*n.*	液化
freeze	[fri:z]	*n.*	冻结
solidify	[sə'lidəˌfai]	*n.*	固化
evaporate	[i'væpəreit]	*n.*	蒸发
condense	[kən'dens]	*n.*	（液化）冷凝
sublimate	[sə'blaimit]	*n.*	升华
deposition	[ˌdepə'ziʃən]	*n.*	淀积
calorimetry	[ˌkælə'rimitri]	*n.*	测热法
specific heat	[spi'sifik hi:t]	*n.*	比热
absorb	[əb'sɔ:b]	*n.*	吸收
intermolecular bond	[ˌintə(:)mə'lekjulə bɔnd]	*n. ph.*	分子间键
latent heat of transformation	['leitənt hi:t ɔv ˌtrænsfə'meiʃən]	*n. ph.*	转化（状态）潜热
latent heat of fusion	['leitənt hi:t ɔv 'fju:ʒən]	*n. ph.*	熔化潜热
latent heat of vaporization	['leitənt hi:t ɔv ˌveipərai'zeiʃən]	*n. ph.*	汽化潜热
coefficient of linear expansion		*n. ph.*	线膨胀系数
coefficient of volume expansion		*n. ph.*	体积膨胀系数
thermal conductivity	['θə:məl ˌkɔndʌk'tiviti]	*n. ph.*	热导率
kinetic theory		*n. ph.*	分子动理论
Avogadro's constant		*n. ph.*	阿伏加德罗常数
Boyle's law		*n. ph.*	玻意耳定律
Charles's law		*n. ph.*	查理定律
Gay—Lussac's law		*n. ph.*	盖—吕萨克定律
ideal gas law		*n. ph.*	理想气体定律
universal gas constant		*n. ph.*	通用气体常数
Mole	[məul]	*n.*	摩尔
periodic table of elements		*n. ph.*	元素周期表
isotope	['aisəˌtəup]	*n.*	同位素
relative atomic mass		*n. ph.*	相对原子质量
relative molecular mass		*n. ph.*	相对分子质量
atom	['ætəm]	*n.*	原子

molecular	[mə'lekjələ]	*n.*	分子
Boltzmann constant		*n. ph.*	博斯曼常数
root-mean-square speed		*n. ph.*	均方根速度
standard temperature and pressure (STP)		*n. ph.*	标准温度及大气压
thermodynamics	[ˌθə:məudai'næmiks]	*n.*	热力学
laws of the thermodynamics	[lɔ:s ɔv ðə ˌθə:məudai'næmiks]	*n. ph.*	热力学定理
P—V diagram	[pi: vi: 'daiəˌgræm]	*n. ph.*	压强—体积图
proportionality constant	[prəˌpɔ:ʃə'næliti 'kɔnstənt]	*n. ph.*	比例系数
molar heat capacity at constant volume		*n. ph.*	等容摩尔热容量
molar heat capacity at constant pressure		*n. ph.*	等压摩尔热容量
internal energy	[in'tə:nəl 'enədʒi]	*n. ph.*	内能
isochoric	['aisəukɔ:rik]	*adj.*	等容的
isochor	['aisəukɔ:]	*n.*	等容线
isobaric	[ˌaisəu'bærik]	*adj.*	等压的
isobar	['aisəubɑ:]	*n.*	等压线
isothermal	[ˌaisəu'θə:məl]	*adj.*	等温的
isotherm	['aisəuθə:m]	*n.*	等温线
adiabat	['ædiəbæt]	*n.*	绝热线
adiabatic	[ˌædiə'bætik]	*adj.*	绝热的
heat engine	[hi:t 'endʒin]	*n. ph.*	热力发动机
internal-combustion engine	[in'tə:nəl kəm'bʌstʃən 'endʒin]	*n. ph.*	内燃机
thermal efficiency	['θə:məl i'fiʃənsi]	*n. ph.*	热效
Carnot cycle		*n. ph.*	卡诺循环
Carnot engine		*n. ph.*	卡诺发动机
expansion	[ik'spænʃən]	*n.*	膨胀
compression	[kəm'preʃən]	*n.*	压缩
Carnot efficiency	[i'fiʃənsi]	*n. ph.*	卡诺效率
absolute zero	['æbsəˌlut 'ziərəu]	*n. ph.*	绝对零度

Electrostatics 静电学

electric charge	[i'lektrik tʃɑ:dʒ]	*n. ph.*	电荷
electromagnetic force	[iˌlektrəumæg'netik fɔ:s]	*n. ph.*	电磁力
induction	[in'dʌkʃən]	*n.*	感应
nucleus	['nju:kliəs]	*n.*	原子核
nucleon	['nju:kliɔn]	*n.*	核子

ionization	[ˌaiənai'zeiʃən]	*n.*	电离
ion	['aiən]	*n.*	离子
elementary charge	[ˌeli'mentəri tʃɑ:dʒ]	*n. ph.*	元电荷
point charge	[pɔint tʃɑ:dʒ]	*n. ph.*	点电荷
unit charge	['ju:nit tʃɑ:dʒ]	*n. ph.*	单位电荷
Coulomb	['ku:lɔm]	*n.*	库仑
Coulomb's law	['ku:lɔms lɔ:]	*n. ph.*	库仑定律
Coulomb's constant	['ku:lɔms 'kɔnstənt]	*n. ph.*	库仑常数
permittivity	[ˌpə:mi'tiviti]	*n.*	介电常数
permittvity of free space	[ˌpə:mi'tiviti ɔv fri: speis]	*n. ph.*	真空介电常数
vacuum	['vækjuəm]	*n.*	真空
electrostatic	[i'lektrəu'stætik]	*n.*	静电的
superposition	[ˌsjupəpə'ziʃən]	*n.*	叠加
electric field	[i'lektrik fi:ld]	*n. ph.*	电场
inverse square law	[in'və:s skwɛə lɔ:]	*n. ph.*	平方反比法则
electric dipole	[i'lektrik 'daipəul]	*n. ph.*	电偶极
conductor	[kən'dʌktə]	*n.*	导体
insulator	['insju:ˌleitə]	*n.*	绝缘体
semiconductor	[ˌsemikən'dʌktə]	*n.*	半导体
superconductor	[ˌsju:pəkən'dʌktə]	*n.*	超导体
electrostatic field	[i'lektrəu'stætik fi:ld]	*n. ph.*	静电场
flux	[flʌks]	*n.*	通量
fluid-volume flux	['flu:id 'vɔlju:m flʌks]	*n. ph.*	流体体积流量
electric flux	[i'lektrik flʌks]	*n. ph.*	电通量
Gauss's law	[gaus lɔ:]	*n. ph.*	高斯定律
Gaussian surface	['gausiən 'sə:fis]	*n. ph.*	高斯面
linear charge density	['liniə tʃɑ:dʒ 'densiti]	*n. ph.*	线性电荷密度
electric potential	[i'lektrik pə'tenʃəl]	*n. ph.*	电势
electric potential difference	[i'lektrik pə'tenʃəl 'difərəns]	*n. ph.*	电势差
electric potential energy	[i'lektrik pə'tenʃəl 'enədʒi]	*n. ph.*	电势能
capacitance	[kə'pæsitəns]	*n.*	电容
equipotential surface	[ˌi:kwipəu'tenʃəl 'sə:fis]	*n. ph.*	等势面
capacitor	[kə'pæsitə]	*n.*	电容器
parallel-plate capacitor	['pærəlel pleit kə'pæsitə]	*n. ph.*	平行板电容器
Farad	['færəd]	*n.*	法拉
parallel capacitor	['pærəlel kə'pæsitə]	*n. ph.*	并联电容
series capacitor	['siəri:z kə'pæsitə]	*n. ph.*	串联电容
dielectric	[ˌdaii'lektrik]	*n.*	电介质

Circuits 电路

(electric) current	[(i'lektrik) 'kʌrənt]	*n. ph.*	电流
direct current	[di'rekt 'kʌrənt]	*n. ph.*	直流电
alternating current	['ɔ:ltə:ˌnitiŋ 'kʌrənt]	*n. ph.*	交流电
drift speed	[drift spi:d]	*n. ph.*	漂移速度
Ampere	['æmˌpiə]	*n.*	安培
voltage	['vəultidʒ]	*n.*	电压
Volt	[vəult]	*n.*	伏特
resistance	[ri'zistəns]	*n.*	电阻
resistivity	[ˌri:zis'tiviti]	*n.*	电阻率
Ohm's law	[əums lɔ:]	*n. ph.*	欧姆定律
electromotive force (emf)	[ilektrəu'məutiv fɔ:s]	*n. ph.*	电动势
joule heat	[dʒu:l hi:t]	*n. ph.*	焦耳热
resistor	[ri'zistə]	*n.*	电阻
positive terminal	['pɔzitiv 'tə:minəl]	*n. ph.*	正极
negative terminal	['negətiv 'tə:minəl]	*n. ph.*	负极
series circuit	['siəri:z 'sə:kit]	*n. ph.*	串联电路
parallel circuit	['pærəlel 'sə:kit]	*n. ph.*	并联电路
internal resistance	[in'tə:nəl ri'zistəns]	*n. ph.*	内阻
terminal voltage	['tə:minəl 'vəultidʒ]	*n. ph.*	路端电压
battery	['bætəri]	*n.*	电池组
cell	[sel]	*n.*	电池
voltmeter	['vəultˌmi:tə]	*n.*	电压（伏特）表
ammeter	['æmitə]	*n.*	电流（安培）表
galvanometer	[ˌgælvə'nɔmitə]	*n.*	检流计
earthed/grounded	[ə:θd / 'graundid]	*n.*	接地的
Kirchhoff's rule	[ru:l]	*n. ph.*	基尔霍夫定律
loop rule	[lu:p ru:l]	*n. ph.*	回圈定则
junction (or node) rule	[ru:l]	*n. ph.*	结（节）点定则
wheatstone bridge	[bridʒ]	*n. ph.*	惠斯登电桥
resistance-capacitance (RC) circuit	[ri'zistəns kə'pæsitəns 'sə:kit]	*n. ph.*	电阻电容电路

Magnetism 磁学

magnetism	['mægniˌtizəm]	*n.*	磁学
magnetic force	[mæg'netik fɔ:s]	*n. ph.*	磁场力
Lorentz force	['lɔ:rənts fɔ:s]	*n. ph.*	洛伦兹力

magnetic field	[mæg'netik fi:ld]	*n. ph.*	磁场
north-seeking pole	[nɔ:θ si:kiŋ pəul]	*n. ph.*	（磁）北极
south-seeking pole	[sauθ si:kiŋ pəul]	*n. ph.*	（磁）南极
magnet	['mægnit]	*n.*	磁铁
bar magnet	[bɑ: 'mægnit]	*n. ph.*	条形磁铁
U-shaped magnet	[ju: ʃeipt 'mægnit]	*n. ph.*	U形磁铁
Right-hand grip rule	['rait 'hænd grip ru:l]	*n. ph.*	右手螺旋定则
Tesla (T)	['teslə]	*n.*	特斯拉
Gauss (G)	[gaus]	*n.*	高斯
helix	['hiliks]	*n.*	螺旋结构
current-carrying wire	['kʌrənt 'kæriiŋ 'waiə]	*n. ph.*	载流导线
permeability of free space	[ˌpə:miəbəl'biliti ɔv fri: speis]	*n. ph.*	真空磁导率
solenoid	['səulinɔid]	*n.*	螺线圈
electromagnet	[iˌlektrəu'mægnit]	*n.*	电磁铁
electric motor	[i'lektrik 'məutə]	*n. ph.*	电动机
Biot—savart law	['bi:əusə'vɑ:lɔ:]	*n. ph.*	比奥—萨伐尔定律
Ampere's law	['æmˌpiəs lɔ:]	*n. ph.*	安培定律
Amperian loop		*n. ph.*	安培圈

Electromagnetic induction 电磁感应

electromagnetic equilibrium		*n. ph.*	电磁平衡
magnetic flux	[mæg'netik flʌks]	*n. ph.*	磁通量
Faraday's law of electromagnetic induction		*n. ph.*	法拉第电磁感应定律
Weber (Wb)	['weibə]	*n.*	韦伯
motional emf	['məuʃənl i: em ef]	*n. ph.*	动生电动势
induced electric field	[in'dju:st i'lektrik fi:ld]	*n. ph.*	磁感应电场
induced current	[in'dju:st 'kʌrənt]	*n. ph.*	感应电流
induced emf	[in'dju:st i: em ef]	*n. ph.*	感生电动势
(electricity) generator	[(ilek'trisiti) 'dʒenəreitə]	*n. ph.*	发电机
inductance	[in'dʌktəns]	*n.*	电感
inductor	[in'dʌktə]	*n.*	电感器
self-induced emf	[self in'dju:st i: em ef]	*n. ph.*	自感电动势
back emf	[bæk i: em ef]	*n. ph.*	反电动势
Henry (H)	['henri]	*n.*	亨利
inductive time constant	[in'dʌktiv taim 'kɔnstənt]	*n. ph.*	感应时间常数

resistor—inductor (RL) circuit	[ri'zistə in'dʌktə 'sə:kit]	*n. ph.*	电阻—电感器电路

Waves 波

mechanical wave	[mi'kænikəl weiv]	*n. ph.*	机械波
traveling (progressive) wave	['trævəliŋ (prə'gresiv) weiv]	*n. ph.*	行波
transverse (traveling) wave	[træns'və:s ('trævəliŋ) weiv]	*n. ph.*	横波
longitudinal (traveling) wave	[ləndʒi'tju:dinl ('trævəliŋ) weiv]	*n. ph.*	纵波
crest	[krest]	*n.*	波峰
trough	[trɔ:f]	*n.*	波谷
wavelength	['weivˌleŋθ]	*n.*	波长
amplitude	['ampliˌtu:d]	*n.*	振幅
period	['piəriəd]	*n.*	周期
frequency	['fri:kwənsi]	*n.*	频率
wave speed	[weiv spi:d]	*n. ph.*	波速
cathode ray oscilloscope	['kæθˌəud rei ɔ'siləskəup]	*n. ph.*	阴极射线示波器
linear mass density	['liniə mæs 'densiti]	*n. ph.*	线质量密度
medium	['midiəm]	*n.*	介质
angular frequency	['æŋgjulə 'fri:kwənsi]	*n. ph.*	角频率
angular wave number	['æŋgjulə weiv 'nʌmbə]	*n. ph.*	角波数
wave number(propagation constant)	[weiv 'nʌmbə (ˌprəpə'geiʃən 'kənstənt)]	*n. ph.*	波数（传播常数）
superposition of waves	[ˌsju:pəpə'ziʃən əv weivz]	*n. ph.*	波的叠加
interfere (overlap)	[ˌintə'fiə (ˌəuvə'læp)]	*n.*	干涉（重叠）
constructive interference	[kən'strʌktiv ˌintə'fiərəns]	*n. ph.*	相长性干涉
destructive interference	[dis'trʌktiv ˌintə'fiərəns]	*n. ph.*	相消性干涉
in phase	[in feiz]	*n. ph.*	同相
out of phase	[aut əv feiz]	*n. ph.*	异相
standing (or stationary) wave	['stændiŋ weiv]	*n. ph.*	驻波
node	[nəud]	*n.*	波节
antinode	['æntinəud]	*n.*	波腹
harmonic (resonant) wavelength	[hɑ:'mənik ('rezənənt) 'weivˌleŋθ]	*n. ph.*	谐波长
harmonic number	[hɑ:'mənik 'nʌmbə]	*n. ph.*	谐波级次

harmonic (resonant) frequency	[hɑ:'mɔnik ('rezənənt) 'fri:kwənsi]	n. ph.	谐波频率
fundamental standing wave	[ˌfʌndə'mentəl 'stændiŋ weiv]	n. ph.	一级（基础）驻波
overtone	['əuvəˌtəun]	n.	泛音
fundamental frequency	[ˌfʌndə'mentəl 'fri:kwənsi]	n. ph.	基频
sound wave	[saund weiv]	n. ph.	声波
compression	[kəm'preʃən]	n.	挤压
rarefaction	[ˌreəri'fækʃən]	n.	稀薄
vibrate	['vaibreit]	v.	振动
diaphragm	['daiəˌfræm]	n.	振动膜
bulk modulus	[bʌlk 'mɔdjuləs]	n. ph.	体积模量
sound level	[saund 'levl]	n. ph.	音级
sound intensity	[saund in'tensiti]	n. ph.	声音强度
Decibel	['desəbəl]	n.	分贝
loudness level	['laudnis 'levl]	n. ph.	响度
threshold of hearing	['θreʃhəuld ɔv 'hiəriŋ]	n. ph.	听觉阈
beat	[bi:t]	n.	颤音
beat frequency	[bi:t 'fri:kwənsi]	n. ph.	差频
resonance	['rezənəns]	n.	共振
Doppler effect	['dɔplə i'fekt]	n. ph.	多普勒效应
Doppler effect for light	['dɔplə i'fekt fɔ: lait]	n. ph.	光的多普勒效应
red shift	[red ʃift]	n. ph.	红移

Optics 光学

electromagnetic spectrum	[iˌlektrəumæg'netik 'spektrəm]	n. ph.	电磁波谱
material medium	[mə'tiəriəl 'mi:djəm]	n. ph.	物质介质
radio wave	['reidiəu weiv]	n. ph.	无线电波
microwave	['maikrəˌweiv]	n. ph.	微波
infrared	['infrə'red]	n.	红外线
visible light	['vizəbl lait]	n. ph.	可见光
ultraviolet	[ˌʌltrə'vaiəlit]	n.	紫外线
X-ray	[eks rei]	n.	X光
Gamma rays	['gæmə reis]	n. ph.	伽马射线
interference	[ˌintə'fiərəns]	n.	干涉
diffraction	[di'frækʃən]	n.	衍射
coherent wave	[kəu'hiərənt weiv]	n. ph.	相干波

phase difference	[feiz 'difərəns]	*n. ph.*	相位差
Young's double-slit interference		*n. ph.*	杨氏双缝干涉
single-slit (or aperture) diffraction		*n. ph.*	单缝衍射
central maximum	['sentrəl 'mæksiməm]	*n. ph.*	中心（最）亮点
light intensity	[lait in'tensiti]	*n. ph.*	光强度
diffraction grating	[di'frækʃən 'greitiŋ]	*n. ph.*	衍射光栅
dark fringe	[dɑ:k frindʒ]	*n. ph.*	暗纹
bright fringe	[brait frindʒ]	*n. ph.*	亮纹
reflection	[ri'flekʃən]	*n.*	反射
refraction	[ri'frækʃən]	*n.*	折射
incident medium	['insidənt 'mi:djəm]	*n. ph.*	入射介质
refraction medium	[ri'frækʃən 'mi:djəm]	*n. ph.*	折射介质
incident light	['insidənt lait]	*n. ph.*	入射光
reflected light	[ri'flektid lait]	*n. ph.*	反射光
refracted (transmitted) light	[ri'fræktid lait]	*n. ph.*	折射光
surface boundary	['sə:fis 'baundəri]	*n. ph.*	边界面
angle of incidence	['æŋgl ɔv 'insidəns]	*n. ph.*	入射角
angle of reflection	['æŋgl ɔv ri'flekʃən]	*n. ph.*	反射角
angle of refraction	['æŋgl ɔv ri'frækʃən]	*n. ph.*	折射角
law of reflection	[lɔ: ɔv ri'flekʃən]	*n. ph.*	反射定律
index of refraction	['indeks ɔv ri'frækʃən]	*n. ph.*	折射率
Snell's law		*n. ph.*	斯奈尔折射定律
dispersion of light	[dis'pə:ʃən ɔv lait]	*n. ph.*	光的色散
prism	['prizəm]	*n.*	棱镜
total internal reflection		*n. ph.*	全内反射
light source	[lait sɔ:s]	*n. ph.*	光源
critical angle	['kritikəl 'æŋgl]	*n. ph.*	临界角
plane mirror	[plein 'mirə]	*n. ph.*	平面镜
real image	['riəl 'imidʒ]	*n. ph.*	实像
virtual image	['və:tjuəl 'imidʒ]	*n. ph.*	虚像
spherical mirror	['sfiərikəl 'mirə]	*n. ph.*	球面镜
radius of curvature	['reidjəs ɔv 'kə:və,tʃuə]	*n. ph.*	曲率半径
focus point	['fəukəs pɔint]	*n. ph.*	焦点
vertex	['və:,teks]	*n.*	顶点
focal length	['fəukəl leŋθ]	*n. ph.*	焦距
paraxial ray	[pæ'ræksiəl rei]	*n. ph.*	近轴光
concave mirror	[kɔn'keiv 'mirə]	*n. ph.*	凹面镜
convex mirror	['kɔn,veks 'mirə]	*n. ph.*	凸面镜

ray tracing	[rei 'treisiŋ]	*n. ph.*	光线跟踪
magnification equation	[ˌmægnəfi'keiʃən i'kweiʃən]	*n. ph.*	放大率等式
upright	['ʌprait]	*adj.*	直立的
inverted	[in'və:tid]	*adj.*	倒立的
thin lens	[θin lenz]	*n. ph.*	薄透镜
converging lens (bi-convex)	[kən'və:dʒiŋ lenz(bai'kɔnveks)]	*n. ph.*	凸透镜
real focus	['riəl 'fəukəs]	*n. ph.*	实焦点
virtual focus	['və:tjuəl 'fəukəs]	*n. ph.*	虚焦点
diverging lens (bi-concave)	[di'və:dʒiŋ lenz(bai'kɔnkeiv)]	*n. ph.*	凹透镜
optical center	['ɔptikəl 'sentə]	*n. ph.*	光心

Atomic and nuclear physics 原子物理与核物理

electromagnetic radiation	[iˌlektrəumæg'netik ˌreidi'eiʃən]	*n. ph.*	电磁辐射
photon	['fəuˌtɔn]	*n.*	光子
photoelectric effect	[ˌfəutəui'lektrik i'fekt]	*n. ph.*	光电效应
quanta	['kwɔntə]	*n.*	量子
photoelectron	[ˌfəutəui'lektrɔn]	*n. ph.*	光电子
bound state	[baund steit]	*n. ph.*	界态
threshold frequency	['θreʃhəuld 'fri:kwənsi]	*n. ph.*	极限频率
work function	[wə:k 'fʌŋkʃən]	*n. ph.*	功函数
Planck's constant	[plɑ:ŋks 'kɔnstənt]	*n. ph.*	普朗克常数
electronvolt (eV)		*n.*	电子伏
Bohr atom model	[bəuə 'ætəm 'mɔdəl]	*n. ph.*	波尔原子模型
atomic spectra	[ə'tɔmik 'spektrə]	*n. ph.*	原子光谱
Rydberg constant	[ridbəg 'kɔnstənt]	*n. ph.*	雷德伯常量
energy level	['enədʒi 'levl]	*n. ph.*	能级
excited state	[ik'saitid steit]	*n. ph.*	激发态
ground state	[graund steit]	*n. ph.*	基态
ionization energy	[ˌaiənai'zeiʃən 'enədʒi]	*n. ph.*	电离能
wave-particle duality	[weiv 'pɑ:tikl dju(:)'æliti]	*n. ph.*	波粒二象性
De Broglie wavelength		*n. ph.*	德布洛意波长
scatter	['skætə]	*n.*	散射
nucleon	['nju:kliɔn]	*n.*	核子
nuclide	['nju:klaid]	*n.*	核素
isotope	['aisəˌtəup]	*n.*	同位素

nucleus	['nju:kliəs]	*n.*	原子核
nuclei	['nju:kliai]	*n.*	原子核（复数）
strong nuclear force		*n. ph.*	强核力
binding energy		*n. ph.*	结合能
atomic mass unit		*n. ph.*	原子质量单位
hydrogen	['haidrədʒən]	*n.*	氢
deuteron	['dju:tərɔn]	*n.*	氘核
deuterium	[dju:'tiəriəm]	*n.*	氘、重氢
tritium	['tritiəm]	*n.*	氚
triton	['traitn]	*n.*	氚核
mass-energy equivalence		*n. ph.*	质能当量
mass defect		*n. ph.*	质量亏损
nuclear fission	['nju:kliə 'fiʃən]	*n. ph.*	核裂变
nuclear fusion	['nju:kliə 'fju:ʒən]	*n. ph.*	核聚变
disintegration energy	[dis,inti'greiʃən 'enədʒi]	*n. ph.*	蜕变能
exothermic (or exoergic)	[,eksəu'θə:mik/,eksəu'ə:dʒik]	*n.*	放热（释放能量）
endothermic (or endoergic)	[,endəu'θə:mik/,endəu'ə:dʒik]	*n.*	吸热（吸收能量）
spontaneous	[spɔn'teiniəs]	*n.*	自发的
electron-neutrino	[i'lektrɔn nju:'trinəu]	*n.*	电子中微子

3 The formulas in AP Physics 1 & 2 AP 物理 1和 2常用公式

Newtonian Mechanics 牛顿力学

$v_f = v_i + at$	$\Delta x = \bar{v}t$ or $\Delta x = \frac{1}{2}(v_i + v_f)t$	$\Delta x = v_i t + \frac{1}{2}at^2$
$\Delta x = v_f t - \frac{1}{2}at^2$	$v_f^2 = v_i^2 + 2a(\Delta x)$	$\sum F = F_{net} = ma$
$F_{static\ friction,\ max} = \mu_{static} N$	$F_{kinetic\ friction,\ max} = \mu_{kinetic} N$	$F_{friction} \leqslant \mu N$
$a_c = \frac{v^2}{r} = \omega r^2 = v\omega$	$F_c = m\frac{v^2}{r} = m\omega r^2 = mv\omega$	$\tau = rF\sin\theta$
$p = mv$	$J = F\Delta t$	$J = \Delta p = mv_f - mv_i$
$W = F\Delta x\cos\theta$	$P = \frac{W}{t}$	$P = Fv\cos\theta$
$T = \frac{1}{f}$	$T_{spring\text{-}block\text{-}oscillator} = 2\pi\sqrt{\frac{m}{k}}$	$T_{pendulum} = 2\pi\sqrt{\frac{l}{g}}$
$F_{spring} = -kx$	$F_G = G\frac{m_1 m_2}{r^2}$	$g = \frac{GM}{r^2}$
$v = \frac{2\pi r}{T}$	$g_{planet} = \frac{Gm_{planet}}{r^2}$	$a_c = \frac{4\pi^2 r}{T^2}$
$F_c = \frac{4\pi^2 mr}{T^2}$	$a_c = \frac{4\pi^2 r}{T^2}$	$PE_G = G\frac{m_1 m_2}{r}$
$v = \sqrt{\frac{Gm}{R}}$	$KE = E_k = \frac{1}{2}mv^2$	$PE_G = E_p = mgh$
$PE_{elastic} = \frac{1}{2}kx^2$	$M.E = PE + KE$	$M.E = mgh + \frac{1}{2}mv^2$
$\theta = \theta_0 + \omega_0 t + \frac{1}{2}\alpha t^2$	$\omega = \omega_0 + \alpha t$	$x = A\cos(2\pi ft)$
$\tau = r_\perp F = rF\sin\theta$	$L = I\omega$	$\Delta L = \tau\Delta t$
$K = \frac{1}{2}I\omega^2$	$U_s = \frac{1}{2}kx^2$	$\rho = \frac{m}{V}$

a = acceleration	F = force	f = frequency
g = acceleration due to gravity	h = height	J = impulse
K = kinetic energy	k = spring constant	l = length
$M.E$ = mechanical energy	m = mass	N = normal force
PE = potential (stored) energy	P = power	p = momentum
r = radius or distance	T = period	t = time
v = velocity or speed	W = work done on a system	x = position
μ = coefficient of friction	θ = angle	τ = torque
ω = angular velocity	α = angular acceleration	ρ = density

Electricity and Magnetism 电磁学

$F=\frac{kQq}{r^2}=\frac{Qq}{4\pi\varepsilon_0 r^2}$	$E=\frac{F_{onq}}{q}$	$E=\frac{kQ}{r^2}=\frac{Q}{4\pi\varepsilon_0 r^2}$
$PE=qV$	$PE=\frac{1}{4\pi\varepsilon_0}\frac{q_1 q_2}{r}$	$E=(-)\frac{V}{d}$
$C=\frac{Q}{V}$	$C=\frac{\varepsilon_0 A}{d}$	$PE_{cap}=\frac{1}{2}QV=\frac{1}{2}CV^2=\frac{Q^2}{2C}$
$I=\frac{\Delta Q}{\Delta t}$	$R=\rho\frac{L}{A}$	$I=\frac{V}{R}$
$P=IV=I^2R=\frac{V^2}{R}$	$R_{series}=R_1+R_2+\cdots+R_n$	$\frac{1}{R_{parallel}}=\frac{1}{R_1}+\frac{1}{R_2}+\cdots+\frac{1}{R_n}$
$C_{parallel}=C_1+C_2+\cdots+C_n$	$\frac{1}{C_{series}}=\frac{1}{C_1}+\frac{1}{C_2}+\cdots+\frac{1}{C_n}$	$F=qvB\sin\theta$
$F=BIL\sin\theta$	$B=\frac{\mu_0 I}{2\pi r}$	$\varphi_B=BA\cos\theta$
$\varepsilon=-\frac{\Delta\varphi}{\Delta t}$	$\varepsilon=BLv$	
A = area	B = magnetic field	C = capacitance
d = distance	E = electric field	ε = electromotive force (emf)
F = force	I = current	l = length
PE = potential (stored) energy	P = power	Q = charge
q = point charge	R = resistance	r = distance or radius
t = time	V = electric potential or potential difference	v = velocity or speed
ρ = resistivity	θ = angle	φ_B = magnetic flux

Wave and Optics 波与几何光学

$v=\lambda f$	$d\sin\theta=m\lambda$	$x\approx\frac{m\lambda L}{d}$
$n=\frac{c}{v}$	$n_1\sin\theta_1=n_2\sin\theta_2$	$\sin\theta_c=\frac{n_2}{n_1}$
$\frac{1}{f}=\frac{1}{d_i}+\frac{1}{d_0}$	$m=-\frac{d_i}{d_0}=\frac{h_i}{h_0}$	$f=\frac{R}{2}$
d = separation or distance	f = frequency or focal length	h = height
L = distance	M = magnification	m = an integer
n = index of refraction	R = radius of curvature	v = velocity or speed
x = position	λ = wavelength	θ = angle

Fluid Mechanics 流体力学

$P = P_0 + \rho g h$	$F_{buoy} = \rho_{fluid} V_{submerged} g$	$P = \frac{F}{A}$
$A_1 v_1 = A_2 v_2$	$P_1 + \rho g y_1 + \frac{1}{2}\rho v_1^2 = P_2 + \rho g y_2 + \frac{1}{2}\rho v_2^2$	$\rho = \frac{m}{V}$
$\frac{V_{submerged}}{V} = \frac{\rho_{object}}{\rho_{fluid}}$		
A = area	F = force	g = acceleration due to gravity
h = depth	m = mass	V = volume
v = velocity or speed	v_e = efflux speed	y = height
ρ = density		

Atomic and Nuclear Physics 原子物理与核物理

$E = hf = \frac{hc}{\lambda}$	$KE_{\max} = hf - \varphi$	$f_0 = \frac{\varphi}{h}$
$E_n = \frac{Z^2}{n^2}E_1$	$p = \frac{h}{\lambda} = \frac{E}{c}$	$\lambda = \frac{h}{p}$
$E = mc^2$		
E = energy	f = frequency	KE = kinetic energy
m = mass	p = momentum	φ = work function
λ = wavelength		

Thermal Physics 热力学

$T(\text{K}) = T(^\circ\text{C}) + 273.15$	$T(^\circ\text{F}) = \frac{9}{5}T(^\circ\text{C}) + 32$	$\Delta l = \alpha l_0 \Delta T$
$\Delta V = \beta V_0 \Delta T$	$H = \frac{Q}{t} = \frac{kA\Delta T}{D}$	$PV = nRT = Nk_B T$
$v_{rms} = \sqrt{\frac{3RT}{M}} = \sqrt{\frac{3k_B T}{m_u}}$	$K_{average} = \frac{3}{2}k_B T$	$W = -P\Delta V$
$Q = mc\Delta T$	$Q = mL$	$\Delta U = Q + W$
$\eta_c = \frac{T_H - T_C}{T_H}$	$\eta = \left\| \frac{W}{Q_H} \right\|$	

续前表

A = area	D = thickness or distance	F = force
H = rate of heat transfer	k = thermal conductivity	L = thickness
l = length	M = molar mass	m = mass
m_u = mass of molecule	N = number of molecules	n = number of moles
Q = heat transferred to a system	T = temperature	t = time
U = internal energy	V = volume	W = work done on a system
K = kelvins	℃ = degrees Celsius	℉ = degrees Fahrenheit
η = efficiency	α = coefficient of linear expansion	β = coefficient of volume expansion

图书在版编目（CIP）数据

AP 物理 1&2/ 赵杰编著 . —北京：中国人民大学出版社，2016.6
ISBN 978-7-300-22843-3

Ⅰ. ①A… Ⅱ. ①赵… Ⅲ. ①力学-高等学校-入学考试-美国-教材 ②电磁学-高等学校-入学考试-美国-教材 Ⅳ. ①04

中国版本图书馆 CIP 数据核字（2016）第 087122 号

AP 物理 1 & 2
赵 杰 编著
AP Wuli 1 & 2

出版发行	中国人民大学出版社		
社 址	北京中关村大街31号	邮政编码	100080
电 话	010-62511242（总编室）		010-62511770（质管部）
	010-82501766（邮购部）		010-62514148（门市部）
	010-62515195（发行公司）		010-62515275（盗版举报）
网 址	http://www.crup.com.cn		
	http://www.1kao.com.cn（中国1考网）		
经 销	新华书店		
印 刷	北京宏伟双华印刷有限公司		
规 格	185 mm × 260 mm 16 开本	版 次	2016 年 6 月第 1 版
印 张	23	印 次	2017 年 6 月第 2 次印刷
字 数	540 000	定 价	59.80 元